Spectrochemical Analysis
by Atomic Absorption

ELEMENTS DETERMINED BY ATOMIC ABSORPTION

I	IIa	IIIa	IVa	Va	VIa	VIIa	VIII			Ib	IIb	IIIb	IVb	Vb	VIb	VIIb	0
H																	He
Li	Be											B	C	N	O	F	Ne
Na	Mg											Al	Si	P	S	Cl	Ar
K	Ca	Sc	Ti	V	Cr	Mn	Fe	Co	Ni	Cu	Zn	Ga	Ge	As	Se	Br	Kr
Rb	Sr	Y	Zr	Nb	Mo		Ru	Rh	Pd	Ag	Cd	In	Sn	Sb	Te	I	Xe
Cs	Ba	La*	Hf	Ta	W	Re	Os	Ir	Pt	Au	Hg	Tl	Pb	Bi	Po	At	Rn
	Ra	Ac	Th	Pa	U												

* Ce Pr Nd Pm Sm Eu Gd Dy Ho Er Tm Yb Lu

Elements determinable in air acetylene flame

Elements requiring nitrous oxide acetylene flame

The remaining elements are not determinable by atomic absorption

Charles Seale-Hayne Library
University of Plymouth
(01752) 588 588
LibraryandITenquiries@plymouth.ac.uk

Spectrochemical Analysis by Atomic Absorption

W. J. PRICE

Pye Unicam Ltd, Cambridge

London · Philadelphia · Rheine

Heyden & Son Ltd., Spectrum House, Hillview Gardens, London NW4 2JQ, UK
Heyden & Son Inc., 247 South 41st Street, Philadelphia, PA 19104, USA
Heyden & Son GmbH, Münsterstrasse 22, 4440 Rheine, West Germany

ISBN 0 85501 455 5

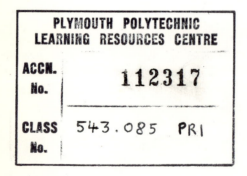
Set by Eta Services (Typesetters) Ltd., Beccles, Suffolk
Printed litho and bound in Great Britain
by W & J Mackay Ltd., Chatham

For all my friends, in many countries,
with the sincere wish that they may find some help
within these pages

Contents

Preface

Since the publication of my previous book *Analytical Atomic Absorption Spectrometry* in 1972, techniques in atomic absorption have progressed in two main directions: improvement in the design and performance of electrothermal atomizers, and the use of microprocessors to take some of the routine operation of the instruments off the hands of the analyst. Methods of flame atomization and the preparation of samples for that purpose have reached a kind of technological plateau and very little of moment in flame AAS is happening. However, the formula of the earlier book—to provide enough theory on which to build up the operational detail as well as to give a self-contained account of most application fields—was well received, and it is now timely to present this new volume which is similarly constructed and also extended to cover the newer aspects of the subject.

The literature and methodology again had to be treated selectively. The unabated torrent of not far short of a thousand published papers every year on AAS not only makes this inevitable just to keep such a text within sensible and economic bounds, but it also places a dire responsibility on the one who presumes to select. The author therefore asks the forgiveness of any reader whose own, or favourite, references do not happen to appear in the bibliography. His objective has been not to create an encyclopaedia, but to present the user with practical ideas and ways of approach, supported wherever possible by experience from within his own and associated laboratories.

Much of the inspiration for the compilation of this second book is derived from colleagues in the laboratory whose interest in the subject seems to be never failing. It also arises from working closely with other enthusiasts in the field, particularly those from both the United Kingdom and overseas who form the Editorial Board of *Annual Reports on Analytical Atomic Spectroscopy*. Without that publication, too, the selective process mentioned above would have been considerably more difficult.

Many of the terms used by the early workers in atomic absorption are now accepted by international agreement—with the notable exception of 'sensitivity' which, as defined and used, was the exact reverse. We have preferred to use the term 'reciprocal sensitivity' for this characteristic which is still often quoted by atomic absorption workers. It has the merit of being the least clumsy of several possibilities.

In the procedures, if acid strengths are not given, the normal commercial 'concentrated' forms are to be understood (e.g. hydrochloric acid sp. gr. 1.18, sulfuric acid sp. gr. 1.98, nitric acid sp. gr. 1.42, perchloric acid sp. gr. 1.54). Dilutions are usually given as percentages, or occasionally as relative volumes, e.g. $1+4$ to avoid the ambiguity of '5 times' or $1+4+2$ to avoid unfamiliar percentage values.

Very sincere thanks are due to my colleague Mr Peter J. Whiteside for reading and making valuable suggestions and additions to the manuscript, and for reading the proofs; also to Dr Ian Bowater who did the thermodynamic calculations on which Figs. 5 and 6 and their associated text is based; and to Dr Lyndon Davies and the management of Pye Unicam for their continued encouragement in this undertaking. Special thanks are due to Barbara, my wife, who typed most of the manuscript and prepared the bibliography and index. Without her help the expeditious completion of the book would not have been possible.

W. J. PRICE
Cambridge, January 1979

Acknowledgements

I gratefully acknowledge permission to reproduce figures and other material from: Mr Colin Watson, Messrs Hopkin and Williams Ltd (Table 7, the organic standard preparation procedure on p. 236 and Table 15); Professor S. D. Brown and the American Chemical Society (Fig. 36); Dr H. Massmann and the editors of *Spectrochimica Acta* (Figs. 25 and 28); Professor T. S. West and the editors of *Analytica Chimica Acta* (Fig. 27); Dr G. F. Kirkbright and Maxwell Scientific Inc., New York (Figs. 17 and 18); Pye Unicam Ltd, Cambridge (Figs. 9, 29, 30, 35, 53–58).

W. J. PRICE

Chapter 1

Introduction

DEFINITION AND HISTORICAL

Atomic Absorption Spectrometry is an analytical method for the determination of elements, based upon the absorption of radiation by free atoms. Atomic fluorescence spectrometry also enables elements to be determined but on the basis of the re-emission of radiant energy absorbed by the free atoms.

Interaction of atoms with various forms of energy results in three very closely related spectroscopic phenomena which may be used for analytical purposes—emission, absorption and fluorescence.

The extreme value of atomic absorption as a versatile laboratory technique—still overshadowing both flame emission and fluorescence which appear to have greater limitations—is evidenced by a rapid annual increase in the amount of published literature from 1960 to 1972, levelling off to a plateau of about 700–1000 papers per annum since 1973.

At the present time, annual sales of atomic absorption spectrometers world wide are believed to be about four and a half thousand, and the annual growth of this market is between five and ten per cent. About one in five new installations is equipped with electrothermal atomizers.

Atomic fluorescence shows few signs of being accepted to this extent, but its possibilities in the field of simultaneous multi-element analysis are still being explored, and this, apart from high sensitivity analyses of a limited number of elements, could be its major use.

No chronicler of this subject fails to point out that the first atomic absorption observations were made in 1802 by Wollaston[826]—these were of the Fraunhofer lines in the solar spectrum—or that Kirchhoff and Bunsen (1860)[403] were first to demonstrate that atomic spectra, used in emission or absorption, could form the basis of a new and highly specific method of analysis. Emission techniques, perhaps because the spectra are more tangible and easier to measure, have become widely used. Emission instrumentation varies in sophistication from simple filter flame photometers to the point where, even in 1950, twenty or thirty elements could be determined in one sample almost in the same number of seconds.

1

Apart from two specialized applications, the identification of some elements in stellar atmospheres and the detection of mercury vapour in laboratory atmospheres, the real potential of atomic absorption was not realized until Walsh's exposition of the subject[769] in 1955. After a few more years of general unconcern, progress became meteoric. Improved instrumentation, more reliable sources of resonance radiation, hotter flames and non-flame atomizers have enabled the technique to be extended to nearly every metallic element in the periodic table. What was thought of as an intriguing method for determining some trace elements is now used in wider concentration ranges with accuracies that compare well with most other accepted techniques.

LITERATURE SOURCES

Several general and topic-orientated textbooks and a monograph on electro-thermal atomization have already been written on atomic absorption spectro-scopy.[1-19] These are listed in the general bibliography.

The present book is written particularly with the practising analytical chemist in mind. Thus, as a considerable part of the published literature is by now either outdated or simply not directly applicable, the bibliography does not pretend to be complete.

Full lists of published papers are compiled regularly elsewhere.[20] Biblio-graphies are also usually included among the atomic absorption instrument manufacturers' literature, and some of the manufacturers include reviews and original papers on the subject in their house journals.[20,21] Apart from the usual sources of abstracts, *Analytical Abstracts*[22] and *Chemical Abstracts*[23] there is a classified source, *Atomic Absorption and Flame Emission Spectroscopy Abstracts*,[24] which aims at complete coverage, at least from 1955, of all possible journals for articles and papers on atomic flame emission, absorption and fluorescence. Reviews appear at regular intervals in the analytical and spectrochemical journals. Some of the more recent critical reviews are included in the general bibliography.[25-28] Biennial literature reviews are included in one or other sections of the annual reviews in *Analytical Chemistry*,[29] and the Chemical Society has been publishing its *Annual Reports on Analytical Atomic Spectro-scopy*[32] from early in 1972.

In the early 1960s, the novelty of atomic absorption resulted in most papers on the subject being published in the spectrochemical, or at least the general analytical, journals. With its acceptance in all fields, articles now appear in the very specialized journals. It is therefore becoming more difficult to trace all references without the help of abstracts services or retrieval systems. The *Annual Reports on Analytical Atomic Spectroscopy* ('*ARAAS*') mentioned above gathers some 1600 abstracts annually from forty correspondents the world over, and is therefore one of the best sources of current information on atomic absorp-tion, emission and fluorescence spectroscopy.

THE PRESENT STATUS OF ATOMIC ABSORPTION

When new physico-chemical analytical techniques are introduced, they appear to acquire, quite quickly, their own particular mythology. This may have the effect of perpetuating certain preconceived notions and, to some extent, hindering the search for the truth. A typical, early example in this field was that chemical interferences are fewer in atomic absorption than in emission. As both techniques depend upon the production of free atoms, they are, of course, the same. But a better subsequent knowledge of flame chemistry has aided understanding of interferences in flame emission, absorption and fluorescence. Other myths of a general nature which, hopefully, have now been refuted are that atomic absorption is a trace technique, that high temperature flames are dangerous and difficult to work with and that the often quoted 'sensitivity' values tell you all you need to know about the performance of a particular instrument. Some more deeply-rooted misconceptions undoubtedly still remain, though Alkemade[46] has started to explode some of these—in particular the often-quoted contention that atomic absorption is superior in sensitivity to atomic emission because the proportion of ground state atoms is higher.

Protagonists of the technique may sometimes give the impression that all other methods of elemental analysis are soon likely to be superseded. Likewise, those with a consuming interest in the use of inductively coupled or other plasmas as an emission source sometimes appear to believe that this, in its turn, has already superseded atomic absorption and is the only technique to be considered. These are also myths, for, though a chemist who has invested a large sum of money in equipment will obviously wish to extract from it as much information as possible, he will also wish to use the technique that offers him most in terms of accuracy, cost and speed. Very often this is atomic absorption, but by no means always.

Comparison with other methods

When atomic absorption is compared with other methods of analysis—spectroscopic or otherwise—at least five capability factors must be taken into account. These are scope of application, ease of sample preparation, sensitivity and/or detection limit, reproducibility and accuracy.

As pointed out by Grant,[293] emission spectrography and spark source mass spectrometry would usually be the methods of choice for qualitative analysis involving metal ions. For quantitative analysis, atomic flame methods, i.e. emission, absorption and fluorescence, come into their own where the sample already is, or is easily converted into, a liquid. Where a number of elements are to be determined together, direct reading emission or X-ray fluorescence multi-channel spectrometers ('polychromators') offer the speediest, though much more expensive, solution. Interest in direct reading emission spectrography has received a boost during recent years through the development of inductively coupled and other kinds of plasma as the atomization and excitation source, particularly for solutions and other liquid samples. Such multi-element instruments may not provide best excitation conditions for all elements simultaneously however, and flame methods still offer excellent precision and accuracy. Sample heterogeneity limits accuracy in techniques where solid specimens are required, but, though this problem is overcome in solution, liquid containers and chemicals used in sample preparation are a cause of contamination. By contrast it is relatively easy to determine several elements sequentially by flame atomic methods, and this usually requires only one dissolution of the sample.

The concentration range where atomic absorption may be applied is from picogrammes and fractions of a part per billion of many cations in solution up to tens of percent of metallic constituents in solid samples. Gravimetric methods with a precision of better than 1 part in 1000 still give higher accuracy in the 10–100% range. However, gravimetry is entirely non-specific and requires complete chemical separations. Titrimetry is usually designed to be specific, but with a precision of about 3 parts per 1000 is hardly better than the best atomic absorption results; it is considerably less versatile though, of course, less expensive. Spectrophotometry usually covers smaller concentration ranges than atomic absorption. It has the advantage that its calibration curves shift less, and as the contents of the measurement cell are static, its precision and accuracy depend critically upon the quality of the instrument and the chemical separations and preparation of the sample—which are often time-consuming. Electrochemical methods compare with atomic absorption only when preliminary separations are not required.

Economic Factors. Economically, atomic absorption falls midway between the very inexpensive instrumentation such as colorimeters, balances, etc. and the very expensive, like automatic, direct reading spectrometers. But, as new innovations are incorporated, relative costs are likely to rise. It is now customary to build in, or make provision for, such refinements as multilamp turrets, data-converting circuitry (so that results can be displayed or printed out direct in concentration), safety interlocks on the gas handling systems, automatic samplers, etc. These may well double the cost of a basic instrument. In the right circumstances this will soon pay for itself in doing the job of several other instruments, and in saving time and labour by avoiding a multiplicity of preparations of one sample for different elements. In every type of laboratory, versatility counts and many analysts have averred that the majority of metal determinations have been appreciably simpler and quicker since the introduction of the atomic absorption spectrometer.

Advantages. Lewis[456] summed up atomic absorption analysis by saying that it has a number of the advantages of an ideal technique: specificity; low limits of detection; many elements can be determined in one solution; there is rarely any 'lapsed time' requirement (as in colour development, drying of precipitates, etc.) in sample preparation; and data output is in a directly readable form. To these we might well add the economic factor, versatility with regard both to types of sample and to concentration ranges and the fact that, though not an 'absolute' method, it can always be made entirely independent.

TERMS AND DEFINITIONS

During the growth period of a new analytical technique, existing terms may acquire a specialized meaning and certain new terms have to be coined. In order that all workers in atomic absorption may communicate with complete unanimity and understanding, a number of terms have been agreed by several international bodies. Most are based upon the definitions proposed by the Atomic Spectroscopy Group of the Society for Analytical Chemistry before the International Conference on Atomic Absorption Spectroscopy, Sheffield in 1969. Further discussions have resulted in the following definitions, many of which have been approved by the International Union of Pure and Applied Chemistry.[354]

General terms

Atomic Absorption Spectroscopy. An analytical method for the determination of elements, based on the absorption of radiation by free atoms.

Atomic Fluorescence Spectroscopy. An analytical method for the determination of elements, based on the re-emission of absorbed radiation by free atoms.

Limit of Detection. The minimum concentration or amount of an element which can be detected with 95% certainty assuming a normal distribution of errors. This is that quantity of the element which gives a reading equal to twice the standard deviation of a series of at least ten determinations at or near blank level.

Sensitivity. Historically, the sensitivity of an element in atomic absorption has been defined as its concentration in solution which produces a change, compared with pure solvent, of 0.0044 absorbance units (1% absorption) in optical transmission at the wavelength of the absorption line employed. This is inconsistent with the definition of sensitivity, which is a direct measure of rate of change of signal with concentration, dx/dC, in other analytical techniques. The term defined above should now be called the *reciprocal sensitivity* or *characteristic concentration*. In this text we use the former.

Terms concerned with the spectral radiation

Characteristic Radiation. Radiation that is specifically emitted or absorbed by free atoms of the analysis element.

Resonance Radiation. Characteristic absorbed radiation that corresponds to the transfer of an electron from the ground state level to a higher energy level in the atom.

Hollow Cathode Lamp. A discharge lamp with a hollow cathode, usually cylindrical, used in atomic spectroscopy to generate characteristic radiation.

Electrodeless Discharge Lamp (EDL). A tube containing the element to be determined in a readily vaporized form and constructed so as to enable a discharge to be produced in the vapour by microwave or radio frequency induction. This discharge is a source of characteristic radiation.

Terms concerned with the processing of the sample in an atomic absorption instrument

Atomic Vapour. A vapour that contains free atoms of the analysis element.

Atomization. The process that converts the analyte element, or its compounds, to an atomic vapour.

Atomizer. The device used to produce a population of free atoms.

Sampling Unit. The part of an atomic absorption spectrometer (often consisting either of a nebulizer, spray chamber, and burner or of an electrothermal atomizer) which accepts the sample and prepares it for atomization.

Nebulization. The process that converts a liquid to a mist.

Nebulizer. A device for the nebulization of a liquid.

Nebulization Efficiency. The ratio of the amount of sample reaching the atomizer to the total amount of sample entering the nebulizer.

Flame Atomization Efficiency. The ratio of the amount of analyte converted to free atoms in the atomizer to the total amount of analyte entering the nebulizer.

Oxidant. The substance, usually a gas, used to oxidize the fuel in a flame.

Fuel. The substance, usually a gas, which is burnt to provide the atomizing flame.

Carrier Gas. The gas used to convey the sample mist to the atomizer.

Spray Chamber. The vessel wherein the sample mist is generated prior to transfer to an atomizer.

Impact Bead. A usually spherical solid bead, placed before the nebulizer orifice, which increases the proportion of small droplets in the nebulized sample solution, thus improving the atomization efficiency.

Flow Spoiler. A device, in a spray chamber, for removing large droplets from a mist.

Direct Injection Burner. A burner in which liquid is nebulized directly into the flame. The flame obtained with such a burner is normally turbulent.

Premix System. A sampling unit in which the fuel, oxidant gas and sample mist are mixed in a spray chamber before entering the flame. Flames obtained using this system are normally laminar.

Semi-premix System. A sampling unit in which either fuel or oxidant gas is added to the sample mist after the spray chamber.

Long Tube Device. A device in which an atomizing flame is directed into a tube lying along the optical axis of an atomic absorption spectrometer.

Long Path Burner. A burner constructed to produce a flame which is extended in one direction at right angles to the direction of movement of the flame gases.

Multislot Burner. A burner with a head containing several parallel slots.

Separated Flame. A flame in which the diffusion combustion zone is so separated from the primary combustion zone as to enable the two zones to be observed independently. Separation may be effected either mechanically or by an inert gas shield.

Observation Height. The vertical distance between the optical axis of the monochromator and the top of the burner.

Burner Angle. The acute angle between the plane of the flame produced by a long path burner and the optical axis of the monochromator.

Electrothermal Atomizer. A device for producing an atomic vapour by ohmic heating, using an electrically conducting tubular cell or filament.

Detectors

Resonance Radiation Detector. A selective detector in which atoms in an atomic vapour are excited by radiation from an external source, and the intensity of the resulting fluorescence radiation is measured.

Terms concerned with the analytical technique

Interference. A general term for an effect which modifies the measured signal produced by a particular concentration of the analysis element.

Depression. An interference that causes a decrease in the measured signal.

Enhancement. An interference that causes an increase in the measured signal.

Matrix Effect. An interference caused by differences between the sample and a standard containing only the analysis element and, where appropriate, a solvent.

Background Absorption or Scattering. Attenuation of the beam of characteristic radiation by wide band molecular absorption or by the light scattering effect of particulate material.

Spectroscopic Buffer. A substance which is part of, or is added to, a sample and which reduces interference effects.

Ionization Buffer. A spectroscopic buffer used to minimize or stabilize the ionization of free atoms of the analysis element.

Releasing Agent. A spectroscopic buffer used to reduce interferences attributable to the formation of involatile compounds in the atomizer.

Comments on the definitions

The relationship between the terms 'reciprocal sensitivity' and 'detection limit' is discussed in Chapter 6 (under Trace Analysis). In this sense it is restricted to describing the capability of an instrument in respect of a specified element. The word *sensitivity* is used in a general non-quantitative sense when describing a method or even an instrument.

The words 'atomizer' and 'atomization' are used in their entirely literal sense:

a thing and a process which turn a substance into atoms. These terms should now never be used in connection with the scent-spray-like device used for producing a mist from a liquid.

The term 'support gas' is still used by different writers to mean either the gas which supports the aerosol, or the gas which supports combustion. To avoid confusion 'carrier gas' and 'oxidant' are proposed instead.

HEALTH AND SAFETY ASPECTS OF ANALYTICAL ATOMIC ABSORPTION

The practice of atomic absorption analysis involves materials and processes, which, if not properly handled or controlled, incur certain hazards to health and safety. It is thus necessary that supervisors and operators in laboratories using atomic absorption be aware of these risks and ensure that all proper precautions are taken.

Such hazards arise in three main areas: the use of gases which form combustible mixtures; the normal hazards associated with electrical apparatus; and the involvement of solvents and samples which themselves are, or readily form compounds which are, toxicologically active or chemically potentially dangerous.

Hazards from use of combustible gases

Perhaps the greatest dangers stem from the use of fuel and oxidant gases, and the fact that the combustion flame is positioned within the equipment. Under certain conditions of incorrect usage, explosions can occur. All atomic absorption instruments should therefore be designed so that, if this happens, there is no direct or indirect danger to the operator, and the effects of the explosion are contained inside the apparatus. Operators should follow manufacturers' instructions for lighting and extinguishing flames carefully, keep the burners clean, and ensure that, if there is a change of solvent, the waste trap is emptied and completely refilled with the new solvent. 'Safe' apparatus is designed so that any one safety device can fail without the whole apparatus becoming unsafe. It is thus essential that, should an explosion occur, it cannot be transmitted back along the fuel gas line. There should be at least two flame traps (blow-back valves) between the possible area of an explosion and the source of acetylene. If an explosion does occur, its force must be held at a reasonable level, and therefore a form of pressure release should be provided in the spray chamber. This must be designed and positioned not only so that minimal damage is caused to the instrument, but also so that the shock wave intensity is limited to $\ngtr 150 \, dB(A)$ in order to avoid injury to the ears of the operator.

Pipes carrying fuel gases must be incapable of leaking near electrical contacts which may arc. Modifications made to fuel gas lines must be properly tested for leaks before use, and pipe joints located in safe and accessible positions. Leak testing should be carried out at pressures at least 50% in excess of those to be applied in normal use. It should also be remembered that solid particles entering the jet of a pressure regulator can cause a slow build-up of pressure in the equipment which it controls, when in fact it should be shut off.

Flexible tubes must be of adequate standard, e.g. they should be reinforced or armoured. Copper piping and fittings must not be employed for carrying acetylene. In many countries special regulations apply to the use of acetylene under pressure. In the UK, for example, this relates to a pressure of more than 9 p.s.i. (0.621 bar) above atmospheric, and before such pressures are used, the Health and Safety Executive should be approached for details and advice. In other countries, local regulations must be ascertained and observed.

In the combustion of nitrous oxide with hydrocarbon fuels, other oxides of nitrogen and water vapour are formed. These combustion products should not be inhaled. Adequate fume extraction (see p. 113) should thus be provided. An acid-laden atmosphere is also detrimental to the equipment itself, and could be the direct cause of an instrument becoming unsafe.

Electrical safety

Electrical hazards are similar to those created by other electrical or electronic instrumentation. Most atomic absorption equipment belongs to Safety Class 1, i.e. it relies on an earth connection for electrical safety. The outer case should therefore be earthed, either directly or through the mains supply lead. Checks on earthing efficiency should be made at least once per annum. It is also standard practice that compartments where electrical hazards exist should only be accessible by the use of a tool, e.g. a screwdriver or spanner. Anyone who makes access to such a section of the instrument should be trained to recognize and avoid these risks. Electrical modifications must be completed in such a way that the equipment remains electrically safe.

At the time of writing there are unfortunately no recognized international standards for laboratory equipment. Some national standards exist, e.g. for general conformity as in the USA, or for certain areas of usage, such as the DHSS standards for clinical laboratories in the UK. The International Electro-technical Commission currently has the matter under active consideration.

Health risks from materials being used

In various analytical methods, the materials being analysed or the solvents employed cause the formation of toxic vapours. This is a further important reason for the provision of efficient fume extractors.

Solvents containing halogens form phosgene or other poisonous gases in the flame. In electrothermal atomizers the solvents themselves are emitted as vapour, and the fume extraction system should itself create an adequate draught as it is not assisted by flame-induced convection currents.

Many elemental vapours, especially those of mercury, lead, cadmium, beryllium and platinum are well known to be exceedingly toxic. Exceptional care must be taken when such elements form part of a sample matrix. But even when they are present in trace or ultratrace quantities, every possible precaution should be taken in the handling and atomization of both the samples and the standard solutions.

Chapter 2

Basic Principles

EMISSION, ABSORPTION AND FLUORESCENCE SPECTRA

The analyst about to use atomic absorption or fluorescence spectroscopy will find it desirable to know something about the way in which atoms react with various forms of energy—particularly light and heat.

Under the conditions in which atomic spectroscopy, emission, absorption and fluorescence, is practised the energy input into the atom population includes thermal, electromagnetic, chemical and even electrical forms of energy. These are converted to light energy by various atomic and electronic processes before being measured. The light energy is manifest in the form of a spectrum, which consists of radiation of a number of discrete wavelengths.

First Balmer and then Rydberg described the spectrum of hydrogen in terms of exact mathematical formulae, which could be extended to other univalent atoms such as the alkali metals. The Rydberg equation, which expresses the wavenumber of a given spectrum line as the difference between two *terms*, one term being constant for a given series of spectrum lines, led to the old quantum theory of atomic spectra in which discrete or quantized energy levels are postulated for the Coulomb forces between the valency electrons and the positive nucleus of the atom. Transition between two such quantized states corresponds to the absorption or emission of energy in the form of electromagnetic radiation, the frequency v of which is determined by Bohr's condition:

$$\Delta E = E_1 - E_2 = hv$$

where E_1 and E_2 are the energies in the initial and final states respectively, and h is Planck's constant.

Both the quantum theory and the more recent wave mechanics theory of Schrödinger have been developed further to correlate atomic structure with atomic spectra, but such detailed knowledge is not necessary for following the processes on which the present subject is based.

An atom is said to be in the ground state when its electrons are at their lowest energy levels. When energy is transferred to a population of such atoms by means

of thermal or electrical excitation, as in the various forms of emission spectroscopy, the transfer takes place by means of collision processes. The amount of
energy transferred may vary considerably from atom to atom, resulting in a
number of different excitation states throughout the population. The subsequent
emission of these different amounts of energy involves not only many higher
energy levels, but also low energy levels of other than the ground state. These
result in radiation of a number of different frequencies, and hence the *emission*
spectrum of any given element may be highly complex.

Theoretically, the reverse process is also possible. That is, if light of any of
these given frequencies is passed through a vapour containing the atoms, it will
be absorbed in performing the process of excitation. However, the proportion
of excited to ground state atoms in a population at a given temperature can be
considered with the aid of the well known Boltzmann relation

$$\frac{N_m}{N_n} = \frac{g_m}{g_n} \exp\left[-\frac{(E_m - E_n)}{kT}\right]$$

where N is the number of atoms in a state n or m, g is the statistical weight for a
particular state and k is the Boltzmann constant. Walsh[769] calculated the ratio
N_m/N_n for a number of common atoms over a range of temperature, see Table 1,
for the case where m refers to the first excited state and n is the ground state.

TABLE 1. Values of N_m/N_n for typical elemental resonance lines

Line, nm	g_m/g_n	N_m/N_n			
		2000 K	3000 K	4000 K	5000 K
Cs 852.1	2	4.44×10^{-4}	7.24×10^{-3}	2.98×10^{-2}	6.82×10^{-2}
Na 589.1	2	9.86×10^{-6}	5.88×10^{-4}	4.44×10^{-3}	1.51×10^{-2}
Ca 422.7	3	1.21×10^{-7}	3.69×10^{-5}	6.03×10^{-4}	3.33×10^{-3}
Zn 213.9	3	7.29×10^{-15}	5.58×10^{-10}	1.48×10^{-7}	4.32×10^{-6}

The very low proportion of atoms in the first excited state, even at temperatures
of 3000 K indicates that absorption of radiation, other than that originating
from a transition involving the ground state, would be very small.

In the sun's spectrum many such absorptions show up as Fraunhofer lines,
but in laboratory experiments, and certainly on the analytical scale, only
absorptions involving the ground state are normally observed. Absorptions
involving the ground state are for this reason known as *resonance* lines.

The absorption spectra of most elements, therefore, when produced under
laboratory conditions, are extremely simple. This accounts for one of the main
advantages of atomic absorption spectra as a means for analysis—there is very
little possibility of coincidence of lines and therefore there are very few examples
of spectral interference.

Atoms excited by absorption of resonance radiation also re-emit the absorbed
energy. This process, by analogy with a similar process known in molecular

spectroscopy, is called atomic fluorescence. The re-emitted energy may, however, be of the same wavelength as the absorbed energy, or it may be of longer wavelength, indicating that an intermediate state is also involved, with the consequent partial loss of energy in some other form.

Direct re-emission, (a) in Fig. 1, is known as resonance fluorescence, and two forms of indirect re-emission, involving fluorescence to an intermediate state (b), and excitation to a state higher than the first excited state (c), are referred to as 'direct line' and 'stepwise' fluorescence respectively. 'Sensitized fluorescence' in which an atom is first excited by transfer of energy from an atom of another element, itself excited by absorption of radiation, is also known.

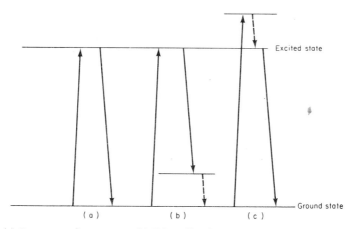

FIG. 1. (a) Resonance fluorescence. (b) Direct line fluorescence. (c) Stepwise fluorescence.

The fluorescence spectrum of a given element, when excitation is by absorption of resonance radiation, is even simpler than the absorption spectrum, and consists of only a very small number of lines.

As the means used for producing atomic vapours always result in the production of heat, all three processes occur simultaneously to some extent, though they can be observed and measured separately by appropriate choice of the experimental conditions.

The very high proportion of ground state to excited state atoms when these are in equilibrium, as shown in Table 1, suggests two particular advantages of atomic absorption and fluorescence measurements over emission. Because absorption and fluorescence are direct measurements of the number of ground state atoms, these would be expected to yield a better sensitivity than emission. This is only one of several factors which must be taken into account when absorption and emission sensitivities are compared, and it is rarely the most important. However, as the temperature coefficient of the number of excited atoms is markedly greater than that of the ground state atoms, absorption and fluorescence

measurements would be expected to be less dependent upon short-term temperature variations in a flame, for example, than emission. This is indeed reflected in
their generally better signal-to-noise ratios.

Alkemade[46] has shown that for a given analysis line the absorption sensitivity
is better than the emission only if the spectral radiance of the lamp source exceeds
that of a black body at the temperature of the flame. Since the sharp line sources
used in atomic absorption are invariably excited by non-thermal means, their
spectral radiance exceeds that of a flame atomizer by many orders of magnitude.
The effective radiation temperature for resonance lines produced in gas discharge
tubes, for example, is about 10 000 K.

A factor which also has practical implications is the wavelength of a resonance
line, which is inversely proportional to its energy ($\Delta E = h\nu = hc/\lambda$). A result of
the Boltzmann relationship (p. 12) is therefore that the ratio N_m/N_n decreases
exponentially with increasing resonance line frequency. Thus it is expected, and
found, that elements whose resonance lines occur at the higher wavelengths are
more sensitive in emission than those whose resonance lines are at low wavelengths. Lithium (670.7 nm) and sodium (589.0 nm), for example, are more
sensitive in emission than copper (324.8 nm) or magnesium (285.2 nm), while
zinc (213.9 nm) is very insensitive indeed in emission.

The converse is not generally true for absorption, for although the absolute
amount of *energy* absorbed is directly frequency dependent, under experimental
conditions one only measures the *proportion* of the incident energy which is
absorbed. The frequency term thus cancels out leaving an expression of the
form given below.

RESONANCE RADIATION

In analytical atomic absorption and atomic fluorescence spectroscopy, one is
dealing largely with the absorption and emission of resonance radiation which
is defined as characteristic radiation of an element corresponding to the transfer
of an electron from the ground state to a higher energy level. A comprehensive
treatise on the interactions between atoms and resonance radiation was compiled
by Mitchell and Zemansky.[520]

The analytical validity of making absorption measurements depends on the
relationship between absorption and the concentration or partial pressure of the
absorbing atoms. This relationship is, according to classical dispersion theory

$$\int K_v \, dv = \frac{\pi e^2}{mC} N_v f$$

where K_v is the absorption coefficient at frequency v, $m =$ electronic mass,
$e =$ electronic charge, and $N =$ number of atoms per cm^3 capable of absorbing
energy in the range v to $v+dv$. The f-value, now called the oscillator strength,
is, for the present purpose, the effective number of free electron oscillators per

atom of the element in question responsible for the absorption effect produced by the incident radiation. From the data in Table 1 it is clear that the value N—the total number of atoms of the element present—can with negligible error be substituted for N_v. (This relationship holds unless the ground state is a multiplet state, in which case the expression is slightly more complex.) Nevertheless, the energy absorption is virtually a direct function of the number of atoms in the absorbing path.

The natural width of an atomic spectral line is of the order of 10^{-5} nm (10^{-4} Å), but this is broadened by Doppler, electric field and pressure effects. Doppler widths vary with the element, the wavelength of the line and the temperature. The sodium 589 nm line for instance, has a Doppler width of 0.0048 nm at 3000 K, but if the line is produced in a flame, the pressure effect may broaden it to about 0.008 nm.

As Walsh pointed out when first seriously examining the potentialities of atomic absorption as an analytical technique,[769] the practical implication of this is that, in order to measure the absorption coefficient of a given line in a white continuum, a spectrograph with a resolution of something like 500 000 would be necessary. This is hardly practicable for the reason that the energy passed by the required small spectral slit-width would be too low to be measured by standard photoelectric methods although limited use has been made of photographic recording. Measurements made over wider band widths than this result in absorption values smaller than the true values, though within certain limits and under controlled conditions these can be put to some analytical use.

However, the absorption coefficient can be measured over an area close to the centre of the absorption line with a sharp-line source which emits lines of smaller half-width than the absorption line itself. If the shape of the absorption line is determined solely by Doppler broadening, then a linear relationship can still be shown to exist between the concentration and the energy absorbed. With a sharp-line source, it becomes unnecessary to use a monochromator bandwidth of the same order as the half-width of the absorption line. A spectrometer is needed simply to isolate the line to be measured from the other lines emitted by the source.

The position is clarified in Figs. 2 and 3. In Fig. 2 the absorption line profile is superimposed on a broader profile, which may represent either a non-sharp emission line, or even, for the purpose of the present argument, the bandpass of a laboratory spectrometer of normal resolution. The proportion of energy actually absorbed is such a small fraction of the total that, at best, the measurement is highly insensitive. When the absorption is superimposed on the sharp emission line, energy in the latter is absorbed over the full width of the emission line and the measured absorption then approaches the theoretical maximum (Fig. 3).

In order to measure the amount of absorption by atoms of a given element, it is necessary to devise an experimental arrangement such as that shown in plan in Fig. 4. A light-measuring device placed beyond (b) is enabled to indicate by difference the amount of energy absorbed by the atomic vapour, while one placed

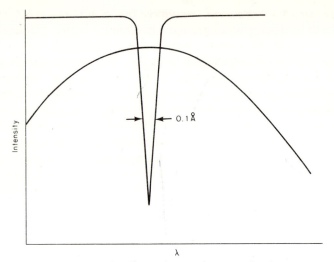

FIG. 2. Absorption line and monochromator bandpass.

at any other position, e.g. at (c), will measure a proportion of the light which is re-emitted as fluorescence.

It is important that the primary source should be a narrow line emitter for absorption purposes, the 'wings' of the spectral profile of a source of wide band-width being unabsorbable light entering the detection system. But the same constraint does not apply to the case of fluorescence, as only the fluorescence radiation itself (plus some scatter) enters the detector. In a rigorous treatment of the subject[823] the fluorescence intensity is shown to be a function of the fluores-

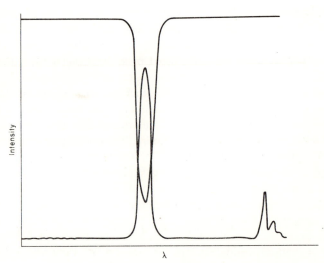

FIG. 3. Absorption line and fine emission line.

cence efficiency, and therefore of the primary source line width, but the effect of increased line width on fluorescence measurements is minor compared with the effect of unabsorbed radiation on absorption measurements.

The Walsh theory of atomic absorption has recently been much refined by another of its early proponents, B. V. L'vov, who discussed the effects of the ultimate width and asymmetry of the source line, its shift and hyperfine structure, on the absorbance values obtained.[486] Hyperfine structure had already been shown by de Galan[264] to be one of the major factors influencing atomic-line profile.

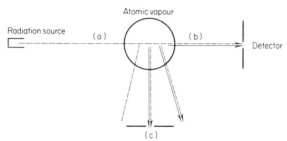

FIG. 4. Illustrating modulated primary beam (a) attenuated by absorption (b) and causing fluorescence at all angles, detected at (c). Note that thermally induced emission also appears at all angles including (b) and (c).

PRODUCTION OF FREE ATOMS IN A FLAME

All the preceding matter referred to the resonance spectra of free, un-ionized atoms. It is on the production of uncombined and un-ionized atoms in an atomic vapour that the success of an atomic absorption or fluorescence analytical procedure depends.

Consider the events that take place when fine droplets of a solution are aspirated into a plasma at a temperature of 2000–3000 K. First, the solvent evaporates leaving small solid particles (referred to by some as 'clotlets'). These particles melt and vaporize. The vapour consists of a mixture of compounds which tend to decompose into individual atoms. Free atoms obviously exist, however transitorily, under these conditions, otherwise there would be no evidence of atomic spectra. The individual atoms absorb energy by collision and become excited or ionized. Taking potassium chloride as an example, the last stages of this process may be written:

atomization: $$KCl \rightleftharpoons K + Cl \tag{1}$$

excitation: $$K + h\nu = K^* \tag{2}$$

ionization: $$K \rightleftharpoons K^+ + e^- \tag{3}$$

and it will be seen that, while the first reaction provides the free atoms required to ensure sensitivity of the method, the second and third reactions tend to remove them. As indicated by the figures in Table 1, however, the effect of excitation is negligible.

Both the molecular dissociation and ionization processes have been investigated thermodynamically to some extent, but it would be difficult and perhaps impossible to give a comprehensive treatment for all the elements likely to be encountered in atomic absorption. Degrees of atomization differ markedly from element to element in a given plasma. It was pointed out by de Galan and Samaey[267] that this factor more than any of the others (oscillator strength, line broadening, etc.) influences the analytical sensitivity. The following simple and partially qualitative approach will at least give some idea of the significance of the atomization and ionization processes, and enable many of the observed effects to be explained.

Diatomic dissociation and subsequent ionization are intrinsically similar processes, and it is possible by means of the Saha equation[659] to calculate degrees of ionization of atoms into ions and electrons for various elements at different temperatures and partial pressures. Some of these for potassium are plotted in Fig. 5, and for calcium in Fig. 6.

For these elements partial pressures of free atoms in the flame of 10^{-6} atm and 10^{-8} atm correspond very approximately to the amounts passing into the flame when a conventional atomic absorption spectrometer is used (at a sample uptake rate of 3 ml min^{-1} and acetylene and air flows of 1.5 and 6 l min^{-1},

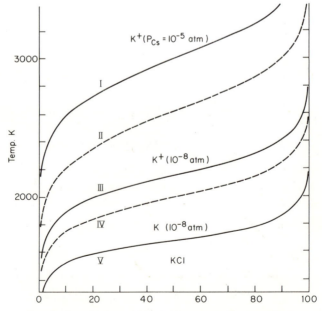

FIG. 5. Atomization and ionization curves for potassium.

respectively) to analyse solutions containing 100 ppm and 1 ppm respectively. The Saha equation is:

$$\log K = -\frac{U}{4.573T} + \frac{5}{2}\log T - 6.49 + \log \frac{g_{K^+} \times g_e}{g_K} \qquad (4)$$

(where K is the equilibrium constant for the ionization process (2) above, the g terms are statistical weights of the species shown as suffixes, U is the ionization potential in cal mol^{-1} and T is the temperature Kelvin) which for potassium reduces to

$$\log K = -\frac{21\,873}{T} + \frac{5}{2}\log T - 6.49 \qquad (5)$$

This can be used to derive values of K for different temperatures in the range 1800–3000 K from which are calculated the degrees of ionization at 10^{-8} and 10^{-6} atm using the equation

$$K = \frac{P_{K^+} \times P_e}{P_K} = \frac{(Px)^2}{P(1-x)} = \frac{Px^2}{(1-x)} \qquad (6)$$

(where P_{K^+}, P_e and P_K are the partial pressures of K$^+$, e and K, P is the total

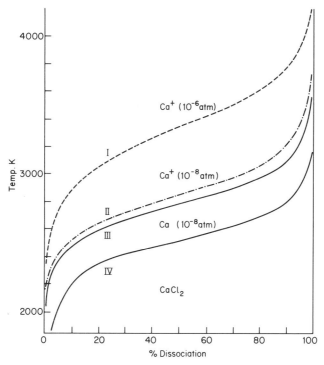

FIG. 6. Atomization and ionization curves for calcium.

pressure of both K species and x is the degree of ionization ($0 \leqslant x \leqslant 1$)) which is then solved for x.

In Fig. 5, curves II and III are degrees of ionization plotted against temperature for 100 ppm and 1 ppm potassium, respectively.

Equilibrium constants for the dissociation of potassium chloride into potassium and chlorine atoms were calculated from known thermodynamic data.[382] The degrees of dissociation are then calculated for $P = 10^{-6}$ and 10^{-8} atm using an equation analogous to (6) above.

The predicted equilibrium constants agree reasonably well with those measured experimentally by Mavrodineanu[500] and Rubeska and Moldan.[4]

The resulting degrees of dissociation are also plotted in Fig. 5, giving curves IV and V for the same two solution concentrations. It will be clear that an intercept drawn horizontally at any temperature will indicate the proportions of the element existing at that temperature and pressure in combined (molecular) form, in free atomic form—the only form in which it is available for atomic absorption measurements—and in ionized form, by the dissociation values at which it cuts a pair of curves for one particular pressure value. These curves are very much idealized. From the pair of curves representing molecular dissociation (V) and ionization (III) for a 1 ppm solution, however, the proportion of free potassium atoms which would be present under such ideal and interference-free conditions can be deduced for any temperature. This is a maximum where the curves show maximum horizontal separation, i.e. at about 1900 K, which is in agreement with the observed fact that the maximum sensitivity for potassium is obtained in an air propane flame giving approximately this temperature. For higher concentrations of potassium, the best sensitivity would be at a higher temperature. This observation is clearly of academic interest only in this connection, but is the basis of one of the contributing causes of calibration curvature.

In Fig. 6, a similar exercise has been done for calcium. The ionization curves I and II for 100 ppm and 1 ppm solutions are obtained from data referring to the singly ionized calcium atom, the Saha equation reducing in this case to

$$\log K_{Ca} = \frac{-6.111 \times 23\,053}{4.573T} + \frac{5}{2}\log T - 6.49 \tag{5a}$$

Curve III is the position of the ionization curve as indicated by some experimental results obtained in the author's laboratory. This is in good agreement with the calculated curve.

Curve IV is a curve drawn from experimental results for a 1 ppm solution of calcium chloride in air acetylene and nitrous oxide acetylene flames and represents the degree of molecular dissociation as a function of temperature. These data suggest a temperature of about 2600 K as being ideal for the measurement of calcium.

The true positions of each of these curves in a given situation depend on all the factors which can affect the dissociation and ionization equilibria. These include the nature of the flame gases, the presence of other cations and anions in the sample solution and the formation of free radicals from the breakdown of the solvent.

The concentration of other ions formed at the temperature of the flame has a direct influence on the vertical positions of the ionization curve and therefore on the horizontal separation between pairs of curves which is a measure of the available free atoms. The curve I in Fig. 5 is the position the potassium ionization curve would assume in the presence of 10^{-5} atm partial pressure ($= 1000$ ppm in solution) of caesium. Caesium is readily ionized and provides an excess of electrons, which suppress the formation of electrons through ionization of potassium. The position of curve I is obtained by substituting the value of the partial pressure of 'caesium electrons' in the potassium ionization equilibrium equations. It is found to be practically the same for both 100 ppm and 1 ppm solutions of potassium.

In this example, caesium acts in the role of an ionization buffer, i.e. it provides an excess of electrons and thus stabilizes the ionization curve at a temperature far enough above that of the flame in use for the effects of other factors on degree of ionization of the analyte element to be minimal.

The importance of this is the much wider temperature range over which a maximum number of free atoms are available between curves V and I, and potassium can now therefore be determined at best sensitivity in higher temperature flames. The practical significance of these curves will also become more apparent in the context of the interference effects to be discussed in Chapter 5.

PRODUCTION OF FREE ATOMS IN AN ELECTROTHERMAL DEVICE

An electrothermal atomizer is generally a small cylindrical furnace or cell which can be raised to a high temperature by resistive (ohmic) heating. In order literally to atomize sample material, such a cell must be capable of being raised to temperatures in excess of 2000 °C for most elements, and up to 3000 °C for the so-called refractory elements.

The great sensitivity of electrothermal atomizers in atomic absorption spectrometry arises from their ability to retain a substantial proportion of the atomized analyte element in the observation zone for a finite period of time. The basic requirement for achieving such sensitivity was first postulated by L'vov:[12] the rate of formation of the free atoms must be equal to or greater than their rate of removal from the optical path, i.e.

$$\frac{dN(t)}{dt} = \left(\frac{dN}{dt}\right)_{in} - \left(\frac{dN}{dt}\right)_{out}$$

At the absorbance maximum,

$$\left(\frac{dN}{dt}\right)_{in} = \left(\frac{dN}{dt}\right)_{out}$$

It would appear that the cylindrical furnace types of atomizer fulfil such a condition better than atomizers based on open filaments, and for this reason the former type has now been adopted almost universally by suppliers and users alike, though in many sizes and configurations.

The mechanism by which free atoms are produced in such a cell depends on a number of factors. These include the compounds still present in the cell at the atomization temperature of the element being measured, the material from which the cell itself is constructed and the atmosphere present in the cell, the rate of increase in temperature and the temperature of operation of the cell.

Atoms are removed from the observation zone by the processes of diffusion, gas expansion and convection, and recombination, and the supply and removal functions combine to give an overall response curve as shown in Fig. 7.

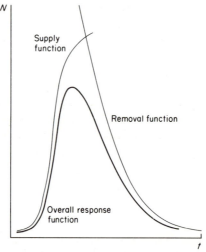

FIG. 7. Response functions and electrothermal atomization.

The atom supply function

The mechanism of atom release from the walls of graphite tubular atomizers has been the subject of several theories.[41,126,714] Fuller[19] has contributed much, and has also reviewed the whole subject of electrothermal atomization in its practical and theoretical aspects. Part of the following has been much condensed from his treatment and part from more recent work by de Galan.[265]

It would seem that, at the point when atomization actually begins to take place, the process should be simpler than in the flame. Various possible reactions have been examined thermodynamically.

Evaporation of metal oxide prior to atomization would occur for some elements whose oxides have high vapour pressure at temperatures where atomization is first observed, e.g. Sb_2O_3. Significant losses of the analyte would be expected to occur under these conditions, and the problem is particularly acute if metal chlorides, which are much more volatile, are present in the atomizer.

Thermal dissociation of metal oxide is another possibility and degrees of dissociation at various temperatures can be calculated from thermodynamic data. Equilibria between the oxygen liberated and the furnace carbon also control the degree of dissociation.

Reduction of the metal oxides by the furnace carbon according to the equation

$$MO(s/l) + C(s) \rightleftharpoons M(g) + CO(g)$$

is the mechanism believed to be relevant for most elements, although theoretically derived data, such as temperatures at which the above equation should proceed, do not always agree with experimentally found 'appearance temperatures'. Notable differences in the results of Campbell and Ottaway are shown by the elements calcium and magnesium which appear at a much lower temperature than predicted by theory.

Carbide formation from the oxide and cuvette material is yet another possibility for some elements, and calculation of the free energy changes involved shows that this occurs at a lower temperature than gaseous atom formation for several elements, in particular, aluminium, calcium and chromium.

In the case of atomizers made from materials other than carbon, Campbell[125] has made thermodynamic calculations from which it is predicted that elements such as lead and cadmium which are readily atomized on carbon are even more readily atomized on a tantalum atomizer. Conversely, elements which atomize only with difficulty on carbon should atomize even less readily on tantalum. These predictions are in general agreement with most practical observations.

It has to be pointed out, however, that atomization is a kinetic process, and two kinds of theoretical treatment have been made on this basis. L'vov[12] was the first to study atomization under conditions of increasing temperature, and Fuller[254] later proposed a theory for isothermal conditions. The latter, which reduces to a simple exponential function, approaches the situation obtaining in modern tubular atomizers which are designed to heat up to atomization temperature in a very short time. It also applies to an isothermal pre-atomization pyrolysis step. L'vov's theory is more applicable to open atomizers and also when the pyrolysis stage has a ramp programme.

Kinetic theories of atomization appear to give more analytically useful information than thermodynamics.[19] For example, the ratios of N(max) to N(integrated) (where N is the number of atoms being measured at a point in time) can be compared to find the conditions under which the peak height measurement or the integrated peak area gives best results. Integration is shown to be better for low temperatures or low atomization rates.

It should also be possible to predict whether a given element can be determined in a particular matrix and to define the best operating temperatures. This approach was also made by Fuller[252] though it is acknowledged that certain simplifications and assumptions are involved: e.g. first-order kinetics were applied and similar kinetics were used for both high and low concentrations.

Kinetic studies have also been extended to the pyrolysis stages of the procedure in an attempt to elucidate some of the interference effects[257] which are well known to be entirely different from those in the flame.

The atom removal function

The sensitivity for a particular element in an electrothermal device must depend on the shape of the total transient absorbance signal observed. The rising slope of this is not equal to the rate of supply and the falling slope to the rate of removal: both increasing and decreasing slopes are functions of both the analyte supply into the furnace cell and the rate of its removal (Fig. 7). Fuller proposed an exponential function[254] for removal of analyte as well as for its production. From a curve-fitting exercise on his transient absorbance signal he concluded that rates of removal of analyte can exceed rate of supply by 3–20 times. De Galan believes[265] that the convolution integral derived by Tessari[581] expresses the conditions for separation of the supply and removal functions better. The rate of removal of analyte atoms is not necessarily proportional to the number of atoms present (i.e. $(\mathrm{d}N/\mathrm{d}t)_{\mathrm{out}} \neq kN_{(t)}$ as assumed by many authors) as there are several disturbing factors:

(i) the analyte is in contact with only a small part of the centre of the furnace, so the response function is at first constant, and then decreases exponentially when diffusion takes place at the ends;

(ii) the analyte is formed during a period of temperature increase and is, therefore, likely to be expelled by the expanding gases. This effect is greater than the diffusivity factor above except for very long furnaces, and is particularly relevant for open atomizers;

(iii) a deliberately introduced flowing inert gas causes convective removal of atomic vapour and enhances the expansion effect (this is reduced by stopping or decreasing the gas flow during the atomize step);

(iv) the atomic vapour may diffuse through the furnace walls (L'vov has shown by use of radioactive tracers[470] that rates of diffusion through the walls of a graphite furnace could be greater than from the open ends of the tube);

(v) the atomic vapour may recombine in the gas phase.

Effects (iv) and (v) can be reduced, e.g. by use of pyrolytic graphite and maintenance of isothermal heating conditions respectively (see Chapter 3). Nevertheless, if the time constants τ_{d}, τ_{e} and τ_{c} for the diffusivity, expansion and gas flow effects, (i), (ii) and (iii), respectively, are approximately exponential, then that of the overall response function will also be a decreasing exponential.

De Galan[265,266] has experimentally isolated the supply function from the removal function. If the removal function is very rapid in comparison with the supply function then the overall response function approximates to the latter. This is achieved in practice by stepping up the flow rate of convective gas until the shape of the recorded signal becomes invariable. This is then the supply function.

From the results of this kind of experiment several important conclusions can be drawn. Firstly, for normal methods of operation, e.g. heating at a rate approaching 1000 K s^{-1}, the supply of analyte vapour lasts about 1 s, and the entire absorbance signal appears within the rising portion of the temperature/time profile. Secondly, the experimentally observed absorbance maximum is lower than the predicted value, and the supply function is not therefore adequately represented as a first order kinetic reaction.

The overall response function is elongated by adsorption and redesorption of the analyte vapour on the walls of the enclosed type of furnace (this might be called a 'chromatographic effect') to an extent which varies from element to element and which also depends on the physico-chemical properties of the graphite surface. This explains why the performances of some tubes vary as the tubes age.

Wall and vapour temperature

While the wall temperature of the furnace is directly measurable the atomic vapour temperature may only be derived by the Schmidt[667a] two line method and the two are not necessarily equal, particularly in larger furnaces.[34] Further work by de Galan shows that the time lag between wall and vapour temperature is usually no more than 5 ms for systems without convecting gas and the temperature difference is usually not more than 5 K. With forced convection, the time lag may be equivalent to ~1.5 cm of travel along the furnace from the point of introduction for the gas to achieve 90% of wall temperature (K). The vapour temperature may consequently be several hundred degrees below that of the furnace walls.

The efficiency of commercial furnaces

It is clear from de Galan's results that the removal function is indeed much faster than the supply function under normal operating conditions. In furnaces with forced convection, atom removal takes about 0.1 s, though in larger furnaces under static conditions, the removal time is about 1 s, i.e. similar to the supply time.

The efficiency of the furnace, defined as its ability to maintain the atomic vapour in the observation zone, i.e. the ratio N_{max}/N_0 (where N_{max} is the maximum number of atoms observed at the absorbance peak and N_0 is the number of atoms actually placed in the furnace) is shown, under such rapid removal conditions, to be equal to τ_R/τ_S, the ratio of removal and supply time constants. If τ_R increases (as in larger furnaces or under static conditions) the ratio tends to unity. N_{max}/N_0 can be calculated from the measured maximum absorbance and known amount of analyte injected, and τ_R/τ_S can be estimated from the observed shape of the overall response function. The two results, which are in reasonable agreement, indicate quite low efficiencies for the types of furnace available commercially. In small furnaces and furnaces operating with forced convective gas flows, only about 5–10% of injected analyte contributes to the maximum observed absorbances. In larger furnaces under static conditions the efficiency may rise to about 50%. Calculations[266] show that 50% efficiency is achieved when $\tau_R = \tau_S = 0.4$ s, when the effective length would be about 3 cm for acceptable diffusivity at an atomization temperature of 1500 K. The overall response signal is then 0.6 s which is manageable by modern signal-handling circuitry and recorders.

Chapter 3

The Formation of Atomic Vapours

THE BASIC ATOMIC ABSORPTION SYSTEM

In order to be able to make measurements in atomic absorption it is necessary to devise an experimental assembly which will convert the material under examination as efficiently as possible to a population of ground state atoms, and then pass resonance radiation of the element to be measured through that population. Ideally, the light-measuring device should 'see' only the wavelength which is being absorbed, as the presence of other radiation lowers the proportion of absorbed radiation, and thus decreases the sensitivity of the measurement.

The basic flame atomic absorption spectrometer is shown in Fig. 8. Light from the source lamp generating a sharp line spectrum characteristic of the desired element passes through the flame into which the sample solution is sprayed as a fine mist. The region of the spectrum in the immediate neighbourhood of the resonance line to be measured is selected by the monochromator. The isolated resonance line falls on to the detector, a photomultiplier, the output of which is amplified and drives a readout device, e.g. a meter, strip chart recorder or, through data processing, to a digital display unit or printer. The intensity of the

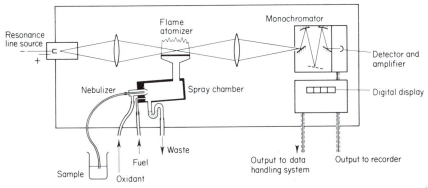

FIG. 8. Practical system for flame atomic absorption spectrometer.

resonance line is measured with and without the sample passing into the flame. The ratio of these readings is a measure of the absorption and therefore of the amount of the element being determined, as will be discussed in detail on p. 120. In non-flame atomic absorption, the flame is replaced by an electrothermal atomizer or other cell in which free atoms can be produced.

[In order to avoid measuring the emission from the excited atoms in the atomizer at the same wavelength, the source lamp intensity is modulated, usually at 50 Hz but sometimes higher, and the amplifier is tuned to the same frequency. Consequently, the continuous component of the radiation signal, originating from the atomizer, is not measured.]

The same or similar components may, in principle, be used to make measurements in flame or furnace emission and atomic fluorescence spectrometry. For emission, the source lamp is removed, and the radiation from the atomizer is modulated at the amplifier frequency. In fluorescence, the source radiation is passed through the flame at an angle, usually a right angle, to the optical axis of the monochromator. The source is again modulated in order to avoid measuring thermally excited emission from the flame.

Instrumentation for atomic spectroscopy—emission, absorption and fluorescence—thus has two parts with essentially different functions, the means whereby the population of ground state atoms is produced from the sample, and the optical system which includes the resonance source, the spectrometer and detector, and which enables the atomic absorption to be measured.

FLAME ATOMIZERS

Combustion flames in premix systems

In spite of the present advanced state of electrothermal atomization, the most convenient, stable and economic source of atomic vapours remains the combustion flame.

Fuel/oxidant mixtures are now commonly used which are safe to handle and produce a range of temperatures from about 2000 to 3000 K. As we have seen, not only is the flame temperature an important parameter in the production of free atoms from a given element but also the chemical effects of radicals and other substances present in the flame. One gas mixture producing a given temperature can have quite different analytical properties from another mixture either of the same or of different gases producing the same temperature.

Fuel gases include propane, hydrogen and acetylene, and oxidants include air and nitrous oxide. Pure oxygen is rarely used as an oxidant as this provides a mixture whose burning velocity is high and difficult to control. However, oxygen may well sometimes be mixed with a monatomic gas such as argon[364] or even helium.[149]

Fuel/oxidant mixtures may be combined from most of the above in ratios which may be stoichiometric, lean or rich. Lean and rich mixtures contain less than and more than the stoichiometric quantity of fuel gas, respectively. The characteristics of some frequently used flames are summarized in Table 2. The flow rates given may not be those that would be shown on flow meters calibrated for air. These are not directly proportional to the quantities of reactants in the stoichiometric flame, as the reading is dependent on the density of the gas, and diffusion of oxygen from the atmosphere affects the flow ratio at which stoichiometry occurs. The indicated flow rates are also likely to depend on the design of the instrument, though real flow rates should not differ widely.

TABLE 2. Characteristics of some pre-mixed gas combustion flames

	Flow rates/l min^{-1}		Approx. temp./K	Expansion factor	Flame speed/ cm s^{-1}
	Fuel	Oxidant			
Air propane					
lean	0.3	8			
stoichiometric	0.3–0.45	8	2200		45[a,c]
rich	0.45	8			
Air acetylene					
lean	1.2	8			
stoichiometric	1.2–1.5	8	2450	1.03	160[a,c]
luminous	1.5–1.7	8			
rich	1.7–2.2	8	2300		
N$_2$O acetylene					
lean	3.5	10			
stoichiometric	3.5–4.5	10	3200	1.64	285[b]
rich	4.5	10			
Air hydrogen stoichiometric	6	8	2300	0.9	320[a,c]
N$_2$O hydrogen stoichiometric	10	10	2900	1.00	380[c]
N$_2$O propane stoichiometric	4	10	2900		250

[a] Gaydon and Wolfard, Ref. 276.   [b] Aldous et al., Ref. 44.   [c] Willis, Ref. 811.

Undoubtedly the most widely used of these fuel/oxidant mixtures is air acetylene as it enables about 30 of the common metals to be determined. The sensitivity of some elements in the hydrogen air flame is not much inferior to their sensitivity in air acetylene (i.e., their degrees of atomization are comparable) but the presence of other substances in the sample solution may cause the formation of stable compounds, and hence interferences are said to be worse. In particular, elements which form stable monoxides give lower sensitivity in hydrogen-based flames.

Air propane mixtures were used extensively in the earlier days of atomic absorption analysis for those elements which were found to be easily atomized. These included the alkali metals, cadmium, copper, lead, silver and zinc. The air propane flame does indeed appear to give a better sensitivity for these elements, though other factors contribute to make the detection limit at least no better, and perhaps somewhat worse, than when air acetylene is used.

Air propane is not widely used at the present time except in some laboratories where the use of acetylene is prohibited. In this case nitrous oxide propane is strongly recommended as it greatly reduces the severe interelement interferences experienced with air propane, and also allows some of the refractory elements to be determined.

Air acetylene mixtures are used successfully for most elements which do not form highly refractory oxides. Calcium, chromium, iron, cobalt, nickel, magnesium, molybdenum, strontium and the noble metals are among the elements normally determined with this flame. The refractory oxide metals may give better sensitivity in a fuel-rich flame—though this on the average is about 150 °C cooler than a stoichiometric flame. Elements which give very low sensitivity in air acetylene flames (and for which this flame analytically is of little use) are those whose dissociation energies for the M—O bond are greater than about 5 eV. Examples are: Al—O, 5.98; Ta—O, 8.4; Ti—O, 6.9; Zr—O, 7.8 eV.

The use of nitrous oxide as the oxidant gas instead of air was suggested by Willis.[810] Its performance under analytical conditions was reported by Amos and Willis.[57] This gas might be expected to provide properties halfway between air and pure oxygen. Mixed with acetylene however, nitrous oxide produces a higher temperature than the equivalent mixture of oxygen and nitrogen, due mainly to the exothermic nature of its decomposition reactions. Temperatures of about 3000 K are produced, not much less than with oxy-acetylene, but the flame propagation is slower. The flame is thus safer to handle than oxy-acetylene.

Nitrous oxide has also been used as oxidant for hydrogen and propane. The nitrous oxide hydrogen flame usually has little to commend it for atomic absorption purposes because of the surprisingly poor sensitivities which it gives. It is highly oxidizing in nature as compared with the nitrous oxide acetylene flame.[144] While it is useful for some elements in overcoming carbide formation, very low sensitivities are given for elements which form refractory oxides. For the very reason that it promotes the formation of refractory oxides, however, the nitrous oxide hydrogen flame is useful for improving the sensitivity of emission band spectra given by metal oxide species. However, this flame has been particularly recommended for the determination of the more readily atomized elements at trace levels when the sample is dissolved in an organic solvent.[461] Benzene, xylene and petroleum can be aspirated into the flame causing very little background signal because the critical C/O ratio (see p. 31) at which luminescence appears is not exceeded as it would be in hydrocarbon flames.

Nitrous oxide propane or butane have been shown[123] to have certain of the interference-reducing properties of the nitrous oxide acetylene flame, while retaining the ease of handling of air acetylene, though sensitivities are again usually disappointing.

There seems little doubt, then, that the two most useful flame mixtures will continue to be air acetylene and nitrous oxide acetylene.

An important study by de Galan and Samaey[267] of the performance of flames based on air or nitrous oxide with hydrogen or acetylene has led to some valuable observations. Best sensitivity nearly always occurs just above the inner cone of the laminar flame. This is usually 0.3–0.5 cm above the burner slot. The air acetylene and nitrous oxide acetylene flames are in thermal equilibrium at this point, whereas the air hydrogen flame attains thermal equilibrium only at about 15 cm above the slot. Better sensitivities might thus be observed at a point in the hydrogen flames which unfortunately is adversely affected by physical instability.

Two main factors arising from the conditions in the flame itself affect the degree of atomization and ultimate sensitivity of a particular metal.

Gas flow rates and thermal expansion of the flame gases after combustion both contribute to the total dilution of the absorbing species in the flame and to the length of time (usually no more than 10^{-4} s) which an absorbing atom actually remains in the radiation beam, and the best sensitivity at a given temperature is given by a flame having the smallest combustion gas/sample volume ratio. At the lower end of the flame temperature scale this is the air propane flame, which explains why this flame gives better sensitivities than air acetylene for elements completely atomized at 2200 K. For hydrogen-based flames the volume of fuel is comparatively greater than for hydrocarbon-based flames. This is part of the explanation of the poorer performance of nitrous oxide hydrogen as compared with nitrous oxide acetylene.

Isothermal degrees of atomization are proportional to the concentration of carbon and carbon-containing species and inversely proportional to the concentration of atomic and molecular oxygen in the flame.[144] The presence of free carbon thus exerts the major effect on the reducing properties of the flame. In a given hydrocarbon flame, as the fuel/oxidant ratio is increased, the concentration of free carbon or small carbon-containing radicals increases up to the point of formation of soot (massive carbon agglomerates which effectively remove the carbon activity). At this transition point the critical carbon/oxygen ratio occurs. For certain elements the sensitivity is observed to increase to a maximum at the critical C/O ratio. The presence of organic solvents increases the critical C/O ratio; thus for metals in solution in organic solvents higher sensitivities are usually found than for aqueous solutions. The critical C/O ratio can be shown to be twice as high for acetylene as for propane, and of course for hydrogen combustion mixtures the effect of carbon activity is absent.

The derivation of flame temperatures and atomization efficiencies has been the subject of much recent research. It is clear that measured flame temperatures depend on a number of factors other than the flame gas ratios themselves, including the method used, the particular experimental assembly etc.[629,630] However, L'vov calculated the temperature values and compositions for a wide range of oxidant/fuel ratios in nitrous oxide acetylene flames. The effects of pressure and presence of water vapour on the atomization efficiencies of a number of refractory elements were then deduced. It was predicted that full dissociation would occur in all monoxide species except those with dissociation energies D_0 in the range 180–200 kcal mol^{-1} (i.e. boron, cerium, hafnium, lanthanum, niobium, praseodymium, tantalum, thorium, uranium and zirconium—all of which have notably low sensitivities in atomic absorption). Low atomization efficiencies for some elements with D_0 between 130 and 150 kcal mol^{-1} were explained by incomplete volatilization of aerosol particles.

The success of nitrous oxide acetylene as an atomizing gas in atomic absorption may therefore be ascribed to the high temperature produced by nitrous oxide as oxidant and the high critical C/O ratio contributed by acetylene.

Nebulizer–burner systems

The prime purpose of these systems is the conversion of the sample solution into the atomic vapour where the absorption measurement is made. In a sense, therefore, this is the heart of the instrument, for the sensitivity of the analysis depends directly upon its correct function and efficiency.

The processes include nebulization (i.e. the conversion of the liquid sample to a mist or aerosol), the selection of mist droplets of the correct size distribution, the mixture of the selected mist with the flame gases and introduction to the burner. These are the processes which occur in pre-mix spray chamber type burner systems. The merits of such systems as opposed to direct injection types have been well established during the past ten years or more, and spray chambers are now almost universally employed in atomic absorption spectrometers, in spite of their comparatively low nebulization efficiency.

A typical system is shown in exploded view in Fig. 9. The sample is drawn into the nebulizer by the low pressure created around the end of the capillary by the flow of the carrier or oxidant gas. The resulting droplets are ejected with the carrier gas into the spray chamber. The design of the chamber is such that droplets with a diameter greater than about 5 μm fall out onto the sides of the chamber and flow to waste. The fuel gas is introduced into the chamber, and also auxiliary carrier gas or oxidant, so that an intimate mixture of sample mist, fuel and oxidant leaves the spray chamber and enters the burner. A laminar flow burner sustains a highly stable flame, and the proportion of sample mist normally introduced has little or no influence over the flame characteristics. Two exceptions to this last statement are when organic solvents (which alter the carbon/oxygen ratio) are employed, and when the volume ratio of liquid sample and combustion gases exceeds a critical ratio of about 1:5000. Despite the complexity of this system, the response time is fast, and only about one second elapses between the introduction of a sample solution and the signal reaching the readout system. A steady reading, corresponding to dynamic equilibrium in the flame, should be reached in 7–10 s.

The particular advantage of the spray chamber system is that homogeneous combustion, ensured by premixing the gases, makes for a nonturbulent or laminar flame. Good stability results together with well defined temperature zones, so that the region of the flame (observation height) where best sensitivity is produced can be selected.

In direct injection burners, the nebulization process takes place in the burner itself. The oxidant, which draws in the sample, is also mixed with the fuel gas in the flame, resulting in non-homogeneous combustion and a high degree of turbulence. Such burners are not well adapted to give the long narrow flames which produce highest sensitivities in atomic absorption.

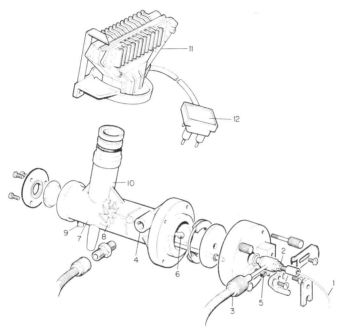

Fig. 9. Sampling unit (from Pye Unicam SP9 series): 1. sample uptake capillary; 2. nebulizer; 3. oxidant inlet; 4. fuel inlet; 5. bead holder; 6. impact bead; 7. spray chamber; 8. flow spoiler; 9. drain; 10. burner stem; 11. burner head; 12. safety interlock.

Pneumatic nebulizer

A nebulizer unit is shown in section in Fig. 10. For a given gas pressure the efficiency and the droplet size distribution given by this device depend almost entirely on the capillary diameter and the relative positions of the end of the capillary and the nose-piece. Best performance may thus be attained by careful adjustment of the position of the capillary, usually by the screw insert which holds it.

Although nebulizer units are constructed robustly to withstand mechanical deformation, they may be damaged by acid attack on the surfaces exposed to the sample solutions. The incidence of acid solutions at high speed with a high proportion of entrained air is well known to be very highly corrosive. Stainless steel will withstand mild acid attack under these conditions but will be rapidly dissolved away by solutions containing 1% ferric ions in hydrochloric–nitric acid. The best nebulizers are therefore fabricated from much more inert material. Platinum–iridium alloy is often used for the central capillary and tantalum, platinum or fluorinated polymer for the nose-piece and annulus.

Nebulizer units are designed to operate at maximum efficiency, i.e. give the largest proportion of fine droplets, for a given oxidant flow rate. If the flow rate is either reduced or increased the efficiency is likely to be impaired, and at higher flow rates the analytical sensitivity will not increase in proportion, but may well decrease. Design parameters include the capillary dimensions and nose-piece geometry.

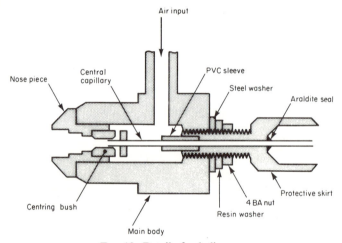

FIG. 10. Detail of nebulizer.

There appears also to be an optimum solution take-up rate for this basic form of nebulizer. If, for a series of nebulizers designed to give maximum efficiency at various take-up rates, the best absorbance is plotted against the take-up rate, a curve envelope of the form shown in Fig. 11 is obtained. The maximum occurs between flow rates of 3 and 6 ml min^{-1} and corresponds to a nebulization efficiency of 10%. Higher absorbance can never be obtained by further adjustment.

These particular values apply to dilute aqueous solutions, but the nebulization rate is also affected by variations in sample viscosity and surface tension and by the equilibrium temperature of the spray inside the spray chamber.

The actual sizes of droplets produced by a nebulizer vary very widely and the diameters range from $<5\,\mu$m to $25\,\mu$m or more. The purpose of the spray chamber is to limit the size of droplets reaching the burner to just those which can be vaporized and atomized in the flame, i.e. of the order of $10\,\mu$m or less, and thus contribute to analytical sensitivity. If a spray chamber prevents vaporizable droplets from entering the flame, analytical sensitivity will be decreased, while if it allows unvaporizable droplets to reach the flame, the flame noise signal will be increased and the flame temperature reduced. The maximum useful population of droplet sizes constitutes about 10% of the total mass of sample nebulized, and this is why this figure also represents the maximum attainable efficiency already quoted.

To improve the nebulization efficiency various ways and means of altering the normal droplet size distribution have been employed. One of these is the application of heat, either to the sample and gases before they enter the spray chamber, or in the spray chamber itself. Heating of the sample produces a minimal effect until the temperature approaches the boiling point. The reduced pressure applied to the sample solution in the nebulizer then results in pre-boiling

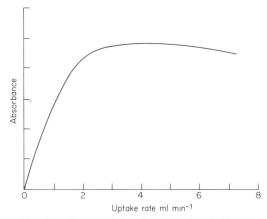

FIG. 11. Characteristics of pneumatic nebulizer.

and erratic nebulization. It has been proved[624] that the heating of the oxidant gas can result in a useful increase in sensitivity, but unless the temperature of the flowing gas is carefully controlled, which is not easy to achieve, the nebulization rate is not constant.

Heating of the carrier gas or of the spray chamber itself results in instability or drift in signal output and there is a more pronounced tendency to carry-over or 'memory' from one sample to the next.

Impact bead and counter-flow nebulizers

Two forms of mechanical device have been introduced into the spray chamber in order to utilize more effectively the momentum of the droplets themselves to produce a comminuting action. In one, a bead or bar is placed close to the orifice of the nebulizer. The droplets, whose speed at this point is near sonic, are fragmented by impact and the mass of material vaporized in the flame can be increased by 50–100%. The material from which the bead or bar is made must necessarily be very inert chemically. A stainless steel bead may be coated with the inert plastic 'penton' or the bead simply moulded from polyvinylidene fluoride.

In a device described by Feldman[225] the oxidant/sample aerosol and fuel nozzles are placed in opposition within the spray chamber (Fig. 12). This also results in a high speed turbulence which produces a larger proportion of the smaller sized droplets. Maximum improvement in nebulization efficiency is obtained when the mass flow rates and velocities of the opposing streams are approximately equal. This produces a region of interaction midway between the nebulizer and fuel jets where the velocity of both streams is effectively reduced to zero and the liquid particles are thus further broken up by the high acceleration forces. Increases in sensitivity of two or three times have been obtained with this system. However, it appears that the maximum improvement occurs fairly sharply at a critical distance between the jets (which should lie on a common axis) and the device is not readily adjusted or maintained in adjustment.

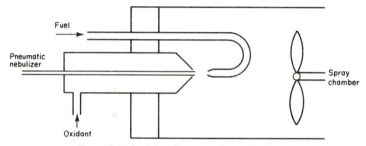

FIG. 12. Principle of counter flow nebulizer.

Ultrasonic nebulizer

A different approach to the problem of producing the largest proportion of volatilizable droplets from a solution is by the use of ultrasonics. While ultrasonic devices can attain higher nebulization efficiency they generally do so at a lower rate of nebulization. This means that they are able to convert 40% or more of small volumes of sample to a mist of useful droplet size, but are not able to produce this mist at a sufficient rate to give higher sensitivities when sample size is not a limiting factor.

Two basic ultrasonic systems are known—liquid coupled and vertical crystal. Ultrasound is produced by the piezo-electric effect on certain crystals, and the crystal vibrations have to be transmitted with the least possible loss of energy to the solution being nebulized. A liquid coupled system was described by Stupar and Dawson.[710] Basically, the sample liquid, see Fig. 13, is held by a diaphragm in contact with a coupling medium (usually water) through which the ultrasonic vibrations are focussed by means of a concave crystal. The mist produced is swept away to the burner by a current of air. Frequencies used were 115 kHz and 70 kHz, and the majority of droplets were in the 17–21 μm diameter range.

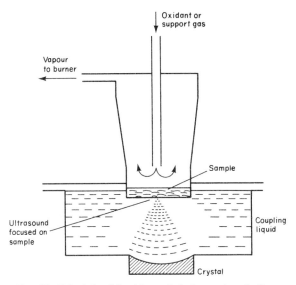

FIG. 13. Principle of liquid coupled ultrasonic nebulizer.

A particular drawback of this device is that, while small samples can be handled, they cannot be fed in continuously as with the standard pneumatic type. To overcome this problem, the vertical crystal nebulizer was devised. Here the sample solution is injected continuously at a constant rate directly on to the vertical face of the vibrating crystal. The crystal must therefore be chemically inert and quartz is best in this respect. Stupar and Dawson[711] have also described the modification of a medical inhalation therapy instrument which employs a metal disc of resonant thickness as the nebulizing surface (Fig. 14). Nebulization efficiency of this system is said to be 50% at a sample flow rate of 0.1 ml min^{-1} and falls off to 24% at 3 ml min^{-1}.

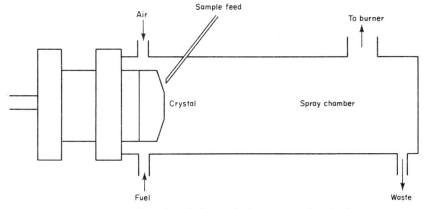

FIG. 14. Principle of vertical crystal direct ultrasonic nebulizer.

These workers' conclusion was that the stability and reproducibility of the absorbance values obtained with the pneumatic nebulizer are better than with an ultrasonic system, the better efficiency of the latter being offset to some extent by its slightly erratic performance. They also suggest that, unless higher frequencies can be employed, e.g. 500 kHz or more, with the corresponding power to generate fine mist with droplet sizes not greater than 5 μm at sample flow rates of up to 5 ml min^{-1}, the sensitivities and even interference effects are likely to be worse than with the pneumatic nebulizer.

Although ultrasonic nebulizers are used in the sample handling sections of some emission spectrometers in which excitation of liquid samples is achieved in an inductively coupled plasma, they appear to offer no advantages over the conventional simple pneumatic nebulizer for the normal continuous mode of uptake of sample solutions in atomic absorption. Indeed, all commercial atomic absorption instruments are still fitted with pneumatic nebulizers. Some special types of application have been described, however, and it is apparent from the literature that some teams of workers are continuing their investigations into this subject.

Denton and Swartz[194] improved the ultrasonic nebulizer for generating high density aerosol dispersions achieving conversion rates of better than 1 ml min^{-1}. Various droplet sizes could be generated by overtone operation. The device was used with an oxyhydrogen flame and was claimed to give significantly better detection limits than a standard nebulizer.

The ultrasonic nebulizer was also employed for the pulse nebulization of small volumes of solution,[715] an earlier claim for enhanced sensitivity having been made for pulsed nebulization of 200 μl of sample.[428] The form of ultrasonic nebulizer more commonly met with in ICP operation, i.e. in a system also comprising a temperature controlled heater and condenser as a desolvation system, was stated to have a sample efficiency of 85% and desolvation efficiency of 72% when operated at optimum temperature and flow rate.[353]

However, the claim that 1–3% salt solutions are handled without burner clogging or memory effects is not particularly likely to interest users of commercial atomic absorption instruments with good pneumatic nebulizers.

Spray chamber

In addition to allowing the selection of sample mist of the wanted droplet size, the spray chamber also allows the sample mist, oxidant and fuel gases to become thoroughly mixed before passing on to the burner, the fuel preferably entering tangentially. Auxiliary oxidant, that is, further oxidant which is required to support the flame but which is over and above that required to make the nebulizer work efficiently, may also be introduced at this point.

The gases are mixed and the larger droplets settled out by imparting a rotary motion to them. Either one or more of the gas inlets and nebulizer are positioned tangentially, or else fixed vanes are inserted as shown in Fig. 9. One tube leads away from the spray chamber to take the mixed gases to the burner and another to remove deposited liquid to waste.

If the nebulizer and spray chamber functions are properly matched, there should be no droplets whatsoever deposited in the burner tube or in the burner itself during the functioning of the instrument. The correctly formed mist is virtually dry and should not condense on any surface. One school of thought even suggests that by the time the mist has reached the burner the solvent has completely evaporated, leaving small suspended solid particles in a partially saturated vapour. This may well be true in view of the surface activity of such small droplets.

The waste tube leads first to a liquid trap, the head of which prevents the gases from escaping and also ensures a constant small excess of pressure in the spray chamber. For combustion mixtures using air as the oxidant a liquid head of about 2 cm is adequate, but for mixtures using nitrous oxide this must be increased to 4 or 5 cm otherwise the back pressure at the smaller burner slot will cause the gas to bubble out with consequent loss of flame stability and risk of explosion.

The size and shape of the spray chamber, within limits, should have little effect on sensitivity provided that it is working at optimum efficiency, and this is borne out in practice. The removal of the flow spoiler vanes, if these are employed, may result in an apparent small increase in sensitivity, but this will almost certainly be accompanied by an increase in flame noise with resulting worsening of detection limits.

It is also important that construction materials within the spray chamber should be easily wetted, particularly those at the end supporting the nebulizer and those forming the flow spoilers. Non-wetting materials like PTFE allow large droplets to reform on the surface, and these dislodge again resulting in a higher degree of flame noise. A suitable material is high temperature (γ) alumina.

As the gases normally passing through the spray chamber constitute an explosive mixture within the confined space, all spray chambers should be provided with a safety device in case the flame burns back from the burner head. This may take the form of a rupturing membrane, or one end of the spray chamber (preferably the one pointing away from the operator!) may simply consist of a push-fit bung. Explosions of nitrous oxide and acetylene mixtures being somewhat more forcible than air acetylene, it is also a wise precaution to place a metal cage around the spray chamber if the construction of this makes it likely to rupture, and restrain the burner head from flying off, if this also is a push-fit, by anchoring with a strong wire loop.

Direct introduction of powdered solids

Few attempts to introduce solid samples directly into the flame in atomic absorption analysis have been successful. Two methods described by Russian workers, however, should be recorded. L'vov and co-workers[384] placed 50 mg of solid sample in a porous graphite capsule which, when closed with graphite powder, was placed horizontally in an air acetylene or nitrous oxide acetylene flame. Atomic vapour then diffused through the graphite and its absorption was measured above the capsule. Because particulate matter cannot diffuse there was no background scatter to be corrected. The detection limit for many elements was between 10 and 100 ppb.

A pneumatic sampling device for direct atomic absorption analysis of powders[278] consisted of a vertical tube, in which the powder was placed, coupled to a 50 Hz vibrator. The 'fluidized' sample was fed into the air inlet nozzle at a rate which was said to be reproducible to 8–10%.

In a study on the introduction of solids into the flame as suspensions of finely ground powder, Willis[812] concluded that only particles of diameter less than 12 μm contribute significantly to the observed absorption. Atomization efficiency is usually lower than in a solution of similar concentration, and is unfortunately variable between sample types making accurate quantitative work impossible.

Burners

The main criteria affecting burner design have been reviewed by Hieftje[318] and by Baker et al.[73] The overriding requirements, particularly for a commercial manufacturer, are the maintenance of a large safety factor and ability to pass solutions of high dissolved solids content without blockage.

In order to avoid corrosion by the usually acidic solutions prepared for atomic absorption analysis, the burners are constructed from stainless steel or titanium. There is a growing tendency for the inner surfaces of the burner shell (see Fig. 9, p. 33) to be coated with an inert plastic material, e.g. polyphenyl sulfide, which is capable of withstanding operating temperatures up to 300 °C.

Burners are designed specifically for the various combustion gas mixtures employed. The major consideration is that the flame propagation velocity should not be greater than the gas velocity through the burner slot, otherwise the flame may flash back down the burner stem and into the spray chamber, with possibly disastrous results. This is often modestly referred to as a 'blow-back'. The flame is stabilized also by what Gaydon and Wolfard[276] called the quenching effects at the walls of the burner slot or holes. This is the cooling effect of the walls on the reacting gases. The quenching distance for a pair of parallel plates and the quenching diameter for a row or pattern of holes is that which will allow a given flame to be just held at the mouth of the burner. Quenching diameter is normally greater than quenching distance.

In addition to the heat-quenching effect of the walls of the burner slot or holes, the latter also exert a drag or friction on the moving gas. Figs. 15 (i) and (ii) represent the velocity v of gas in the burner orifices, in which the distances between the walls are a_1 and a_2 where $a_1 > a_2$. Thus the total gas flow, which is proportional to $\int_0^a v \, da$, the area under the curve, is not proportional to a.

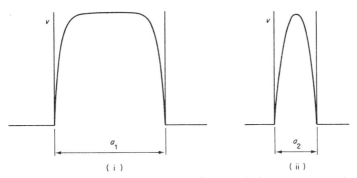

FIG. 15. (i) Gas velocity across wide burner slot. (ii) Gas velocity across narrow burner slot.

Hence flow rate is not proportional to slot area, and the smaller the area, the lower is the flow rate in proportion.

The effective burner mouth thus depends on the actual area and the shape and size of the holes or slots. In particular, for two burners of the same area, one consisting of holes will have a smaller effective area than a single slot; one consisting of two or more narrow slots will have a smaller effective area than a single wide slot; and a burner consisting of a long narrow slot will have a smaller effective area than one half as long but twice as wide.

The general form of burners for atomic absorption provides laminar flow and a long path length flame. In a laminar flow burner the direction of flow is generally perpendicular to the flame front. In turbulent flow, the direction is not well defined. Although it may appear that the longer the path length, the higher the absorbance will be, this is true only insofar as the effective slot area allows a given gas flow rate to be maintained. Higher absorbances are therefore

TABLE 3. Compatibility of burners and gas mixtures

Burner	Slot length	Gas mixture supported
Air acetylene	10 cm	air acetylene
		air hydrogen
		argon hydrogen (see p. 46)
Air propane	10 cm	air propane
Multislot ('Boling burner')	10 cm	air acetylene (with auxiliary air)
		air propane
		air butane
		argon hydrogen (see p. 46)
N_2O acetylene	5 cm	N_2O acetylene
		air acetylene
		N_2O propane
		N_2O hydrogen
		air hydrogen
		argon hydrogen

produced by limiting the slot width, so that the absorbing species are concentrated in the narrow light envelope. This principle is illustrated by the direct substitution of a 5 cm path length 'nitrous oxide' burner for a 10 cm air acetylene burner for an air acetylene flame. The absorbance values are reduced by only about 10% due to the effect of the smaller slot area on the flow rates.

While it is always advisable to use the correct burner for a given gas mixture, it is sometimes possible to use a burner for a slower mixture than that for which it was designed, but it is imperative that it not be used for a *faster* burning mixture, otherwise a blow-back will occur.

In Table 3 the most common types of burner presently used with atomic absorption spectrometers are listed together with the flames they will support. In general, Meker-type emission burners may only be used for the flame they are intended for. It is, of course, a wise precaution to consult the manufacturers of any burner before using it for a different flame from that intended.

Burners for use with solutions having high total dissolved solids

A limitation of most single slot burners, or burners where the slot is replaced by a single row of circular holes, is that the maximum concentration of dissolved solids in the samples to be atomized may be of the order of only 2%. This applies particularly to burners with slots of small dimensions, e.g. air acetylene and nitrous oxide acetylene. Solid solute tends to build up on the underside of the jaws more or less quickly, producing a drift in the flame characteristics.

Burner blockage was greatly reduced in the air acetylene flame by Boling[107] using a triple-slot burner and a higher throughput of oxidant gas. Multislot burners provide a smaller quenching distance than would a single slot burner of the same total slot area, but need a greater gas flow rate than standard single slot burners, the amount of oxidant required usually being more than that which can pass through the nebulizer. An auxiliary supply of air, bypassing the nebulizer, is then introduced into the spray chamber, mixed and passed to the burner. Multislot burners can usually allow up to 12% of dissolved solids to pass evenly to the flame without fear of build-up on the burner jaws.

For some elements, the sensitivities produced in the multislot burner are worse than in the single slot and for some other elements they are better. Two opposing effects appear to be in operation. The species being measured is effectively diluted by the higher volumes of combusting gases and the increase in equivalent slot width, but this can be more than compensated for by the increase in flame temperature of the central slot which is protected by the two outer slots. This applies particularly to those elements which are incompletely atomized at the temperature of the single slot air acetylene flame or whose degree of atomization is easily decreased in the presence of entrained air.

Since the introduction of the multislot burner, however, the design of long pathlength burners has been much refined, and a single slot burner can now approach the multislot in both performance and ability to handle more concentrated sample solutions (Pye Limited, UK Patent No. 1420599).

Blockage in burners whose jaws are of simple square cross-section is shown to be due to the turbulence induced at the lower edge of such jaws. The greater the total gas flow rate, the greater the turbulence, and the greater (or sooner) the consequent deposition of salts leading to blockage.[73] This is obviated by streamlining or 'flaring' the jaws with a curved surface of suitable radius, extending the whole length of the flame slot (Fig. 16). Salt deposits tend to be nucleated by machining imperfections on the jaw surfaces, and these should therefore be finished to a high standard and polished to a mirror-like surface.

(a)   (b)

FIG. 16. Jaw sections of modern burner (a) basic; (b) possible variation.

Such burners should be readily dismantled for cleaning, as some kinds of deposit are not readily removed by washing.

An additional problem with a nitrous oxide acetylene flame is that hard carbon deposits, which glow brightly at the temperature of the flame, build up along the slot when a fuel-rich flame is used. Such a deposit can affect the analytical results through flame instability and eventually cause a safety hazard. In the past it has been the practice to dislodge the deposit, without extinguishing the flame, with a (preferably old) screwdriver or similar implement. This is an unsafe habit as glowing particles may be pushed through the burner slot with violent results. The flame should be switched back to air acetylene or extinguished completely before the carbon is removed.

Burners designed to prevent this nuisance have a thin raised edge to the slot, also shown in Fig. 16. This enables extra air to be entrained at the base of the flame, effectively oxidizing the carbon before it is deposited. It may also raise the temperature of the slot, but this is not particularly desirable and a suitable operating temperature can be maintained by the incorporation of cooling fins.

Flames in normal use

Air Propane. This flame is now rapidly losing favour, and many of the interference effects reported in the early literature were simply a result of its comparatively low temperature. It is much less stiff than the air acetylene flame and is liable to distortion by draughts or 'flicker'. It is therefore not recommended for instruments where the flame compartment is not enclosed. For those elements which do show an improved sensitivity, this is obtained only when the flame is completely non-luminous. Under these conditions the flow rate of propane may be as low as 300 or 400 cm^3 min^{-1} with some instruments where the real air flow rate is 8 l min^{-1}.

The lean air propane flame has a very small blue inner cone which may tend to lift off the burner head. The stoichiometric flame is stable and just not luminous, while the rich flame is luminous and usually tends to be 'floppy'.

Air propane flames may be better sustained on a multislot burner, particularly if organic solvents are in use. In some countries propane may not be available—only propane butane mixtures or butane gas. Butane has roughly the same analytical characteristics as propane, but the flame will be successfully maintained only on a multislot burner.

Air Acetylene. Perhaps the most generally useful flame of all, this can be supported on the nitrous oxide acetylene and multislot burners in addition to its own burner. For most elements, best sensitivity is obtained with the stoichiometric flame, though others, notably chromium, molybdenum and tin, show best sensitivity when the flame is very rich and luminous. Under such conditions, increased noise, originating from the flame, occurs in the output signal.

The four types of air acetylene flame listed in Table 2 can be distinguished visually. The lean flame is very stiff with small inner blue cones. The stoichiometric flame is stiff with blue cones 2–4 mm in height, but is just not luminous. The blue cones can still be distinguished in the luminous flame, but tend to disappear in the rich flame, which is luminous almost to the point where sooty smoke appears from the top.

It may be useful to remember that when acetylene is taken from cylinders or tanks, these are filled with kieselguhr and acetone to dissolve the acetylene. The cylinder must therefore always be used with the valve uppermost, and it is wisest to discard the last 20% of the acetylene capacity as this is usually contaminated with acetone and the flame characteristics become different.

—*Nitrous Oxide Acetylene.* Although this flame can be burned on a 10 cm burner slot, the optimum length for most metals is 5 cm, as this incurs less cooling effect and improves sensitivity. The burner slot size is thus usually 5 cm × 0.45 mm. The burner top itself is massive and either water- or air-cooled in order to avoid the possibilities of pre-ignition and blow-back through excessive heat at the burner slot.

The flame produced in this standard type of burner is characterized by three distinct regions: the primary reaction zone, about 2–3 mm high and whitish-blue in colour; the red interconal zone which takes the appearance of a red feather (by which description it is usually known) varying in height from 0–30 mm depending on the oxidant/fuel ratio; and a blue secondary diffusion zone. Kirkbright et al.[406] have shown that the red feather, where the highest temperature is to be found, gives strong CN and NH band emission, and have suggested that the high degree of atomization of some elements in this flame is at least in part due to the more complex reactions occurring with these species and not simply to the action of atomic or incandescent carbon as is often assumed for the air acetylene flame.

In the lean flame, the red feather is less than 10 mm in height, but this increases to 10–20 mm in the stoichiometric flame, which is just not luminous. In the luminous flame, the red feather again increases in height, and the top is generally lost in the brilliant luminosity.

The high gas flow rates have in the past caused operators to show some reluctance to ignite this flame directly with a taper or lighter. It is important to remember that blow backs will not occur as long as the flow rate through the burner slot is greater than the flame propagation rate. The critical flow rate for this gas mixture with the slot size given above is about 2.5 l min^{-1}. With individual gas flows of 10 and 3 l min^{-1} of oxidant and fuel respectively there is a very large safety margin. Indeed there is still a safety margin if either of the gases fails completely. Most commercial instruments, however, are provided with self-ignition systems, which light the flame with air acetylene. With a manually or automatically operated valve, the oxidant can then be changed to nitrous oxide, the fuel flow rate being increased at the same time. This system also enables rapid interchange between the flames when several elements are to be determined, the nitrous oxide acetylene burner always remaining in position.

Nitrous Oxide Propane. Although this flame is not now considered to be one of the more important in atomic absorption analysis, it has some useful features and a few details are therefore recorded in Table 2. Precautions advised for nitrous oxide acetylene should be observed. The flame is easily controlled and there is no tendency to carbon deposition along the edges of the burner slot.

Nitrous Oxide Hydrogen. This is an oxidizing flame which tends to be more useful for elements which can be determined by their molecular, especially oxide, emission spectra. It is used, for example, for determining boron at 518 nm.[796] It can provide a carbon-free flame which improves sensitivity for some carbide-forming elements (boron may be determined in this flame by atomic absorption at 249.7 nm) or which may be used in conjunction with organic solvents which otherwise give a smoky flame.

Air Hydrogen. The premixed air hydrogen flame is of limited use, hydrogen generally being the fuel for the diffusion flames described in the following section. It provides best sensitivity for the determination of tin,[129] however, and a hydrogen-rich flame also improves sensitivity for selenium.[545]

Diffusion flames

The high degree of absorption of radiation of low wavelength by hydrocarbon flames has led to some applications of flames based upon hydrogen. These are much more transparent at low wavelengths. In diffusion flames, the fuel is surface-mixed with ambient air, and not pre-mixed with oxidant. Such flames are usually cooler than the corresponding pre-mixed flames,[117] but allow a very safe, non-explosive system to be devised.

Hydrogen diffusion flames are supported on a standard air acetylene, nitrous oxide acetylene or multislot burner and the hydrogen is diluted with an inert gas such as nitrogen or argon to stiffen the flame. When this type of flame is used with standard atomic absorption equipment no modification is required as the diluent gas may be used for nebulization. The apparatus may be set up as follows: the nitrogen or argon supply is connected to the 'oxidant' inlet of the instrument and the hydrogen supply to the fuel inlet. Extra care must be taken to ensure that the spray chamber and associated tubes are leak free because of the high diffusibility of hydrogen gas. The spray chamber is flushed for a few seconds with the hydrogen before ignition, the hydrogen flow is adjusted to about $1.5\,l\,min^{-1}$ and then ignited. After ignition, the diluent gas is turned on. A convenient flow rate for argon is $5\,l\,min^{-1}$ if an air acetylene burner head is being used, but $6\,l\,min^{-1}$ or more will be required with a multislot burner. If necessary the 'auxiliary oxidant supply' system can be utilized to increase the argon flow rate to the burner. Higher flow rates can also be used for the hydrogen.

The type of flame described here was first used by Kahn.[377] Menis and Rains[510] used a total consumption burner with the same gas mixture.

The use of argon has one particular advantage over nitrogen. Because of its lower heat capacity, the temperature of the flame is higher, and better sensitivity figures generally result. Incidentally the quenching effect is lower than that of nitrogen, and thus it is a much better flame to use in atomic fluorescence. A disadvantage is its relatively higher cost.

Separated flames

The laminar premixed flames described so far always exhibit three distinct reaction zones. After the preheat zone, these are: (i) the primary reaction zone; (ii) the interconal zone, which is small in stoichiometric flames but increases as the gas mixture becomes richer in fuel. The interconal zone is the most productive of free atoms for those elements which form stable monoxides; (iii) the secondary reaction, or diffusion zone, sometimes called the 'plume'. These were separated as long ago as 1891 by Teclu.[727] More recently Mavrodineanu[500] has made a study of the molecular reactions taking place in each zone. In the primary zone of the air acetylene flame inside the flame and above the blue region of unburnt gases, the products mainly formed are carbon monoxide, hydrogen and water. The secondary zone, a cooler outer mantle, is the diffusion flame in which these gases are burnt in the atmospheric oxygen to carbon dioxide and water.

Kirkbright and West[413] have examined the properties of the primary and interconal zones of the air acetylene and nitrous oxide acetylene flames for use in atomic absorption and fluorescence. Separation of the secondary zone was achieved mechanically in the first instance by supporting a silica tube on the burner stem with its open top about 7–10 cm above the circular Meker head. Lean gas mixtures tend to produce turbulent flames with this arrangement, and rich mixtures lead to the deposition of carbon on the walls of the tube, making observation uncertain. A better system is to surround the flame with a curtain of inert gas such as nitrogen or argon. The means whereby this is achieved is illustrated in Fig. 17. Although this shows a circular Meker burner, the same kind of device can be placed around a long path atomic absorption burner. The potential value of separated flames can be deduced from Fig. 18 in which the background emission from unseparated (A) and separated (B) air acetylene flames are superimposed, the intensity ordinate for the unseparated flame being on a scale 1/15 of that of the separated flame.

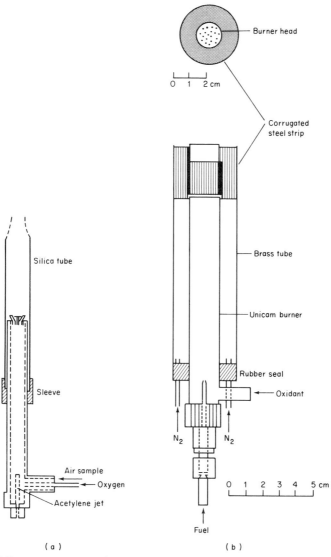

FIG. 17. (a) Burner arrangement for separated air acetylene flame employing silica separator tube. (b) Burner arrangement for separated air acetylene flame employing nitrogen shielding gas.

The OH and CO band emission in the interconal zone is at least two orders of magnitude lower in the separated flame. Such an effect will clearly help to improve signal/background ratios in atomic emission and fluorescence, and also in absorption, though more indirectly. This is because the very intense band emissions increase the photomultiplier shot noise (see p. 118). In addition they may saturate the detector or contribute markedly to flame noise by their component of amplifier modulation frequency.

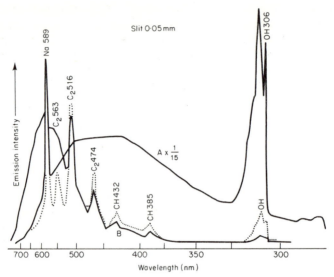

FIG. 18. Background emission from unseparated (A) and separated (B) air acetylene flame. The ordinate of A should be multiplied by 15 to compare absolute intensities. The effect of decreased fuel flow rate on emission from separated flame is shown by broken line.

Separation of the diffusion zone of the nitrous oxide acetylene flame was described by Kirkbright, Semb and West[408] and considerable improvements for a number of elements in flame emission spectrometric measurements are quoted. Indeed, detection limits for aluminium, beryllium, molybdenum, titanium and vanadium appear to be as good as or better than those expected by atomic absorption.

In atomic absorption, separation of the air acetylene flame would seem to have the most benefit for the determination of elements whose resonance lines occur at low wavelengths, especially arsenic and selenium, where the measurements are made below 200 nm. At such wavelengths, absorption by flame gases in the separated flame is much less than in the normal flame and so better sensitivities and detection limits are obtained. Separation of the nitrous oxide flame maintains the reducing properties of the interconal zone flame gases by preventing the diffusion of atmospheric oxygen. Better sensitivities are thus obtained for some of the elements, such as aluminium, boron and silicon, which form highly refractory oxides.

The possible advantages of using separated flames in analytical atomic spectroscopy were discussed in some detail by Cresser and Keliher.[162]

Discrete volume nebulization

For the majority of applications of atomic absorption where flame atomization is adequate, the continuous nebulization achieved in the nebulizer/spray chamber systems previously described is suitable. If the volume of sample solution actually available is very small, however, e.g. less than 0.5–1.0 ml, the system may not have time to attain equilibrium and produce a steady reading on meter or digital display.

The discrete volume method of nebulization (also variously known as the 'direct injection', 'aliquot' or 'pulse nebulization' method) enables a sensible reading to be obtained using a small volume of solution, e.g. 50 or 100 μl.[671]

The sample uptake capillary from the nebulizer is made to terminate in the point of a small plastic conical funnel as shown in Fig. 19. A standard micro-

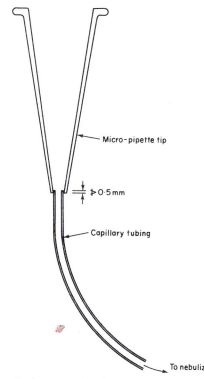

FIG. 19. Accessory for discrete volume nebulization.

pipette tip is ideal for this purpose. It will usually be necessary to trim a millimeter or so off the end of the tip until its internal diameter is fractionally smaller than the outside diameter of the nebulizer tubing in order to ensure a good seal. The tube is pushed into the micropipette only about 0.5 mm to avoid forming a liquid trap which could cause cross contamination. The tip is then mounted vertically on a suitable clip on the front of the atomic absorption spectrophotometer.

The instrument is set up in the normal way for flame analysis except that the spray chamber and burner are allowed to run dry when no sample is being passed.

The analysis is carried out by transferring 50 or 100 μl aliquots of the sample into the tip by means of a precision micropipette. The readout is obtained in the form of sharp peaks on the recorder trace. The peak-reading facilities incorporated in newer instruments can also be utilized.

Sensitivities obtained by this technique are only marginally lower than those given by conventional continuous aspiration. In instruments where the dissolved solids handling capacity of the nebulizer or burner is limited, e.g. to less than 5%, discrete volume nebulization allows considerably more concentrated solutions to be employed, e.g. up to 12–15%. Detection limits depend on the overall reproducibility attained, which should normally be of the order of 2%.

OTHER SYSTEMS BASED ON FLAMES

The main virtues of the nebulizer/spray chamber are stability and, in conjunction with the appropriate gas mixture and burner, very high reproducibility and atomization efficiency. In a conventional burner, however, any one atom remains in the light path for a very short period of time. Sensitivities could be improved if the atomic vapour could be constrained to remain longer in the resonance beam and also if the sample could be introduced without an inefficient nebulization process, or indeed without any pretreatment involving dilution at all.

A number of devices which take advantage of one or both of these ideas have been described by individuals and some are available commercially.

Long tube absorption cells

Although flame gas-flow rates cannot be appreciably decreased to improve the dwell time of atoms in the observation zone (these are dictated by flame propagation velocities) the combustion products themselves, including free atoms, can be directed along a tube, the axis of which coincides with the optical axis of the spectrometer. The success of a device of this nature depends upon the mean lifetime of free atoms. This, in turn, varies with the element, the temperature and composition of the flame gases.

A simple tube system in which the flame from a total consumption burner is directed in at one end, and the combustion products emerge from the other was described by Fuwa and Vallee.[262] The sensitivity is not proportional to the tube length for most elements as their times of transit exceed the lifetimes of the free atoms. It can, however, be improved to some extent by proper choice of the tube material and by heating the tube with a second burner or an electrical heating jacket. The tube material should be capable of withstanding considerable thermal shock and must not be corroded by the combustion products.

Silica, alumina and ceramic tubes up to nearly 100 cm in length have been used. Such tubes, especially silica, act as light guides because of the reflectivity of the internal surface. Sensitivity may therefore fall off as material from the flame condenses. Condensation also causes memory effects, i.e. the absorption falls off only gradually after the sample has ceased to be aspirated. A further difficulty is that background absorption with tubes of this type is considerable, probably because of the presence of more molecular absorbing species and the greater effect of scattering in a long absorption cell.

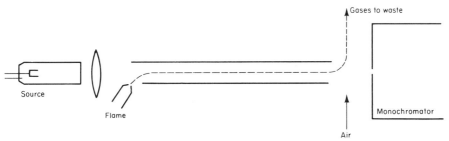

FIG. 20. Long tube absorption cell.

A successful long tube device[656] consists of an alumina tube 45 cm long, with one end cut at a slight angle so that the flame can enter when the burner is slightly tilted (Fig. 20). The burner may be a total consumption type, or simply the 'emission' burner from a spray chamber system. Best results have been obtained with an air hydrogen flame. It is desirable to be able to heat the tube up independently to about 600 °C for some elements. A stream of air must be arranged to blow the combustion products to an exhaust system.

The best sensitivities for all possible elements are achieved under fuel-rich conditions. These occur when the burner is so positioned and the flow rates of hydrogen and air adjusted so that the primary and secondary reaction zones are at the entrance and exit to the tube respectively, leaving the interconal zone to occupy the tube. The sensitivity enhancement then improves with increase in dissociation energy of the metal–oxide bond.

Improvements in sensitivity by a factor of approximately 50 over standard flames may be expected for some elements, e.g. silver 0.002, gold 0.01, bismuth 0.025, cadmium 0.001, copper 0.0075, manganese 0.01, lead 0.02, zinc 0.0015, tin 0.04 (all in ppm per 1 % absorption) and others. The operation and results given by this system are described in more detail by Rubeska and Moldan.[654,655] These workers also mention that some elements which form stable oxides give improved sensitivity in the tube, which under the conditions mentioned protects them to a great extent from atmospheric oxygen.

Complete protection from atmospheric oxygen was obtained by the use of a long absorption tube with separated flames[319] (p. 46). A long side arm was provided on the flame separator, along which the partly combusted gases were drawn. An optical window was provided at each end, and one tube was surrounded by a furnace which maintained its temperature at 1100 °C. Combustion gases were air acetylene with auxiliary hydrogen. Sensitivities using this separated flame system were at least comparable with the other long tube systems, but there was less background, less noise and less deposition of material in the tube.

Flame adapters

The long tubes described above suffer from the disadvantage that they cannot be accommodated in the burner position of a standard atomic absorption spectrometer. A device which uses the existing long path burner both to atomize and maintain the element in the free atomic state and which can also be made to handle small solid samples directly is the tube flame adapter. As developed by White[794] this consists of a nickel tube about 10–12 cm long and 1 cm diameter supported in the optical axis with the burner positioned about 2 cm below. A hole 5 mm diameter is made midway along the wall of the tube nearest the burner (see Fig. 21). The flame is lit, heating the tube, and the sample is introduced on a platinum loop or a miniature crucible (the Delves cup)[187] between the flame and the hole. The sample becomes partially atomized and the atoms are retained in the optical path by the tube for considerably longer than with a conventional burner.

Only comparatively cool flames such as air propane or the very lean air acetylene flame could be used with this device as first conceived. This limited its

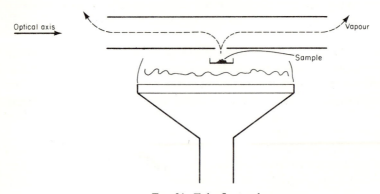

FIG. 21. Tube flame adapter.

use to measuring those elements which are atomized at low temperatures, such as lead, cadmium and thallium. Very good sensitivity is in fact obtained for lead, e.g. less than 10 μg/100 ml of whole blood, but since reproducibilities are not usually of the highest order its most common use has been as a readily applied screening method.

In more recent commercial versions of this device the nickel tube is replaced by a ceramic tube allowing the use of hotter flames. It has even been possible to use a nitrous oxide acetylene flame[374] in conjunction with a silicon carbide absorption tube and molybdenum cups sheathed by nitrogen to prevent oxidation. Absolute detection limits were quoted for some fifteen elements ranging from 0.01 ng, 0.08 ng and 0.1 ng for cadmium, silver and lead to 10 ng and 20 ng for nickel and chromium respectively.

The sampling boat

Somewhat simpler in concept, but based on the principle of evaporating the sample completely and quickly in the flame in order to record a high narrow absorption peak, is the sampling boat.

The boat itself as first described by Kahn *et al.*[376] is about 5 cm in length and a millimeter or two wide and deep. It is made of tantalum because of the resistance to heat and good conductivity of this metal. A multislot burner is preferred as it heats the boat more uniformly and produces a smooth flow of the combustion gases around it. A simple device is made to enable the boat to be loaded about ten centimetres away from the flame, then pushed close to the flame to dry the sample, and finally placed in the correct position within the flame.

If the sample is aqueous, it must be completely dried before it enters the flame, otherwise some is lost by forceful evaporation. Many samples can nevertheless be prepared as solutions, with standards made in the same way. Solid samples, particularly those containing organic matter, can either be previously wet-ashed, or, more conveniently with the very small samples for which the technique is designed, dry-ashed in the boat itself close to the flame.

Although the air acetylene flame may be used, the sample itself can never become as hot as the flame and the method is therefore limited to elements which are readily atomized such as arsenic, cadmium, indium, lead, mercury, selenium, silver and thallium. Fortunately, these are also the common toxic metals and this method has been used[748] to detect small traces of some of them.

The absorption response curve reaches its peak in 1 to 2 s depending on the element, and returns to the baseline after a total time of about 4 s. The instrumental electronics and, in the absence of an integrating circuit, the chart recorder, must therefore have a rapid response and the area under the response curve should be summed or integrated.

Detection limits depend upon the type and volume of sample used. A typical boat, capable of holding 1 ml of solution before it is dried off, could give a detection limit of 0.0001 μg ml^{-1} for silver or cadmium—some fifty or more times better than the conventional method.

HYDRIDE GENERATION AND REDUCTION METHODS

The chemical properties of some elements are such that special methods can be employed both to separate them effectively from the main sample matrix before their introduction into the light path of the atomic absorption spectrophotometer, and to convert them into an atomic vapour once they are there. These elements include antimony, arsenic, bismuth, germanium, tin, selenium, tellurium, and even lead which readily, or comparatively readily form volatile hydrides. The hydride is generated by chemical reduction of the sample, and is then entrained in a current of argon and led into the observation zone where it is decomposed by heat to form the atomic vapour. A number of methods in use are based on this principle, but they differ in the ways both the reduction and the atomization are carried out.

Some of the earlier methods used a conventional zinc/acid reduction, together with some kind of collecting device for the hydrides produced. Holak[324] collected arsine in a U-tube in a nitrogen trap and Manning used a rubber balloon for the same purpose.[479] Titanium III chloride/hydrochloric acid and magnesium/zinc were used to extend the method to bismuth, antimony and tellurium, and sodium borohydride was used for germanium.[594,595] For some elements, particularly tin, lead and tellurium, the hydride formation is comparatively slow and hence a collection vessel may be necessary. Other workers[741] have obtained adequate sensitivity without this facility. Sodium borohydride is now generally used as the reductant, the sample having been brought into solution, after suitable wet oxidation if necessary, and brought up to 1 M with hydrochloric acid.

Much of the literature on the hydride generation method fails to mention the possibility of interferences by other elements present in the sample. Such elements usually influence the rate or extent of the hydride formation, though if a cool flame is used as the atomizing medium there is also a possibility of sensitivity reduction by the rapid formation of other compounds in the flame. A. E. Smith[691] has made a comprehensive review and study of interferences in the determination of seven of the elements (not lead). Fleming and Ide[237] investigated interferences in the method as applied to steels.

The generated hydrides are decomposed in the optical path in one of three ways. The simplest is direct introduction into a cool flame, e.g. argon/hydrogen. In instruments provided with an auxiliary oxidant inlet this is simply made to pass first via the reduction vessel where the hydride is to be formed. Argon is connected to this line and hydrogen to the fuel inlet. The normal oxidant supply is also replaced by argon. The device is represented in principle in Fig. 22. If the auxiliary inlet is not available, a reaction vessel bypass should be fitted in order to maintain flame stoichiometry.[236]

Alternatively, as in the method of Thompson and Thomerson,[741] the hydrides, entrained in the argon stream, are passed through a silica tube which is coaxial with the resonance beam and which is heated by an air acetylene flame supported on a multislot or other wide-path burner. Better sensitivities are claimed for this method, presumably because of the longer residence time of the atomic vapour in the resonance beam. A special tube was designed in which end windows were replaced by streams of nitrogen which induced the hydrogen to ignite away from the actual light path.

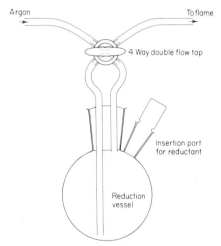

FIG. 22. Principle of basic hydride generator.

In a third method of atomization[148] the hydrides are passed into a similar long-path silica cell, e.g. 15 cm long and 2.5 cm diameter, in this case heated electrically.

When atomization is carried out in cells of this kind it would seem to be important to use the longest possible heated pathlength, but also the smallest possible diameter consistent with not interfering with the light beam or affecting energy throughput. Thompson and Thomerson's cell was, for example, 17 cm long but only 0.8 cm in diameter.

Experimental procedure

The sample solution, prepared by a wet oxidation procedure suitable for the particular samples in hand, is made up to 1 M in hydrochloric acid. It has been noted that the response to arsenic, bismuth, germanium, antimony, selenium and tellurium is not very dependent on acid concentration in the range 1–4 M when sodium borohydride is the reductant,[741] though for tellurium the response increases somewhat through this range. The sensitivity of tin decreases markedly when the acid concentration is greater than 0.7 M. For lead both the acid and sodium borohydride concentrations are critical. Because of this fact and its relatively poor sensitivity the determination of lead by hydride formation should be considered only in exceptional circumstances.

For all the above elements except lead, the concentration of sodium borohydride also is not very critical. 2 ml of 1 % m/v sodium borohydride for a 1 ml volume of sample solution is recommended. For greater volumes of sample, e.g. waters up to 20 ml with arsenic or selenium up to 0.2 ppm, sodium borohydride pellets, 98 % (Aldrich Chemical Co. Ltd., Gillingham, Dorset, UK, or Ventron Corporation, Alfa Products Div., Danvers, Mass., USA) are extremely convenient. If difficulties are experienced in the preparation of stable solutions of sodium borohydride, it has been suggested[416] that one pellet of potassium hydroxide be added for each 100 ml of 2 % m/v sodium borohydride solution and the cloudy liquid vacuum-filtered through a 0.45 μm membrane. The solution may then be used for three weeks.

Mercury

Mercury is unique among metallic elements in that it not only has an appreciable vapour pressure at ambient temperatures but the vapour is stable and monatomic. Hence, mercury produced in elemental form by reduction of its compounds can be entrained in a stream of inert gas, or even in air, and measured by the atomic absorption of the cold vapour without the need of either flame or flameless atomizers. It was indeed determined in this way[77,829] long before the advent of the atomic absorption techniques which are the principal subject of this book. Correction for broad band molecular absorption with a continuum source was also anticipated.[76]

In current methods, samples are oxidized, often with sulfuric acid and permanganate, so that organic matter is destroyed and mercury is converted to mercuric sulfate. Excess permanganate is removed with hydroxylamine hydrochloride and mercury vapour is generated by the action of stannous chloride. It may be collected prior to measurement by amalgamation on gold[754] or copper[115] or simply entrained in an air stream and passed directly into an absorption cell.[308] Since 1968 very many variations of the cold vapour method have been published. These variations arise from procedures of releasing mercury from chemical bonding in the original sample and devices for trapping or collecting the mercury vapour prior to its being transferred to the absorption cell. In some methods all the collected or generated mercury is allowed to pass through the absorption cell and thence to extraction, resulting in a sharp absorption peak. In others the vapour is retained in a circulating system (Fig. 23) so that the absorbance rises to an almost noise-free plateau.

Methods for mercury determination have recently been reviewed by Ure.[755]

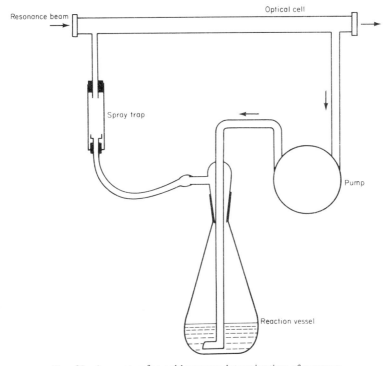

Fɪɢ. 23. Apparatus for cold vapour determination of mercury.

Further details of the methods and individual applications are given elsewhere in this book.

Apparatus and Method. In the simple recirculating system of Fig. 23, air is circulated with a small diaphragm pump (Charles Austen Capex Mark II fitted with polypropylene inlet and outlet pipes) at about 1 l min^{-1} through, in the following order:

(i) The reaction vessel. This should have a volume convenient to the sample preparation procedures to be used, e.g. a 150 ml conical flask is suggested for the tissue analysis method on page 276 but a large volume of air above the reaction solutions is undesirable, and this should be fitted with a B24 ground glass socket to take a Drechsel bottle head.

(ii) A spray trap consisting of cotton wool and a porosity G2 sinter (some workers prefer a magnesium perchlorate drying tube).

(iii) The absorption cell with longest path length and minimum volume possible, consistent with transmission of sufficient energy to the detector. A length of 15 cm and diameter of 0.75 cm would be suitable for many atomic absorption spectrophotometers. Windows should be made of ultraviolet transmitting glass. Alternatively a standard 10 cm ultraviolet spectrophotometric gas cell can be used but this may give a sensitivity lower by a factor of about 2 than that which would otherwise be possible. The burner itself acts as a convenient support for the cell which can be fixed either with adhesive tape or with a specially made bracket.

(iv) Thence back through the pump.

The atomic absorption spectrophotometer is set up to make absorbance readings on the mercury line 253.7 nm.

The clear prepared sample solution, containing 0–300 ng of mercury as mercuric sulfate is transferred to the reaction vessel and the reductant is added. This may be 2 ml of 10% stannous chloride solution or an equivalent amount of sodium borohydride solution.[639] The Drechsel bottle head is immediately inserted and the circulating pump switched on. The mercury is then liberated in monatomic vapour form and is swept through the cell. The absorbance value rises to a plateau, reaching a steady value in about 1 min. When this constant value has been obtained and recorded, the reaction vessel is disconnected from the system and the vapours allowed to escape through the extraction hood. When the absorbance returns to zero the next sample or standard can be reduced and inserted.

Sensitivity and Accuracy. Under the conditions described, an absorbance of unity should be obtained with about 1 μg of mercury in the system. In the absence of a flame, high scale expansion factors should be possible and the detection limit should be of the order of 1 ng. The repeatability of the measurements depends directly on the constancy of the total volume of recirculating air. Thus flask volumes and prepared solution volumes should also be the same during a particular analysis run. A single reaction vessel for all the sample solutions is not usually convenient.

Another consideration is that the mercury in the vapour phase is in equilibrium with that in the liquid. The larger the volume of liquid the less will be the concentration of mercury in the vapour phase. If the liquid volume is variable in a system of constant total volume, these two sources of error do, to some extent, cancel. For maximum sensitivity, however, both total volume and liquid volume should be as small as possible. With a 50 ml reaction flask, for example, the absolute detection limit would be improved by a factor of two or more.

Interferences. Ions normally precipitating mercury from solution, e.g. iodide, tend to depress or suppress the evolution of elemental mercury, though these are unlikely to remain after the sample has been wet oxidized. Most metal ions up to 100 ppm do not interfere.

A rapid method for mercury determinations using the cold vapour principle and an ingenious dual bubbler system for sample reduction has been described by Simpson and Nickless.[686]

Combined function

A logical development from the foregoing is a reduction apparatus suitable for the formation either of the arsenic group hydrides or of mercury vapour. Such an equipment, made by Berghof in West Germany, is constructed entirely of PTFE and allows all parameters, e.g. inert gas flow, tube temperature, pump and stirrer speeds, to be optimally adjusted.

NON-FLAME ATOMIZATION

Combustion flames, though cheap to produce, stable in operation and, depending on the gas mixture used, able to give a wide range of temperatures, nevertheless have certain serious disadvantages. Chief of these is that the atomic vapour always contains other highly reactive species. It is therefore not possible to predict with any certainty exactly how a given mixture of elements may respond in absorption or indeed how a non-absorbing species may affect or interfere with the elements to be measured. Many attempts have been made to produce the atomic vapour in a completely neutral or unreactive medium, and various electrical methods have been proposed to introduce the necessary amount of heat energy into the system. Almost without exception the initial cost of the electrical equipment required for these methods is comparatively high, as in order to avoid the effects of selective volatilization at the atomization stage large amounts of energy must be dissipated in a time which is short compared with the total time of measurement.

Non-flame methods fall into two main categories, resistively heated devices and electrically induced plasmas.

ELECTROTHERMAL ATOMIZERS

The forerunner of all electrothermal atomization devices as we know them today was the furnace of A. S. King[395] of the Mount Wilson Observatory (1908). This was a graphite tube heated to about 3000 °C in an atmosphere of hydrogen (generated from zinc and sulfuric acid!). The neutral spectra of a number of elements were studied, sodium, calcium, iron, copper, mercury and caesium at that time and progressing to tungsten and rhenium in 1932. R. B. King used a wound resistance furnace of alundum at 1400 °C to determine absolute *f*-values of copper and cadmium[396] and iron.[397]

The use of a graphite furnace for generating atoms as the absorbing medium in analytical atomic absorption spectroscopy was first described by L'vov.[465] In his first furnace the sample to be analysed was placed on an electrode which was introduced into the horizontal graphite tube through a hole in the underside. This furnace was 5–10 cm in length and 3 mm in diameter. Although the tube was heated, the atomization temperature was attained by striking an arc between the sample electrode and the graphite furnace tube. In later models[466] the sample was atomized by passing a current through the narrowed section of the electrode to the furnace tube, which was itself heated to prevent rapid recondensation of the atomic vapour (see Fig. 24). This device was placed in a purged chamber and a batch of prepared sample electrodes placed on a turntable. It is interesting to note that L'vov applied simultaneous background correction using a deuterium continuum in what might now be called a dual channel spectrometer. He also used inert metal liners in his earlier furnaces and pyrolytic graphite later. He emphasized particularly that his graphite tube acted as an atom cell rather than as a vaporizer.

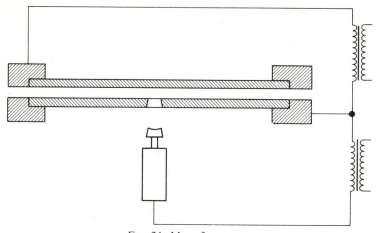

FIG. 24. L'vov furnace.

The use of this type of atomizer as a serious tool in analytical atomic absorption does not appear to have been considered until 1968 when both Woodriff[828] and Massmann[491] published work on enclosed graphite furnaces which may be looked upon as directly developed from L'vov's ideas, though simplified in that their source of heat was their own electrical resistance.

Since that time many electrothermal atomizers have been described in the literature and a number of these have been made available commercially. They fall into three main groups:

tubular graphite atomizers of the Massmann or 'mini-Massmann' type;
open filament atomizers of graphite or metal;
vertical crucible or 'well' furnaces.

Graphite tube atomizers

In Massmann's furnace[491] and the commercially available form first described by Manning and Fernandez[482] the tube dimensions are approximately 5 cm length and 6.5 mm diameter (Fig. 25). A 2 mm hole is placed halfway along for the introduction of liquid samples of 2–200 μl volume. The tube is supported between water-cooled steel end-cones which make the electrical connections to a low voltage supply and which are insulated from the mounting. The whole device is held within a metal casing which contains an inert atmosphere, e.g. argon, to prevent oxidation of the tube during operation. The operating cycle of this and the subsequent models based upon it consists of three principal steps. First, the drying step to remove the solvent would be at 100 °C if the solvent, as is usual, were water. The ashing or pyrolysis step at a higher temperature is to remove as much as possible of the organic or other matrix material. The atomization step raises the mineralized residue to the temperature at which the element to be measured is converted to an atomic vapour by one of the processes described on p. 22.

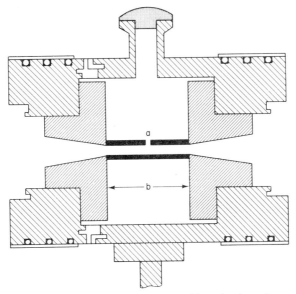

FIG. 25. The Massmann Cuvette: a, graphite tube; b, end cones.

Woodriff's furnace is of more massive construction and is operated at a constant temperature, i.e. without the temperature programme described above. Liquid samples are introduced in already nebulized form in a stream of inert gas, and solid samples are placed into a small graphite boat which is pushed into the atomizer. Some claims are still being made for the superiority of performance of this design[827] but because of its large size and lack of versatility in operation it has not received the same general acceptance as smaller furnaces.

All commercially available tubular atomizers are based on Massmann's original principle of ohmic heating. They differ from each other principally in the dimensions of the graphite tube, though there are also considerable practical differences in the design and functions provided by the power supply and in general operating convenience.

The original Perkin-Elmer heated graphite atomizer 'HGA70' had tube dimensions similar to Massmann's. A few years later the dimensions were reduced in the HGA74 to 2 cm in length and 3 mm diameter. Varian produced a 'mini-Massmann' version of their carbon rod atomizer in which the internal dimensions of the furnace were 5 mm long and 1.5 mm diameter. This device would accept only 2 μl of sample solution.

If the furnace is too large it will take too long to attain its highest operating temperatures and may give rise to greater relative background absorption effects. Nevertheless, large furnaces accept larger sample volumes and are therefore capable of better analytical precision and relative sensitivity. Very small furnaces tend to have better absorption efficiencies because of their small volumes, as the focus of the radiation beam occupies a larger proportion of their measuring volume. Because of this and their shorter atomization (temperature rise) times, they generally have better absolute sensitivities. This is offset by the limitation in sample volume which causes worse relative sensitivity and accuracy. This is also borne out by the performance of a metal micro-tube atomizer (25 mm long and 1.5 mm in diameter) reported by Ohta and Suzuki.[559,560]

Cross-section of the standard graphite tube.   Cross-section of the 'Profile' graphite tube.

Fig. 26.

Most present-day furnaces are of a size which is a compromise between these extremes. The standard graphite tube used in the Pye Unicam electrothermal atomizer is 5 cm long and 8 mm in diameter. The central (cylindrical) portion of their Profile tube, designed to conform to the shape of the radiation beam, is 3 cm long and 5 mm in diameter (Fig. 26).

Open filament atomizers

The carbon filament atomizer of West and Williams[792] received considerable interest in 1969 and following years, as this type of device appeared to offer considerable possibilities for both atomic absorption and atomic fluorescence measurements. The design was essentially simple, consisting of a carbon rod 40 mm long, and between 1 and 2 mm in diameter supported between two stainless steel electrodes which made connection to the low voltage supply.

The atomizer was enclosed in a Pyrex glass chamber through which flowed a stream of argon. Silica windows allowed the light from the radiation source to pass just above the surface of the rod for atomic absorption, and a side arm at 90° to the optical axis of the spectrometer was provided for the source radiation in atomic fluorescence. Power requirements for this atomizer were lower than for a furnace, 0.5 kW instead of several kilowatts. The system was further simplified[43] by replacing the glass chamber with a vertical flow of argon (Fig. 27). The useful sample capacity of filament atomizers is usually not more than 5 μl.

In further developments of this kind of system, the carbon rod has been replaced by a platinum or tungsten loop[118] and a tantalum boat[197] or ribbon (Barnes Engineering) which would hold 20 μl of solution. More recently, some experimentation has been done with filaments of graphite braid,[528] the advantage of which seems to be that a larger volume of sample solution can be absorbed in the braid than can be placed on the flat surface of the other open atomizers.

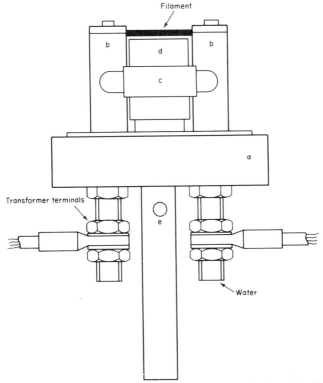

FIG. 27. Rod atomizer as used by West: a, base; b, electrodes; c, cooling water link; d, laminar flow box; e, shield gas inlet.

In operation the open atomizers require a similar temperature programme, viz. drying, ashing and atomizing steps, to the tubular atomizers.

In theory, because the atomic vapour formed above a filament-type atomizer is not constrained to remain in the observation zone as for a tubular atomizer, the sensitivity of such a device would be expected to be lower. In practice, however, the time constant of the atom supply function is extremely fast due to the low thermal mass. This gives a high initial density of free atoms with consequent good analytical sensitivity. On the other hand, because of the small sample handling capability, relative detection limits are not appreciably improved.

Vertical crucible furnace atomizers

Massmann[491] was the first to describe a vertically orientated graphite atomizer (Fig. 28). This was used specifically for atomic fluorescence measurements, the primary source radiation entering the crucible vertically downwards through the open top and the fluorescence being viewed above the sample through a slit in the wall.

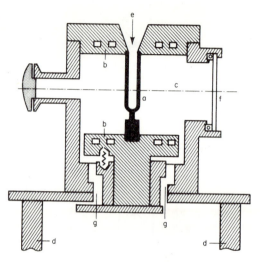

FIG. 28. Argon chamber with graphite cuvette for atomic fluorescence analysis: a, graphite cuvette; b, steel holders; c, optical axis; d, mounting holder; e, incident radiation; f, quartz window for viewing fluorescence radiation; g, insulation.

A vertical furnace has also been used by Headridge and his group[66] for work on solid metallic samples. This furnace is of the constant temperature type and is heated inductively at radio frequency. During one analytical run, a number of samples are introduced consecutively through an air-tight valve, the number being limited only by the amount of unvaporized sample residue remaining in the well of the furnace. Good results were obtained for traces of a number of comparatively volatile elements in steel and other alloys.[63]

For convenience in automatic operation the vertical configuration was chosen by George[277] in a monitoring system for trace metal concentrations in river water. Lead, cadmium and zinc were measured, with sensitivities of 48 pg, 2.5 pg and 1.5 pg respectively. The monitor operates unattended over a period of one week, continuously diluting and measuring samples and standards. Operation of the furnace, furnace loading and handling system and control unit is based on the use of programmable read-only memories.

Atomizer materials

An ideal atomizer will be constructed from a material which is chemically inert, has good thermal and electrical conductivity, can be obtained in a high state of purity, is easily machined, has low porosity and has a low expansion coefficient and high melting point.

Graphite most nearly satisfies these conditions, though some metals, particularly tungsten and tantalum have also been considered and used either as a tube material, or as a liner for a graphite tube.[71,254] Graphite is made in a process in which amorphous carbon is heated resistively to about 3000 °C. At normal pressures it sublimes at about 3500 °C. This sets the upper temperature limit for graphite atomizers. Graphite is, of course, oxidized at temperatures much lower than this in contact with air, and graphite atomizers are always operated in an atmosphere of nitrogen or argon.

Normal graphite is naturally porous and its porosity seems to increase after it has been maintained at high temperatures, e.g. as during its life as an atomizer. Another disadvantage of graphite as an atomizer material is that some elements readily form stable carbides in contact with graphite at elevated temperatures.

The form of graphite known as pyrolytic graphite overcomes both of these problems to a large extent.[12,467] Pyrolytic graphite can be formed as a coating on a graphite substrate by maintaining it in an inert atmosphere containing a small proportion of methane or other hydrocarbon, at a temperature of up to 2500 °C. A pyrolytic layer produced in this way usually cannot exceed 1 mm in thickness.

Micrographs show that the structure of the coating is crystalline and conical, growing from points on the substrate surface.[795] The coating is of high purity as impurities are not deposited, and has the required linear expansion, mechanical strength, thermal conductivity and resistivity. Its permeability is too low to be measured.

In analytical use, the reduction in permeability of the pyrolytic graphite reduces the loss of atomic vapour by diffusion through the furnace walls and improves residence time. The pyrolytic form is also less inclined to form carbides with vanadium, molybdenum, titanium, tungsten, silicon, etc. and such elements may therefore be measured with greater sensitivity.

Alternatively, pyrolytic coatings may be produced on graphite tubes *in situ* in the electrothermal atomizer by bleeding methane into the inert gas supply to the tube.[152,480,740] A 10% methane/90% argon mixture is available commercially and is used to give a 1% methane mixture at the tube, which is held at a temperature of 2000 °C. Alternatively, the 1% methane mixture may be passed during the analysis. A tube treated in this way, while for many purposes better than an untreated tube, is likely to be inferior in use to the commercially produced item, as the temperature gradient which must necessarily result from ohmic heating in these circumstances leads to a coating of variable thickness and quality.

Power supply for electrothermal atomizers

For the majority of purposes the sample is raised to its atomization temperature through the sequential steps of drying, ashing or pyrolysis followed by the atomization step proper. It is usual to add a fourth step, 'tube clean', in which any solid material remaining in the tube after the atomization step is removed by volatilization at the highest attainable temperature. A fifth step may also be provided in which the efficiency or otherwise of the fourth step may be monitored by repeating a reading at the atomization temperature.

A typical graphite tube atomizer would have a resistance of about 10 milliohms. When a low voltage, up to 7 volts, for example, is applied to the ends of such a tube a current of several hundred ampères flows, resulting in a rapid temperature rise.

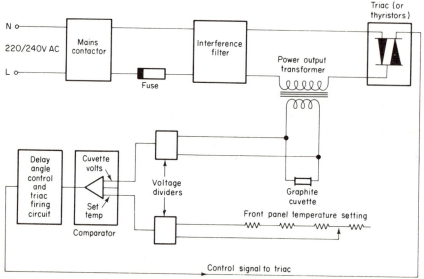

FIG. 29. Simplified block diagram of a phase-angle controlled power supply.

The function of the power supply is to provide a low voltage high power waveform which may be controlled to give the required temperature for the required time for each of the steps described.

Figure 29 shows a simplified block diagram of a basic power supply employing a phase angle control system. Power is fed directly from the mains via a triac or thyristor control to the power output transformer. The secondary of this transformer supplies the low voltage, usually not more than 10 V, to the graphite atomizer. This voltage is controlled for individual steps, the programme, by monitoring and feedback to a comparator circuit. This compares the actual voltage with that set on a front panel temperature selector and causes the triac to conduct only over the part of the mains AC waveform which will produce the desired temperature. An example of the switching of an AC waveform is shown in Fig. 30, the power to the furnace tube clearly depending on the delay in each cycle before conduction is initiated.

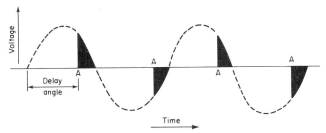

FIG. 30. Switching of AC waveform.

Control of tube temperature

In this kind of power supply system it is necessary to control the temperature actually attained at the steady state. This is not easily achieved because factors such as variations in tube dimensions (even within a batch), drift in physical properties of the graphite with number of previous firings, variations in impurity levels etc., can have a large effect on the final temperature.

In a routine analytical situation it is also more important that temperatures achieved should be reproducible for a given setting than that the operator should know the precise temperature. This applies both to atomization and to ashing temperatures if meaningful calibrations and final results are to be obtained.

In atomizing systems where the temperature performance achieved—its actual value and its reproducibility—depends directly on the electrical parameter controlled, Whiteside has shown that it is more important to stabilize applied voltage than either current or power.[795]

Direct monitoring of the tube temperature with feedback control of voltage supplied to the atomizer has been achieved in several different ways. A solid state controller[159] and an inexpensive monitor using a photovoltaic cell have been used for this purpose[777] and Mullins[539] described a system based on a tungsten resistance thermometer in contact with the atomizer tube. The latter was said to allow temperatures to be accurately maintained over the whole of the temperature range and to enable higher rates of heating to be employed. The temperature rise curve naturally lags behind the voltage rise curve, but with control the voltage can 'overshoot' and be reduced at the point when the desired temperature is reached. Chakrabarti[712] believes, however, that a thermocouple is not only upper temperature limited, but suffers thermal lag, resulting also in temperature overshoot. Nevertheless, a consequence of fast heating rates is that a higher density of atomic vapour is formed transitorily, giving improved analytical sensitivity.

Lundgren[462] designed an infrared sensor to operate with a fibre optic and in the commercially produced 'HGA2200' an optical sensor provides the continuous temperature monitoring which allows a fast heating cycle with its attendant advantages mentioned above.

MISCELLANEOUS ATOMIZERS

Plasma torch

Induction coupled plasmas are able to volatilize the most refractory materials as an electronic temperature of something like 16 000 K is produced at the centre of the discharge. The temperature decreases along the tail flame and it should in principle be possible to select that part in which conditions are optimum for a given element. Wendt and Fassel[787] showed that a number of highly refractory metals including niobium and tungsten would give sensitivities in the part per million range when solutions containing their salts were nebulized ultrasonically in a stream of argon. The performance was thus at least equivalent to a nitrous oxide acetylene flame. It was also shown that there was no interference of aluminium or phosphorus in the determination of calcium.

While this atom source would seem to have much to commend it—easy sampling, direct sampling of powdered solids, freedom from chemical interferences—the main disadvantages are the very high initial cost, instability ('flicker') in the tail flame and the high degree of ionization of most elements in more stable but hotter parts of the discharge. The induction coupled plasma has been much developed in recent years as an emission source.

Cathodic sputtering

A demountable hollow cathode-like device, into which metallic samples could be introduced, was described by Gatehouse and Walsh.[275] The samples had to be machined to form an open-ended cylinder, and positioned carefully in the optical beam in the sputtering chamber. This was sealed, pumped down and filled with argon at the correct pressure. The discharge was then switched on with the sample as the cathode. The absorption of the sputtered atom cloud was measured.

This system did not show particularly high sensitivity, and preferential sputtering of certain elements may be found to give rise to some lack of reproducibility. Nevertheless, silver and phosphorus were determined in copper, and silicon was determined in aluminium and in steel.[770] Later it was shown by the same workers that precisions of the order of $\pm 1\%$ and detection limits down to 10 μg/g could be obtained for many elements.[291] There is also the possibility of measuring carbon, phosphorus and sulfur by atomic fluorescence in the same cell.[773] The main advantage of this technique may be in the analysis of thin coatings and alloys,[55] or for analysis of metals that would otherwise require complex dissolution procedures. Best indications of present potential are probably those given by Gough.[290]

Instead of machining the cathode from the metal sample as in the work referred to above, the sample solution can, of course, be evaporated on the inner wall of a suitable hollow cathode. Both aluminium[285] and graphite[355] have been used as cathode material for this method. Nanogram amounts of a number of the more volatile elements may be measured by evaporating the sample on small aluminium or graphite discs which are subsequently transferred to a brass cathode.[269] The method has been extended to the measurement of iodine at fractions of a microgram using the line at 183.0 nm by Kirkbright and Wilson.[415]

Lasers

Lasers have been investigated for two possible functions in atomic absorption spectroscopy, as a substitute for conventional atomic spectral lamps and as a means of producing the atomic vapour.

Commercial solid- or gas-state lasers are most unlikely to be used in the former role as coincidence of a laser line with an atomic resonance line would be entirely fortuitous. 'Tunable' lasers of various types are now becoming available, however, though as spectroscopic sources they still have severe disadvantages from the points of view of wavelength coverage and cost. An ion-laser-pumped continuous wave dye laser[296] is said to exhibit better wavelength accuracy and stability than the pulsed lasers used by most investigators hitherto, and also not to suffer from pulse-to-pulse variations. The power available with a laser seems to make it a particularly attractive source for atomic fluorescence measurements.

A number of workers have reported the successful use of lasers for atomization of samples both for atomic absorption and for atomic fluorescence. A particular advantage of the laser in this respect is that a very large amount of energy can be concentrated into a small area and so very small regions of a larger sample can be atomized without their being removed from the main sample. Such a laser-microprobe/atomic absorption assembly was described by Mossotti et al.[533] This required a fast response system to measure the absorption in the transient atomic vapour.

It seems, however, that laser atomization may suffer from inefficient atom production. With a pulsed CO_2 laser with 0.1 J pulses, a width of less than 1 μs and a repetition rate of 5 Hz,[494] metal samples had to be contained in a graphite furnace at a temperature just below that required for continuous perceptible atom production. This problem has also been overcome by use of a repetition rate of 100 kHz.[615] Recent examples of the microprobe approach are the raster analysis done by Neumann[422] and the vapour transfer method of Kantor.[380]

In Kantor's system the vapours produced by a laser microprobe are led to a conventional chamber/burner system for AA measurement.

Chapter 4

Instrumental Functions

THE OPTICAL SYSTEM

As will be seen from Fig. 8 and from the more detailed optical layout of Fig. 35 the optical system of an atomic absorption spectrometer has four main constituent parts. These are:

(i) the primary radiation source;

(ii) the pre-slit optics by means of which the source radiation is first focused in the centre of the atomizer position, then brought to a second focus at the entrance slit of the monochromator;

(iii) the monochromator;

(iv) the detector.

The parts are discussed separately but they must also be considered as an integral unit for, in a properly designed instrument, best performance can only be obtained by careful matching of all optical parameters.

Primary radiation source

In Chapter 2, Resonance Radiation, it was shown that the primary radiation source must emit a sharp resonance line spectrum. Ideally a half-width of 0.001 nm should be provided. Appreciable absorption, though not the maximum possible, is still obtained with emission lines of 0.01 nm half-width. Thermally excited spectra, and arc and spark spectra, contain lines that are broadened by Doppler and pressure effects. Arc and spark spectra are also subject to considerable electrical and magnetic field broadening. They are therefore unsuitable as the primary radiation source in atomic absorption.

Vapour discharge tubes were the first sources to be used, as they are readily available for some of the volatile common metallic elements. Undoubtedly the best source is the sealed-off hollow cathode lamp, which has undergone considerable development and improvement in intensity and reliability since atomic absorption became popular. More recently, microwave and radio frequency excited electrodeless discharges have been introduced commercially.

70

Vapour discharge tubes

Vapour discharge lamps consist of a glass or silica tube containing an inert gas at a pressure of several torr and a quantity of the metal whose spectrum is to be produced. Oxide-coated electrodes are sealed in. They can be run on AC or DC. If AC they are controlled with a variable transformer and ammeter. When first switched on a gas discharge takes place which warms and vaporizes the sealed-in metal. The metal vapour then takes over the discharge and the radiation consists almost entirely of the metal spectrum. Such lamps should be run at currents considerably below those recommended by the manufacturers in order to minimize self-reversal of the resonance lines. For example, a current of less than 0.5 A is desirable instead of the normal rating of 1 A or more.

This type of lamp is available for sodium, potassium, zinc, cadmium, mercury and thallium, and these are manufactured by both Osram and Philips. Elenbaas and Riemans[212] have given details about their construction and operation. Lamps which do not have a quartz outer envelope may require a hole to be cut in the protective glass envelope or the envelope removed altogether in order to pass the resonance lines in the ultraviolet.

Vapour discharge lamps have the advantage of much higher intensity than hollow cathode lamps. This means that better ultimate signal to noise ratios should be obtained, and therefore better detection limits, but at the required low currents, these lamps tend to be more unstable, and the advantage is to a large extent ruled out. In atomic fluorescence measurements, however, where the sensitivity is a direct function of source intensity and where excitation line widths are less important, much better analytical performance is achieved.

In atomic absorption, modern hollow cathode lamps are undoubtedly better from the point of view of stability and line width for cadmium, zinc, thallium and even mercury. Good hollow cathode lamps are now also available for sodium and potassium, and the use of vapour discharge lamps appears to be dying out.

Hollow cathode lamps

The hollow cathode discharge has been known for many years to spectroscopists as a fine line source, and indeed one capable of producing spectra where the fine structure could be studied.[746] It has also been used as the excitation source for microsamples in emission spectroscopy. In these applications the source is, of necessity, demountable, and indeed in the earliest atomic absorption work it was also used in demountable form.

Demountable Hollow Cathode Lamps. These allow the cathodes to be changed at will, and hence one lamp system can be used for any element (or combination of elements if suitable alloy cathodes are available). After each change of cathode, however, the lamp has to be purged with fill gas and then pumped down to the operating gas pressure. A refinement in some systems is that the fill gas is kept circulating through a cleaning-up section. In such cases the cathode never becomes poisoned, and the atom cloud which normally forms in front of the cathode, increasing the self-absorption of radiation, is effectively removed. Stable high performance is therefore possible. A demountable hollow cathode lamp system is, however, costly to set up as it requires high vacuum techniques, and the time taken to change from one element to another is too long to be considered in routine analysis. A system ideally suited to research requirements was described by Elwell and Gidley,[1] and a commercial system is available from Barnes.

For the above reasons, therefore, it is the sealed-off type of hollow cathode lamp which is in common use as the primary radiation source at the present time. The hollow cathode discharge is in fact a low pressure discharge ('Geissler' gas discharge tube) with a special geometry. If the cathode is made hollow in shape, and the gas pressure reduced to the correct value, the discharge takes place entirely in the hollow cathode, when the radiation emitted becomes rich in the spectrum of the cathode material in addition to the spectrum of the fill gas. The pressure is somewhat higher than for a gas discharge tube.

Typically, the internal diameter of the cathode would be 3–5 mm, the fill gas argon or neon and the operating pressure between 4 and 10 torr. This is contained within a glass envelope through which the cathode and a tungsten wire anode are sealed. The whole is energized with a potential of about 300 V and currents of 4 mA up to 50 mA or more may be passed, depending on the element being excited.

The choice of fill gas depends primarily upon two factors. Firstly, the discharge emission lines of the gas itself must not coincide with the resonance lines of the elements to be measured. Secondly, the relative ionization potentials of the fill gas and cathode metal must be taken into account. The ionization potential of neon is much higher than that of argon and many metals and hence will tend to increase the proportion of 'spark' lines produced, as excitation is caused by collisions of the second kind. Neon is therefore normally used with elements of high ionization potential. Other considerations are that argon, being heavier, has a more efficient sputtering action and also 'cleans up' (adsorbs on sputtered metal films) less quickly; on the other hand, the spectrum of neon is much less rich in lines than that of argon and hence neon is probably better when the lamps are used with instruments of lower dispersion.

Sealed-off Hollow Cathode Lamps. The earliest sealed-off lamps consisted of a glass tube through which the electrodes were sealed, with an optical window of glass or silica (depending on the wavelength of the wanted resonance line) attached, usually with a thermosetting resin or vacuum wax. The cathode normally had an internal diameter of 10 mm. The components of this device had to be cleaned up by baking before finally pumping down to operating pressure. Because of the wax sealing, very efficient cleaning was not possible, and the fill gas and cathode became poisoned more or less quickly. This led to a reputation of short life for early sealed-off lamps, but this has fortunately now been dispelled.

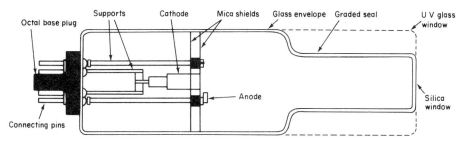

FIG. 31. Diagram of modern hollow cathode source.

The move towards all-fused construction and modified geometry over the past ten years has now led to a lamp which provides much higher intensity over a long useful life. Most manufacturers guarantee their lamps for 500 hours and they usually run for at least twice this length of time. No more than 2 or 3% of reputable lamps fail under this guarantee, such is their present-day reliability.

An internal cathode diameter of 2 mm is now much more common (Fig. 31) as this concentrates the energy of discharge on a smaller area and thus produces a much higher intensity. Also, the energy appears to be dissipated to a greater extent in the resonance lines, giving these a higher intensity in relation to the rest of the metal spectrum and the gas lines. This clearly results in better analytical performance. A mica shield helps to prevent outward spread of the discharge, and for the same reason some lamps are provided with a ring anode. However, the anode shape does not appear to be of major importance. A higher gas pressure is used—up to 10 torr—and this both helps to maintain the discharge inside the cup and takes longer to be 'cleaned up' in use. Figure 32 shows the relationship between operating current and gas pressure for hollow cathode lamps in general. This shows that this modern type of lamp can be run to advantage at a lower

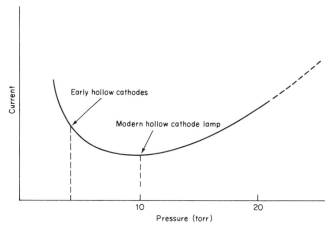

FIG. 32. Pressure/current characteristic of hollow cathode sources.

current than the corresponding early lamp. There are thus several factors contributing to improving the length of working life and analytical performance.

In the all-fused construction of these lamps, borosilicate glass may be used for the body and optical window where the resonance line has a wavelength higher than 400 nm. Between 400 and 250 nm the window can be made of 'ultraviolet transmitting' glass. Below 250 nm the transmission of ultraviolet glass falls off steeply, e.g. 70% at 240 nm, and a silica window, attached to the body by a graded seal, must be used instead.

For hollow cathode lamps, many manufacturers now appear to have adopted a maximum outside diameter of 37 mm.

Boosted Output Lamps. This is a type of hollow cathode lamp, introduced by Sullivan and Walsh,[720] in which the cloud of atoms normally formed in front of the cathode is itself excited by a secondary discharge. In this way these atoms are made to *contribute* to the intensity of the resonance radiation rather than diminish it by absorption. The secondary discharge is of a low voltage but higher current (300–400 mA). The auxiliary electrodes are almost entirely enclosed in glass sleeves but openings are provided to position the discharge correctly.

Such lamps gave improved signal to noise ratios by virtue of their increased intensity, and also straightened calibration curves because the resonance lines were less self-absorbed before reaching the flame. Their higher degree of complexity, however, seemed to make them less reliable and shorter lived, and as they appear to have no real advantage over the lamps described above, they are no longer commercially available.

Sullivan[719] has recently reported improvements in demountable hollow cathode lamps based on the boosted output design, pointing out their potential as a source for atomic fluorescence measurements. On the other hand, Falk[221] has shown that the avoidance of self-absorption in the hollow cathode lamp could not improve the output intensity of a resonance line by as much as an order of magnitude. The advantages of the boosted output type of lamp over the usual modern hollow cathode design in atomic absorption analysis are thus open to doubt.

Slotted Cathodes. In order to increase the proportion of light from the hollow cathode lamp entering the monochromator, i.e. the total energy throughput, it is possible in designing the optics to match the tangential astigmatic image of the cathode, formed after off-axis reflection from a concave mirror, to the shape of the monochromator entrance slit. In instruments with lens-only optics, this image is not formed, and in order to achieve the same effect lamps with 'slotted cathodes' have been devised by one manufacturer.

Multi-element Lamps. The use of multi-element lamps, i.e. lamps whose cathodes contain more than one element, has been advocated, partly on the grounds of economy and partly for their convenience.

It is difficult to prove their economy, because when the lamp fails, one has lost all the elements. Also, for some reason which is not entirely apparent, such lamps usually cost more than those for a single element. It is clearly not economical to use a multi-element lamp when a single element only is being determined as this is equivalent to wasting the life of the other elements.

Experience has shown that the different rates at which elements sputter cause them to lose their intensity and be lost in turn. The alloy from which the cathode is made should therefore contain the elements in concentrations which are proportional to their rates of sputtering under the conditions in which the lamp is to be used. Alternatively, cathodes formed by powder metallurgy or by discrete rings of individual metals have been used with varying degrees of success. There is thus a considerable limitation on the elements which can be brought together in a single cathode.

The convenience of a multi-element lamp is without doubt, particularly where the spectrometer has no lamp 'turret'. It would be an advantage where a routine analysis for some three or four elements is regularly carried out and those elements can be obtained together in one cathode. 'Warm-up' and change-over times can then be reduced. The resonance line intensities of individual elements are somewhat diminished as a result of the elements being brought together in one cathode, however, and overall performance is therefore poorer.

Hollow Cathode Lamp Currents. The supply to a hollow cathode lamp should be either current-stabilized or limited with a series resistor, as the current is almost independent of the applied voltage. A given lamp will operate over a fairly wide range of currents, but it is desirable to choose the best operating current for the particular lamp and analysis in hand. Some manufacturers label lamps with 'maximum' currents and others with 'maximum operating' currents. A maximum current should under no circumstances be exceeded as irreparable damage may be done. High currents, and currents above 'maximum operating' values will result in excessive line broadening. It is therefore nearly always desirable to run a lamp at a current value less than that given on the envelope. Many users prefer to run the lamp at the lowest current that gives adequate stability and freedom from intensity drifts. This improves the sensitivity because of minimal line broadening, and increases lamp life. With modern instruments it should be unnecessary to run lamps at higher currents with the object of reducing the amplifier gain to lower the electronic noise levels.

It is desirable that the lamp 'noise' should be less than 0.2% and intensity drift less than 5% per hour. For a double-beam instrument in which variations in intensity are to a large extent compensated, or for a single beam instrument with an automatic zeroing control, the intensity drift limitation can be somewhat relaxed.

Lamps need a short warm up period before use, the length of which depends on the type of lamp and the element. Periods of between 5 and 20 minutes are normal. During this time the lamp intensity drifts to its equilibrium value, sometimes gradually upwards, sometimes going first above and then slowly downwards. During this time, also, the atom cloud mentioned earlier attains its equilibrium density so that self-absorption of resonance radiation, and hence the line-shape absorptivity coefficient of the line, also show drift towards a stable value. These effects are discussed again in relation to single-beam and double-beam optical systems.

Electrodeless discharge lamps

The main disadvantage of hollow cathode lamps, viz. comparatively low resonance line intensity, may be overcome for the easily vaporizable elements by use of electrodeless discharge lamps. Such sources have also been used for studies in spectral structure for some years and more recently as intense line sources for both atomic absorption and fluorescence.[168–171,821]

The electrodeless discharge tube takes the form of a sealed quartz tube, 3–8 cm in length, 1 cm or somewhat less in diameter, containing a few milligrams of a metal or volatile metal salt. The tube is filled with an inert gas at a pressure of a few torr which starts the discharge and helps maintain it with collisions of the second kind. A microwave- or radio-frequency electromagnetic field provides the excitation energy through a wave-guide cavity or coil.

Microwave EDLs. For much of the published work up to the present time, the power source has been one of the comparatively inexpensive medical diathermy units (Electromedical Supplies Ltd.) which provide power at about 2450 MHz up to 200 W. These can be connected to various wave-guide cavities and antennae.

Detailed instructions on the preparation of microwave electrodeless discharge tubes have been published by Dagnall and West.[172] The material placed in the tube to generate the discharge must have a vapour pressure of about 1 mm at 200–400 °C in order to function with the microwave power unit and cavities mentioned. This is the factor which dictates whether the metallic element itself or whether its chloride or iodide is to be used. In the case of the iodide, this may well be best formed by using a little of the metal and a small excess of elemental iodine. The inclusion of one or two milligrams of mercury, or even just saturated mercury vapour, is said to improve the reliability of 'striking' of the lamp by acting as a carrier of the initial discharge. It also tends to improve lamp life by preventing adsorption of the main element on the quartz tube.

The thermally uniform nature of the discharge and even a skin effect (concentrating the discharge near the tube walls) ensure that there is little or no self-reversal of the emitted resonance lines. However, the tube must be small enough to allow this condition to obtain without movement of vapour from cooler parts of the tube itself. There is thus an optimum size of tube for each element, which depends upon the vapour pressure of the material used and the nature of the cavity. It follows that a tube made to contain two elements can only be successful if the metals or compounds used have similar vapour pressures at the working temperature of the tube.

The cavities employed should be tunable and air-cooled. A reflected power meter may be used with some advantage to assist in tuning the cavity and to ensure that the reflected power itself is not sufficient to damage the magnetron. If the power is less than 75 W, however, visual tuning may be adequate.

Lamp temperature is the most important parameter controlling spectrum intensity[121] because vaporization within the lamp is mainly thermally induced. Therefore microwave EDLs must normally be thermostatted in order to make them sufficiently stable in output intensity for analytical use.

When correctly prepared and run, electrodeless discharge tubes should emit only the principal lines in the spectrum of the main element together with the iodine line 206.2 nm and several of the strongest mercury lines (assuming that the metal iodide is the active substance and that mercury has been introduced either deliberately or incidentally by way of the vacuum pump). Lines may, of course, also originate from impurities in the substances used. Resonance line intensity of the wanted elements may well be ten to one hundred times greater than in hollow cathode lamps.

Tubes have been prepared for about fifty elements, but of these arsenic, antimony, bismuth, selenium and tellurium are certainly the best. This is a happy coincidence because, probably for the same reasons of elemental volatility, these five also have a reputation for making the worst hollow cathode lamps. Electrodeless discharge lamps are thus a very useful tool in the hands of a research analyst, who can construct them as the need dictates, thus saving considerable expenditure on hollow cathode lamps.

Radiofrequency EDLs. Several companies have made radiofrequency lamp systems available in recent years. Although light intensity from these may be lower than from microwave lamps, they have proved to give better short and long term stability without the need for thermostatting. It is generally found, too, that operating conditions are more easily established and reproducible from day to day. They are therefore particularly suitable for routine applications in atomic absorption, especially for the volatile elements which do not make good hollow cathodes,[79] and for the elements having lines at lower wavelengths.

The radio frequency commonly employed is 27.12 MHz (within the model control, USA Citizens' and scientific investigation bands).

RF lamps are recommended for arsenic, selenium, antimony, tellurium and phosphorus and are also available for about ten other elements.

In recent work Novak and Browner showed[557] that radiofrequency-pulsed electrodeless discharge lamps have much higher mean output intensities with shorter, e.g. <0.8 ms, pulses. In the cases of some elements the intensities can then be *better* than those of microwave lamps.

Pre-slit optics

The Optical Aperture. A major constraint in the design of the atomic absorption system is the long narrow flame or electrothermal atomizer. As much light from the source as possible must pass through the atomizer, otherwise lower absorption (and lower sensitivities) than the maximum possible will be experienced. Ideally, the atomizer should fit the light envelope like a 'minimum volume' infrared gas absorption cell. Attempts to achieve this with a flame were not particularly successful, but in electrothermal atomization a recently described profiled furnace tube[605] has been shown to give improved sensitivity over the straight cylindrical type.

The smaller diameters of hollow cathodes, e.g. 3 mm as already mentioned, which give better intensity and a smaller focus, are most suitable for use with furnaces. The magnification of the image at the focus in the atomizer is usually about $1 \times$. Careful design of the optics can then provide a better resonance line/ furnace emission intensity ratio at the detector, possibly with use of a field stop at the entrance slit of the monochromator. The fact that the source is not a point but has finite dimensions tends, to some extent, to improve the situation in the flame because the light envelope though conical at the ends is more cylindrical at the centre.

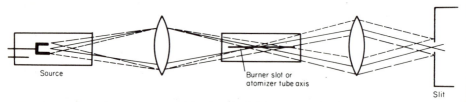

FIG. 33. Pre-slit optics, single beam, plan view.

Because the source is focused on to the monochromator slit, more energy will pass into the monochromator as the source image width (or diameter if, as is usual, the source is circular) approaches the slit width. A small source is thus preferred, and the small cathodes of modern hollow cathode lamps are also advantageous in this respect.

A typical pre-slit optical diagram is given in Fig. 33 with the width of the light envelope much exaggerated. The long narrow flame is seen to influence the angle of acceptance at the monochromator slit, dictating the optical aperture of the monochromator and also of the source lamp itself. A narrow acceptance angle means that, for the purposes of atomic absorption measurements, a large monochromator aperture is unnecessary, and may even be an embarrassment as stray light levels, i.e. light not subjected to the absorption process in the flame, are then greater.

The design of multi-purpose instruments—which are to be used for flame emission or fluorescence as well as absorption—must therefore include a compromise, since in these the detection limit in emission or fluorescence can be improved by making the flame fill a larger optical aperture.

It thus turns out that a monochromator aperture of $f.10$ such as is normally used with ultraviolet spectrometers is suitable for atomic absorption requirements. A good flame emission spectrometer should have an aperture of $f.4$ or thereabouts, and an instrument to be used for measuring flame fluorescence should have the same. Such a monochromator can be quite expensive but instruments designed to perform emission and/or fluorescence measurements as well as absorption usually compromise with an aperture of not smaller than $f.7$.

In some early home made atomic absorption instruments the pre-slit optics consisted of two slit-shaped diaphragms, one at each end of the burner. The source lamp was placed as close to one of these as possible, and the monochromator slit close to the other. Such a system helps to reduce the amount of stray or unabsorbed light from the lamp falling on the detector, but it also seriously limits the total energy throughput. This results in good reciprocal sensitivity values but poor detection limits.

Single and Double Beam Systems. The pre-slit optical system shown in Fig. 33 is typical of a single-beam instrument. As in any other single beam ultraviolet spectrometer, the light falling on the detector is proportional to the *transmission* of the 'sample' in the optical path, and it is thus necessary to make a reading with and without the sample in order to obtain the *absorbance* which by Beer's law (see p. 120) is nearly a linear function of the concentration. Accurate absorbance values thus require a primary source of stable intensity. Stability in this context should, ideally, be absolute over the whole period during which measurements are taken so that the need to run reagent blanks or other reference solutions at frequent intervals is obviated. While modern hollow cathode lamps and good electrodeless discharges certainly emit at the same intensity over long periods, no source is entirely without some intensity drift.

The effects of source variation can be overcome to a very large extent by employing double beam optics. In the most usual form of this device, the beam falls onto a rotating sector mirror before passing through the flame. This directs the beam alternately through the flame and along a path by-passing the flame at a frequency corresponding to the amplifier lock-in frequency which may be 50 Hz or higher. Past the flame, the beams are recombined with a half-silvered mirror as shown in Fig. 34. At the detector, the output signals corresponding to each beam are divided, amplified separately and compared in a bridge circuit.

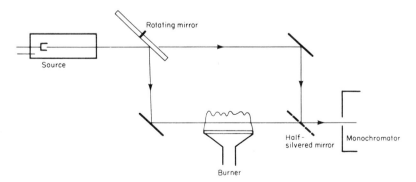

FIG. 34. Basic double beam optical system.

The out-of-balance signal is then compensated electronically and converted to absorbance.

A double beam atomic absorption spectrometer does not have all the advantages imputed to a double beam 'ultraviolet and visible' spectrometer. In the latter the reference beam passes through a cell which can be made to contain everything except the actual species being measured. Thus the true absorbance is obtained directly. In the atomic absorption instrument the reference beam does not pass through the flame. If it passed through a second flame, flames being *dynamic* systems, neither the chemical nor the noise characteristics could be guaranteed to be even similar and certainly not identical. The normal double beam system, therefore, corrects only for variations in primary source intensity and detector response. It does not correct for any spurious absorption or scatter in the flame, nor will it allow for variations in the absorptivity coefficient of the primary radiation which, as has been noted, occur particularly during the initial warm-up period of the primary source.

A further disadvantage of double beam optics is that of necessity at least 50% of the incident energy is wasted. Otherwise the beams cannot be recombined without modulating the flame emission.

Zeeman Line Splitting. Some recently introduced instruments in which Zeeman splitting of either the emission source line or the analyte absorption line is incorporated are, in the above sense, true double beam instruments. In these instruments, both measured components of the resonance beam pass along exactly the same path through the atomizer. More detail will be given in the sections on Background Correction (see p. 86).

Modulated Sample Input. An interesting way of retaining most of the advantages of double beam systems is to modulate the aerosol supply to the flame. This was achieved with two matched nebulizers and a fluid-switching technique.[64] By dispensing with a spray chamber, modulation frequencies of up to 70 Hz were possible. With such a device, true background and flame noise effects are overcome by spraying the sample through one nebulizer and the blank solution through the other.

Monochromator

Resolution. With the ultimate resolution of the system depending on the spectral bandwidth of the resonance line emitted by the primary radiation source, the function of the monochromator is to isolate that resonance line from non-absorbing lines situated close to it in the source spectrum. Such lines may originate from the cathode metal or the lamp fill gas. The monochromator resolution should therefore be the best that will separate lines in the most complicated source spectrum to be employed, but as this will depend on the analysis to be undertaken, and to a great extent on the lamps to be used, no hard and fast rule can be given.

A monochromator of high resolution is not required when the alkali metals and perhaps calcium and magnesium are to be determined in biological samples, but for analysing metallic alloys, particularly those containing nickel and chromium, better resolving power is required. The problem is accentuated when source lamps emitting the spectra of more than one element are employed. A general purpose monochromator should thus be capable of separating two lines 0.1 nm apart or less when operating at minimum effective slit width. The better the quality of the monochromator, the smaller the slit width will be before further narrowing has no more effect on resolution. In very good instruments the minimum spectral band pass may be about 0.01 nm.

A further function of the monochromator is to isolate the measured resonance line from molecular emission and other background continua which originate in the flame. To prevent all radiation from the flame reaching and 'saturating' the detector, the monochromator is always placed in the optical path after the sample cell, and not before, as in conventional ultraviolet spectrophotometry.

With the aperture dictated by the limitation of the pre-slit optics, the maximum effective slit height is the diameter of the source image at the primary slits. Longer slits would simply allow unwanted radiation to enter the monochromator during absorption measurements.

Prism or Grating. The remaining choice lies between a prism or grating dispersing element. A glass prism is unlikely to be considered as this will not pass ultraviolet light. The essential difference in performance between prisms and gratings is that the dispersion given by prisms is high in the ultraviolet but decreases rapidly with increasing wavelength. For gratings it is substantially constant throughout the spectrum, and depends on the number of grooves per unit width, the spectral order and the focal length of the collimator.

Prisms can therefore be quite useful in atomic absorption work as the majority of resonance lines occur in the ultraviolet region. At the other end of the spectrum, the alkali metal resonance lines are separated both from each other and from lines of other elements, though gas lines produced in hollow cathode lamps may cause stray light effects when these elements are determined in absorption. In general, however, all the light is transmitted by a prism in a single 'order' so the energy concentrated in a particular line is higher and there is less stray light and other spurious reflections such as unwanted orders or ghosts.

Prisms fail mainly on their lack of resolution in the range 240 nm upwards, particularly up to about 450 nm, within which region many important resonance lines of elements with complex spectra occur. Gratings are therefore becoming far more widely used, as good replicas are no longer expensive. Holographic gratings, a more recent development, are now also able to be blazed economically and reproducibly. As light energy passed falls off rapidly in the higher orders, gratings blazed to give a maximum diffraction at a particular wavelength in either the first or second order will usually be adopted. In some instruments two gratings may be used to cover the whole range, e.g. 180–400 nm and 400–860 nm, to maintain good resolution in the important lower wavelength region.

It is interesting to note that the widespread use of electrothermal atomizers has called for more ingenuity in the design of monochromators. The emission from the furnace walls is seen over a wider cone than the light from the spectral source. The monochromator aperture is therefore chosen, e.g. by limiting grating area, to be no greater than the hollow cathode image area in order to reject the emission image from the furnace tube. In a mirror-type monochromator, the emission radiation and spectral source radiation can also be effectively separated by the use of baffles at a sagittal focus following off-axis reflection from a concave mirror.

Background corrector

An essential part of an atomic absorption system for many applications, particularly for electrothermal atomization methods, is a device for correcting the atomic absorption signal for non-specific molecular absorption or light scattering effects. Both of these attenuate the light beam, giving an apparently increased absorption signal. Such spurious signals can be corrected for under the continuous measurement conditions offered by the flame by making the total absorption measurement and the background measurement at different points in time as will be described on p. 139. A simultaneous background corrector is nevertheless very convenient in flame work, but absolutely essential in electrothermal atomization.

Such a corrector requires the incorporation of continuum sources covering the ultraviolet region and, in more expensive instruments, the visible. The spectral bandwidth of the monochromator is large compared with the width of the absorption line (see Fig. 2) and an absorption reading taken over this bandwidth using a continuum source contains only a negligible component due to absorption by the absorption resonance line at the same monochromator wavelength setting. Such an absorbance value represents the non-specific absorbance and scatter (the 'background') and it can be measured and subtracted from the total absorbance measured by the atomic emission resonance line.

This is achieved in a double beam spectrometer by positioning the continuum source (usually a deuterium arc lamp or a hollow cathode lamp containing deuterium gas for the ultraviolet and a tungsten halogen lamp for the visible) with respect to the chopping mirror (see Fig. 35) so that the latter reflects the hollow cathode lamp (atomic source) radiation and deuterium lamp (continuum source) radiation along the sample beam in alternate pulses.

The instrument then measures the total absorbance in sample channel electronics, and the background absorbance in the reference electronics. The latter is subtracted from the former by the output circuitry, and the reading given is the corrected atomic absorption. It is to be noted that in this mode many double beam spectrometers operate essentially in the single beam mode, because the reference light beam is inoperative.

A single beam spectrometer is therefore readily designed to give background correction by incorporation of the continuum source and chopper together with a 'double beam' readout system. This is sometimes referred to as a 'quasi double-beam' or dual channel system.

In some recent instruments, true double beam operation with background correction is also possible. This system corrects for intensity drifts in both source lamps as well as electronic drifts, whereas the single beam/background correction system can correct only for detector response drift. Such a system incorporates four electronic channels and one detector. As there is usually at least one static beam splitter in each optical channel, relative noise levels must be adversely affected.

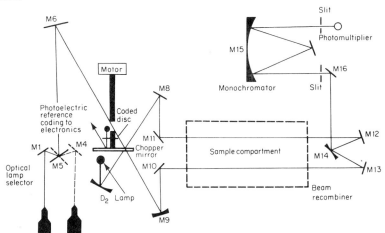

FIG. 35. Optical diagram of double beam AAS with background corrector.

Static beam splitters necessarily lose 50% of incident light energy. A rotary chopper in conjunction with pulsed lamps overcomes this problem but is more expensive. Furthermore, the lamp pulse must be shorter than the chopping frequency otherwise the chopper would partially cross the beam, affecting the beam shape and thus, particularly in electrothermal atomization, the efficiency of background correction.

Phasing problems of this nature with pulsed lamps are overcome by using a coding disc on the chopper (see optical diagram Fig. 35) but the difficulties of using this system both for beam splitting and recombining are likely to be great, hence the usual practice of incorporating a static device at one of these points in the optical path.

Because the atomic absorption and the scatter or background signals in electrothermal atomization are produced by different species in the furnace tube, these species are not expected to form a homogeneous smoke or vapour. Indeed they may give their maximum effects at somewhat different points in space and time.

It is worthwhile considering at this point how to ensure that the background corrector operates at best possible efficiency. This requires attention to the chopping frequency, optical alignment and amplifier offset voltages.

The rate at which a cycle of pulses occurs, i.e. one pulse from each lamp, the chopping frequency, is of great importance when the corrector is working with an electrothermal atomizer. As already noted, during the atomization step, the atomic population increases quickly to a maximum, then falls more slowly. The background absorption behaves similarly, though not identically. The accuracy of correction therefore depends on the time interval between successive pulses relative to the rate of change of both background and total absorption. It can be shown that at 50 Hz chopping frequency, a maximum background correction accuracy of 99% can be obtained for a scatter peak whose rise time is half a second. At 15 Hz the possible accuracy falls to 95%.

In order to ensure that correction is applied only to the volume within the cell where the atomic absorption is being measured, it is essential that the continuum beam be accurately aligned with the atomic beam within the atomizer. Directions for achieving this are given in Chapter 7.

It will also be appreciated that high background absorbance or scatter will leave only a small signal level in the reference channel of the instrument's read-out electronics. The signal in the sample channel will be even smaller. The final result is therefore the difference between two small signals, and analytical accuracy depends on the accuracy with which these signals are processed in the data handling section of the instrument. The decoder, integrator and logarithmic amplifier circuits must be designed to enable small offset voltages, i.e. signals present in the absence of light, to be eliminated. Furthermore, the time constants of the circuits in the two channels must be the same.

Zeeman effect background correction

A fundamentally different approach to background correction has recently become of interest. The Zeeman effect occurs when an atomic vapour which is absorbing or emitting resonance radiation is subjected to a magnetic field of several kilogauss. This splits the spectrum lines into a number of components. In the simplest cases, exemplified by cadmium, mercury and zinc, the principal resonance line is replaced by a π component with half the original intensity situated at the original wavelength together with two σ lines, each of a quarter of the original intensity, displaced by equal wavelength intervals higher and lower than the original line (see Fig. 36). For most other elements, the picture is more complicated because the π and σ lines are themselves split into further components. This effect can be used in background correction by applying the magnetic field either to the atomized sample or to the source of resonance radiation. An example of the first of these is shown in Fig. 37.

In addition to being split by the magnetic field, the π components of the spectrum line become polarized in a plane parallel to the magnetic field and the σ lines perpendicularly. In the system shown, therefore, the light from the

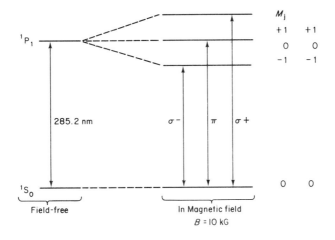

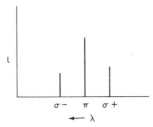

FIG. 36. Normal Zeeman effect in magnesium.

source is polarized alternately, by means of a rotating polarizer, in planes perpendicular and parallel to the magnetic field applied to the atom cell, before it actually passes through the atom cell. With the Zeeman effect only light polarized in a parallel plane is absorbed by the π components, i.e. undergoes atomic absorption as well as attenuation by background absorption and scatter. Light polarized in a perpendicular plane suffers no atomic absorption, only attenuation by background and scatter. The perpendicular and parallel components are detected in the usual way and their corresponding photocurrents

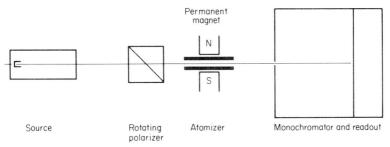

FIG. 37. Analyte-shifted Zeeman background correction.

amplified and separated by a phase-sensitive amplifier as in a normal double beam or dual beam system.

If the magnetic field is applied to the light source, instead of to the atom cell, the rotating polarizer allows the π and σ components to pass through the atom cell. The π component, being at the resonance wavelength, is subject to both atomic and background absorption, while the σ component, being at a slightly different wavelength, suffers only background absorption.

The two beams are detected and their corresponding signals separated as before. In other systems the magnetic field itself may be modulated. As with the other methods of background correction the difference in absorbance between the two signals is the true atomic absorption. The advantage of Zeeman effect background correction is that only a single standard light source is required, e.g. a hollow cathode lamp, which gives a single beam of two components which are already matched for intensity. It follows that there are no beam alignment or intensity matching problems as there can be with systems involving separate atomic and continuum sources.

It will also be clear that an atomic absorption spectrophotometer with Zeeman effect background correction has all the attributes of a true double beam spectrophotometer, i.e. it will automatically correct for variations in source intensity and detector response drift. It is the only system in which the reference beam path is identical with the sample beam path in every respect except for absorption by the analyte.

There are, however, certain practical problems in operating these systems. It would be extremely difficult to apply a high enough field strength to a normal long path flame for purely dimensional reasons. Magnetic fields, when applied to the atom cell, have therefore been used largely with miniature electrothermal devices. Also for dimensional reasons, it is equally difficult to apply the magnetic field to a hollow cathode lamp. In cases where this has been done the lamps become difficult to strike. The magnetic field has, however, been successfully applied to specially small electrodeless discharge lamps.

Several instruments have been described,[182,198,302] one commercial,[419] and reviews of the application of Zeeman effect in atomic absorption have been published.[30,31,550]

In recent work, Dawson[294] has shown that a background absorbance of 2.0 could be corrected to better than 0.005 absorbance unit.

The main interest in this technique is likely to be in electrothermal atomization because of the stringent requirements this imposes on background correction. It is possible, though, that the variability of the splitting pattern from element to element may result in its application being less than universal.

Detectors

The use of photographic plates and gas-filled photocells is now of historical and limited research interest only. Photomultiplier tubes are in almost universal use.

The wide spectrum range to be covered (from the arsenic line at 193.7 nm to the caesium line at 852.1 nm) by a general purpose analytical instrument poses some problems of sensitivity, particularly towards the higher wavelengths.

The spectral sensitivity of photomultipliers depends primarily upon the photosensitive material used to coat the cathode. These materials are usually alloys of alkali metals with antimony, bismuth and/or silver. Most of these materials provide adequate output at wavelengths down to 190 nm provided the envelope material has adequate transmission. Caesium–antimony cathodes operate well up to 500 nm and the output falls rather steeply to around 760 nm, where there may or may not be sufficient sensitivity to detect the potassium and rubidium lines (766 nm and 780 nm). Photomultipliers of this type may well have to be selected for adequate performance. Trialkali cathodes, antimony–sodium–potassium–caesium, respond well enough up to 850 nm and can usually be relied upon if caesium is to be detected. This type of photomultiplier is now manufactured in a side-window version which is only slightly inferior to the end-window tubes hitherto available and costing about the same as caesium–antimony tubes. More recently gallium arsenide cathodes have become available and their response at high wavelengths is a considerable improvement upon that of other types.

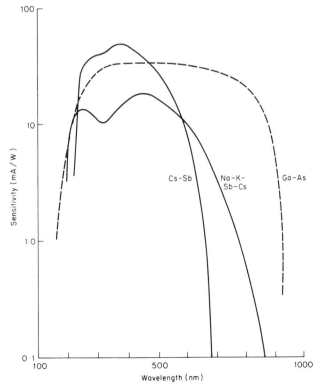

FIG. 38. Photomultiplier sensitivity curves.

Typical spectral sensitivity curves for photomultipliers most used in atomic absorption, the Cs–Sb and trialkali (NaKSbCs) types are given in Fig. 38. Ga–Sb photomultipliers, though giving response over a remarkably wide wavelength range are still extremely expensive. Principal manufacturers include Centronics, EMI, Hamamatsu, Philips and RCA.

Important characteristics of a particular photomultiplier from the point of view of atomic absorption are its dark current and noise. An amplification factor within the tube itself of 10^6 or greater is commonly achieved by increasing the voltage applied over the dynode system. The gain g of a photomultiplier increases exponentially with the interdynode voltage V. The relationship is

$$g = kV^{0.7n}$$

where n is the number of dynodes. It is therefore most important that the high voltage supply should be highly stabilized, e.g. to 0.05% or better, in single beam instruments. Increasing interdynode voltage, however, results in an increase in dark current and the noise component of the dark current (known as 'dark noise'). Also increased by increasing dynode voltage is the 'shot' noise, that is the actual statistical variations in output caused by the electron showers generated between dynodes. Shot noise is proportional to the square root of the intensity of the radiation falling on the photocathode. To a point, therefore, photomultipliers allow a relatively large almost noiseless gain factor. The range of hollow cathode lamp intensity and monochromator slit width requires a dynamic range of at least 100–$1000 \times$ in the detector/amplifier system. This is best provided by the ability to vary the dynode voltage which can be controlled indirectly by a front panel mounted potentiometer, usually labelled 'gain'.

READ-OUT SYSTEMS

The component sections of an atomic absorption spectrometer have been discussed in some detail, as they are common to virtually all available instruments. The way in which the instrumental readings are presented probably most distinguishes one instrument from another. It is therefore of interest to know what the readout section of the instrument can be expected to do, but beyond the scope of this book to explain the many ways of doing it.

Modulation

The signal received by the detector consists of the resonance radiation (attenuated or not by absorption in the atomizer) and emission from the atomizer. Flame emission may include both resonance emission of the wavelength at which absorption is being measured, and molecular band emission and scatter from small particles in the flame, both of which are likely to extend beyond the monochromator bandpass. These problems are accentuated with electrothermal atomizers. Only the resonance radiation originating from the source lamp is wanted. Other radiation falling on the detector diminishes the value of the absorbance that can be recorded. The lamp output is therefore 'coded' by modulation and the post-detector amplifier is tuned to the same modulation frequency, thus effectively preventing the continuous signal from the atomizer from being recorded. The source can be modulated either by using an AC supply current or by interposing a synchronous chopper in the beam before the atomizer. It should be remembered however, that the total light signal still falls on the detector and that if the unwanted part is very high the photomultiplier can become 'saturated', thus giving a difference signal which is not proportional to the absorption being measured. Even before this happens a substantial contribution is made to shot noise, which, it will be remembered, is dependent on the total incident light intensity. A further problem arises if the continuous signal contains a high noise component, such as would be the case with a highly emissive or luminous flame. The noise will almost certainly contain a contributing frequency the same as the modulation frequency. There will thus be a noise breakthrough which will obscure the wanted signal. This is impossible to remove electronically and can only be overcome by modifying the flame conditions.

Various modulation frequencies have been employed. Most convenient from the electronic point of view is normal mains frequency, 50 or 60 Hz. In general, any frequency above about 40 Hz is satisfactory, as the normal flame noise component of such frequencies is low. Frequencies below 15 Hz are definitely unsuitable as there is appreciable noise breakthrough even under the best flame conditions. Such frequencies are also unsuitable for signal handling for electrothermal atomizers. In some instruments modulation frequencies of 300 Hz or 285 Hz have been used (the latter to avoid mains harmonics) as there is said to be less noise at higher frequencies. In a double beam spectrometer the beam is modulated and switched by a rotating beam switch mirror, the supply to the source being DC or pulsed.

Absorbance conversion

At any given instrumental setting the photomultiplier output is proportional to the transmission of the flame. A simple linear amplifier with an output feeding a meter thus provides readings on a linear transmission scale. A contiguous absorbance scale can easily be provided but since $A = \log 1/T$ this is much compressed above absorbance readings of 0.5 and virtually unreadable above 1.0. The errors incurred by a linear transmission scale are discussed in more detail on pp. 157 and 159. Commercially available instruments therefore include a circuit for converting the output to be linear in absorbance.

The absorbance conversion can be achieved electronically in several different ways. A logarithmic amplifier can be employed or the logarithmic decay of voltage appearing across a discharging condenser may be utilized.

Meter and recorder readout

Unlike the liquid cell in a spectrophotometer, the flame is a dynamic entity and the sample passes through it in a quantized fashion. Source lamps may also introduce their own short term fluctuations and these together may contribute a noise factor over and above the purely electronic one. A meter will, as faithfully as its inertia permits, attempt to follow this noise and will present to the observer anything from a barely perceptible flicker under the best conditions to a considerable irregular waver under the worst, the latter making a sensible reading impossible. The worst conditions referred to will exist when a high amplifier gain or scale expansion is used to measure a trace element in a part of the spectrum, for example where emission noise is contributing to the measured signal. An operator watching the meter needle for a minute or more could not possibly state with confidence its average position.

A certain amount of electronic damping is permissible in order to reduce the noise amplitude, but if the damping is excessive the meter becomes sluggish and readings are again meaningless. The damping time constant in the meter circuit should therefore not be greater than 2 s.

The principal advantage of using a chart recorder instead of a meter is that the pen movements—including noise and drift from all sources—are recorded and a more accurate estimate of an average position is possible. The recorder thus contributes very positively to the accuracy of the measurement. To ensure meaningful traces in flame atomization, the full scale response time should be not more than 1 s, and the damping time constant between 1 and 5 s.

When the absorption signal is in transient form, as in electrothermal atomization and the non-continuous flame systems, the recorder response must be faster and consistent with an amplifier time constant of ~ 0.2 s (the time constant is the time taken to reach $(1 - 1/e) = 64\%$ of the final reading, which is reached as a rising exponential). Full scale deflection must thus be reached in no more than 0.5 s. Such a recorder/amplifier system probably follows the absorbance signal given by most types of furnace atomizer in use at present, but may not be fast enough for open atomizers or for furnaces of faster response should these become available. This is an important reason why peak height detection and peak area measuring circuitry, which is less time-constant limited, has been developed.

Digital presentation

The modern practice in scientific instruments is to present the results digitally. This avoids errors in scale readings either on the meter or recorder through parallax, mis-interpolation between scale divisions, etc. Most instruments are equipped with digital display, and inexpensive digital display units are still available which can be plugged into certain recorder outlets.

Input Signal Integration. When digital *printout* of the concentration is desired, the noise component is unacceptable because the printer operates at a fixed moment of time and will not necessarily print out the noise- or drift-averaged result. The signal presented to the printer must have been either averaged or integrated over a predetermined period of time.

This process may consist of taking 10 or 100 instantaneous readings over perhaps 10 s during the sample nebulization period. The average of these is then passed through the whole process and printed out as concentration. More commonly, integration continuously sums and normalizes the absorption signal over a selected period of time. This appears to give the more correct final result.

Scale Expansion. These facilities are useful in atomic absorption instruments for accurate work at both high and low concentrations. The reason for the latter is easy to understand. Scale expansion simply involves an electrical expansion of the presented signal by a chosen factor which is continuously variable. All the signal, including noise, is expanded, and the expanded signal is therefore not more selective than the normal signal. Only the noise originating from the readout device itself is not expanded and this should be negligible with the possible exception of a chart recorder. Scale expansion considerably facilitates the reading of small absorbances, however, and is particularly valuable when the noise level is 0.5% absorption (0.002 absorbance units) or less.

Real detection limits are often lower than reciprocal sensitivity figures, as already defined, by a factor of 5 or 10, and thus concentrations close to the detection limit are only measurable by scale expansion means. Scale expansion is always used in instruments which are capable of reading out directly in concentration. Atomic absorption instruments usually have a separate scale expansion control. This should be completely independent of the amplifier gain and zero control and should simply increase the signal fed to the readout device like the gain control on an independent external chart recorder. Also, any scale-expandable outlet to a recorder should be independent of the digital display.

Electronic linearization and concentration readout

A further error-incurring step, that of calculating concentration, still remains. This is not merely a matter of multiplying by a factor (the absorptivity coefficient) because in atomic absorption, Beer's Law, which states that the absorbance of an absorbing species is proportional to its concentration, is strictly true only at low absorbance values. Deviations are usually apparent at absorbances above 0.5, the calibration curve bending towards the concentration axis. Reasons for this are discussed under Calibration (p. 121). It is therefore convenient to include circuitry in the readout system whereby this curve function can be straightened, so that the absorptivity factor can be applied in order to obtain a direct readout in concentration.

Calibration graph curvature usually tends towards an asymptote of reduced slope and nearly always ultimately to an asymptote of zero slope. Manual curve straighteners are of two kinds: those which assume that curvature begins at a certain absorbance value and that the curvature is of a defined shape, and those that assume that all graphs are curved (however slightly at low absorbances) and tend towards a horizontal asymptote whose vertical co-ordinate is the stray light level beyond which no further absorption can take place. The latter system is based on the stray light equation (see p. 158). Neither is entirely correct, but both are capable of extending the linear range to a degree where factorization for concentration is worthwhile. The first type requires three controls, slope, onset and curvature, while the second dispenses with an onset control.

Linearization and concentration readout facilities in the form described are invariably used in conjunction with signal integration and digital display. This ensures both a steady reading on which to carry out the arithmetic, and also a noise-free readout for each sample.

Microprocessor assisted curve correction and concentration readout

While the electronic linearization and concentration readout facilities clearly improve both accuracy and convenience over the old manual and graphic procedures, much more comprehensive facilities are made available by the introduction of microprocessor systems. The setting up of the calibration is also much faster with these systems. Two main types are known at present, viz. inbuilt microprocessor data handling and use of an on-line bench-top programmable calculator.

Inbuilt Microprocessors. These may, in some instruments, be limited to the types (or even to the one type) of correction programme foreseen and provided by the manufacturer of the instrument. In this case it would provide an advantage in speed and in flexibility of presentation of calibration solutions to the instrument. The procedure is usually to aspirate a blank and up to 5 standard solutions, entering the standard values by keyboard. The calibration is then automatically established. It is expected that many of the facilities and the extra versatility of external on-line calculators will be included with internal microprocessor systems in future.

On-line Data Handling Systems with a programmable desk-top calculator are by far the most versatile at the present time for use with a single instrument. They enable a programme of calibration and curve correction which is particularly appropriate to a given problem or situation to be selected. The programmes may be provided as software by the instrument manufacturer, or they can be devised and written by the user himself with some practice.

The programmes usually available include those based upon the stray-light equation, or upon a general polynomial relationship between concentration and absorbance of the type $C = p + qA + rA^2 + sA^3$. Other kinds of relationship can also be provided.

A particular advantage of microprocessor assisted curve-fitting is that a number of readings can be taken for each calibration point, and the final operation is automatically carried out on the stored means.

Calibration can be carried out by any of the normal methods, which are described in more detail in the section on Calibration (p. 120), including the following:

Direct Comparison with synthetic matrix-matched standards, the method employed hitherto in most atomic absorption methods.

Slope-corrected Direct Calibration in which the overall calibration is done with synthetic standards, but the final slope is adjusted with a standard reference material of the same matrix as samples. This eliminates matrix interferences and also offsets some types of systematic error in sample preparation.

Method of Standard Additions as described on p. 125 can also be programmed thus avoiding the drawing of the calibration for each individual sample.

On-line calculators can also be programmed to give additional information, the reciprocal sensitivity of a particular determination for example, and the standard deviation of a series of measurements and hence the detection limit.

On-line computers, which can be programmed to provide all the functions mentioned above, would be used in large laboratories for the collation and formated presentation of data from a number of instruments and techniques.

The use of microprocessors or computers in instrument control

The applications of microprocessors so far mentioned have been limited to the processing of the raw data obtained from the measuring circuits of atomic absorption spectrophotometers and their conversion and presentation in intelligible form.

The human operator is involved only in setting up the instrument itself and presenting to it the calibration standards, at the same time entering the standard information into the processing system, followed by the samples. With an 'autosampler' calibration itself is done entirely automatically, provided the data system is programmed for the order of blanks, samples and standards. Clear advantages of this are the saving of the operator's time and the avoidance of calculations and report-writing and possible attendant errors.

A further aim in microprocessor or computer control is the automatic selection of all instrument parameters for any given analysis in a routine or control laboratory. It would be desirable to be able to do this with an accuracy and speed which a good human operator might find difficult to match. This is an area which is currently receiving much attention.

The entering of instrument parameters via a keyboard by the operator himself is an interesting, but up to now not particularly useful, exercise. It can and frequently does take longer than setting up in the traditional way. The problem of automatically selecting the absorption wavelengths with the required degree of accuracy is a particularly difficult one in this respect though in principle several possible solutions are known. Newstead's method[547] is to scan the region accurately and select the maximum intensity within the region with a peak-seeking device, having driven the monochromator to the near vicinity of the wanted line. In a commercially available unit, the line is approached always from the high wavelength side, the region scanned, the wavelength setting of maximum intensity being recorded and the monochromator driven back to that setting.

The difference between these two procedures is one of degree rather than of kind, but it is believed the former is probably more precise.

The technology is now available for all other instrument functions to be selected both on main instrument and peripheral devices, e.g. gas control box, electrothermal atomizer or autosampler. Programmes for instrument setting and data processing would be stored, for example, on magnetic cards, one card for each analysis. Although, as already indicated, the actual speed of analysis may not be vastly improved, the advantages lie in the better reliability and setting accuracy obtainable and in the possibility of more efficient use of the time of a skilled analyst.

Chapter 5

Requirements in Instruments for Practical Analysis

ATOMIC ABSORPTION SPECTROMETERS

With a large number of commercial atomic absorption spectrometers to choose from at the present time, the chemist about to embark on the technique may have some difficulty in deciding which features are important as far as his particular requirements are concerned. There is probably no perfect instrument and if there were it would be extremely expensive. Furthermore, manufacturers' specifications change at frequent intervals. It is not proposed therefore to review instruments individually, but to point out some of the features which are desirable in a good atomic absorption spectrometer for general analytical work, and to list the manufacturers in the Appendix.

Sampling unit for flame atomization

It should be a simple matter to introduce the sample into the burner, and hence the nebulizer take-up capillary usually extends outside the instrument. This also enables sample presentation devices (p. 101) to be connected, thus making the instrument virtually automatic particularly if concentration printout facilities are also available.

The nebulizer, spray chamber and particularly the burner, though readily accessible for dismantling and cleaning, should be enclosed. This prevents draughts from disturbing the flame and avoids damage both to the operator's eyes from ultraviolet radiation and to his person should a blow back occur. The flame should be viewable through an ultraviolet cut-off filter. The spray chamber should be constructed so that damage to itself is minimal and does not occur to the rest of the instrument in the event of a blow back.

The nebulizer should be constructed so that the sample only comes in contact with inert plastic or an inert metal. The nebulizer capillary is usually made from platinum iridium alloy, and the body of the nebulizer made in tantalum or coated with an inert plastic. Stainless steel nebulizers are sometimes used, but these are attacked by most strong acids in aerosol form and are soon irreparably corroded even by solutions containing ferric chloride. Impact beads should be made from inert plastic, not glass.

Automatic ignition of the flame is not simply a luxury. It is extremely useful, and inspires confidence in a reluctant operator, particularly if interlinked with the gas controls so that the flame can only be ignited with the correct burner and under the correct conditions. Should any supply fail, flow sensing devices in the gas lines can be made to shut the instrument down in the correct manner for which air and power must be made available. Although automatic control to this extent is more expensive, it ought to be considered seriously if unskilled or junior staff are employed, but it must be proven to be entirely reliable, as a single failure would negate its whole purpose.

The spray chamber and burner system should be designed to run as long as possible without dismantling and cleaning. In practical terms this must imply at least the duration of one analysis run, but preferably at least a whole working day. Dismantling should be quick and easy.

The burner, and particularly the burner jaws, should be massively constructed, to prevent both overheating (which could cause a blow back) and deformation (leading to variations in atomization characteristics). The burner may be pre-aligned in a horizontal sense but should be capable of adjustment both for height and angle. Burner heads for different gases should be easily interchangeable. A simple push-on fit is best.

It should also be convenient and quick to change from normal flame operation with a burner to atomization with a furnace or other electrothermal device, a hydride kit, a mercury kit or Delves' microsampling accessory. Such accessories can be designed to fit in place of a burner head.

Flow controls and meters are essential for both oxidant and fuel gases. It is an advantage to have a third flow control and meter to be used either for auxiliary oxidant or for more sophisticated combustion mixtures such as hydrogen/oxygen/argon, etc. which are used in research in atomic absorption and fluorescence and may be more generally required for routine determination of some individual elements.

It is convenient to be able to connect all gases used, e.g. air, nitrous oxide, acetylene and propane or hydrogen, to the instrument and to select the required pair by valves on the instrument.

Electrothermal sampling unit

As cylindrical graphite tubes of one size or another constitute by far the largest proportion of electrothermal atomizers in general use, the following remarks apply most particularly to them.

The graphite tube itself should have a low thermal mass to ensure firstly that heating rate is fast enough so that a high density of free atoms is obtained and secondly that cooling is fast enough to avoid undue loss of time between firings. Heating should be typically at the rate of about 1000 K s^{-1} initially.

The bore of the tube should be wide enough to prevent attenuation of the atomic radiation beam, and to allow separation of unwanted light emitted from the hot graphite tube from the main beam by one of the means mentioned on p. 82.

Because the radiation beam tends to be almost conical in the vicinity of the focus at the centre of the tube and of increasing diameter as it extends towards the ends of the tube and beyond, a tube profiled to the approximate shape of the light beam[249,605] makes more efficient use of the beam.

The tube should provide an isothermal environment within the measuring volume so that molecular recombination and condensation towards the ends of the tube are avoided. Such condensation occurring during the ashing phase can cause revolatilization at the higher temperature atomization step, thus increasing the interference from non-specific|scatter or absorption. In some furnace designs this effect is prevented by forcing a current of inert gas in through the ends of the tube and out through the sampling hole during the drying and ashing stages (referred to earlier as forced convection).

The graphite tube should be designed so that the sample can be easily and conveniently introduced. A micropipette delivering accurately reproducible volumes or an automatic injection accessory is essential for solutions. The tube should be able to accept solution volumes of up to 50 μl and retain it without spillage in the part of the tube that will attain the highest temperatures during the heating programme which will follow.

The atomizer head (containing the tube) should be readily installed and interchangeable with burners. It should also be a simple and relatively quick operation to install a new atomizer tube and to ensure that it is optically aligned.

Radiation source

Turrets containing up to six positions are usually provided for hollow cathode lamps and perhaps vapour discharge lamps. These may rotate or slide to bring the required lamp into position. It is probably better to leave the lamps stationary but to select the one required by a rotating mirror, though this may introduce two further reflecting surfaces in the optical path. This is certainly better for some bismuth and lead lamps which operate with molten cathodes, and is also more convenient to operate. Some turret systems hold all or some of the lamps not actually in use on a warm-up current which is lower than the operating current. This may save lamp-life marginally, but does not save time as an equilibrium intensity still has to be attained.

All lamps of whatever type require some warm up time, and thus it is useful to be able to switch the lamps on independently at the operating current.

The current supplied to the lamp should be stabilized to 0.1 % or better, and the lamp intensity drift should be better than 5 % per hour, particularly if it is to be used in conjunction with a single beam instrument.

Modulation of the lamp intensity may be electronic or mechanical. Even with mechanical chopping, however, electronic switching of the lamp so that it is 'off' during the dark period of the cycle has the advantage that it can run at a higher current and intensity during the 'on' period of the cycle, thus improving the overall intensity/energy ratio and thus the signal to noise ratio of the actual measurement.

Electrodeless discharge lamps are less easily modulated electrically than hollow cathodes. Radio frequency induced lamps are better than microwave in this respect, and if a power supply separate from the main instrument is to be used it should be possible to use one where the modulation frequency and phase can be controlled from the main instrument. A suitable signal can be found by tapping into the main power supply but is much more easily and conveniently obtained from the unused supply to the hollow cathode lamp which the EDL replaces.

Optics

The merits of single and double beam optics have been discussed. If stability of better than 5% per hour is essential, e.g. in quality control, monitoring etc., double beam ought to be chosen. Double beam instruments also give better precision for work of the highest analytical accuracy.

The monochromator must cover the range from 185 to 860 nm to include all metallic element resonance lines. The detector may not, for economic reasons, be sensitive over all of this range, but it should be capable of being replaced by one that is, with minimum modification to the instrument.

Gratings are now used in practically all commercial atomic absorption spectrophotometers. A prism monochromator can only be an advantage if there is to be some emphasis on the determination of metalloids—arsenic, selenium and tellurium—and a few metals—zinc and cadmium. In this respect the most important parameter is the resolving power of the monochromator, which varies considerably with wavelength for a prism but is virtually constant for the grating. For most purposes a resolving power of about 0.1 nm and a dispersion of 3 nm mm^{-1} are adequate.

The focusing optics, both pre-slit and monochromator, may use either lenses or mirrors or both. Each has certain advantages and disadvantages. Lens systems tend to be cheaper, to have a better energy throughput and to favour the 'optical bench' type of instrument design which has mechanical simplicity and stability. However, the principal disadvantage is that the positions of foci depend upon wavelength, and wavelength compensation may have to be introduced, particularly where focusing is important, e.g. with the use of small diameter electrothermal atomizers.

The advantages of mirror optics include invariability of foci and the ability to 'fold' the optical system, making for smaller and more compact instruments. On the other hand, mirror reflectivity is low compared with lens transmission, and mirrors require special coatings to preserve their initial reflectivity values. Also, mirror coatings are particularly liable to attack by the corrosive vapours so often present in analytical laboratories.

The manufacturers' choice and that of the user too is therefore a compromise between these things. Compactness may be very important to a particular user, and the advantages of a non-rotating lamp turret are had at the cost of using mirrors. Best performance/cost ratios are undoubtedly obtained with lenses. A good compromise would be to use a lens outside the monochromator to focus the atomic radiation on the atomizer, followed by mirrors in the monochromator itself.

Optical accessories

There is little need for a wavelength scanning device in routine analysis, but it is often useful in investigational work, e.g. to check interferences from emission lines near the resonance line or molecular emission from the flame.

Many applications, particularly those involving electrothermal atomization, benefit from the use of a background correction system to avoid errors caused by continuous absorption or scatter (see pp. 82 and 138).

The ability of an atomic absorption instrument to make measurements in emission has been acknowledged for some time, and it is easy to incorporate the necessary facilities in the instrument. It is equally easy to 'design in' the possibility of rearranging the optical system so that fluorescence measurements can be made but perhaps not so easy to modify an instrument that has not been so designed.

Double channel spectrometers

Several double channel spectrometers (as distinct from double beam) are available commercially. These allow simultaneous measurement at two wavelengths either in absorption or emission. This requires two separate optical systems—primary source, monochromator, detector—and only the atomizer is common to both. The data-handling sections are usually composite, so that if the channels are designated A and B at least three forms of read-out can be computed.

A and B Simultaneously. This could be, for example, the determination of two elements, with the object of halving the time and/or volume of sample normally required for the analysis. This would be particularly valuable in the solid sampling methods described in the next section when only minute pieces of sample are available. Emission and absorption could be used simultaneously if required.

$A - B$. The intensity—or absorbance—difference function may be used to obtain background corrections directly, the one channel being tuned to the element line being measured, and the other to a conveniently near point where the background is of similar strength.

Alternatively this mode can be used to correct for spectral interference, either in the flame by emission, or in the source lamp by absorption by tuning the B channel to the wavelength of the interference and balancing the outputs so that $A - B = 0$ for a sample which does not contain the element being determined.

A/B. This function is used in procedures involving internal standardization (see p. 126). Mixed emission/absorption measurements can be made if required.[225]

The value of double channel spectrometers for background correction is more questionable now that single beam instruments are readily equipped with simultaneous background correction. Internal standard methods are now perhaps not as useful for overcoming variations in performance of one part or another of the instrument as for avoiding the need for the very accurate weighing out of samples which contain their own internal standard element, e.g. steels and other alloys.[375]

In this kind of analysis the ratios of, for example, copper, manganese or molybdenum to iron are independent both of sample weight taken and of subsequent dilution and therefore there would be no need to spend time carrying out these operations with the usual analytical accuracy. Probably the most important use of the double channel spectrometer is that of a two-element analyser.

Automation

Atomic absorption techniques as we know them at the present time are not ideally suited to automation to the extent where the operator can happily leave equipment performing analyses unattended for many hours at a time.

Though double beam optics overcome the effects of variations in resonance source intensity and detector and amplifier response, it is still undesirable to leave flames running unattended for long periods of time without a flame control accessory. Electrothermal atomizers are safer in this respect but, until their analytical accuracy can be said to be as good as that attainable in the flame, probably useful only in a more limited range of applications.

In considering automation, different analysts may have different processes in mind, and it may be useful to consider the question in relation to each of the various operations that make up an atomic absorption analysis: (i) sample preparation, (ii) serial dilution, (iii) presentation to the instrument, (iv) multi-element analysis and (v) data handling.

Sample Preparation. The process of preparing a master solution from solid samples is usually the most time-consuming step in an atomic absorption analysis. This has not yet been successfully automated in any automatic chemistry system. It is not inconceivable that weighing out, dissolution, digestion, filtration and making up to volume could all be carried out without human intervention, but the mechanical problems would be considerable. Perhaps the nearest approach is the automation of the Kjeldahl digestion of plant or other biological materials. Here a portion of the prepared solution would be suitable for atomic absorption analysis after serial dilution.

Metals and some other substances in an already comminuted form might also be dissolved in similar equipment.

Serial Dilution. Equipment is available from laboratory supply houses (e.g. Griffin 'Diluspence', Fison automatic and Mettler processor-controlled diluters) by means of which predetermined volumes of any number of samples may be accurately diluted with another liquid, e.g. solvent, buffer solution etc. The use of this equipment saves a considerable proportion of the time that is often spent in pipetting and making up to volume.

Presentation to the Instrument. Accessories are available for most atomic absorption spectrometers enabling the operator to load a number of already prepared and diluted samples, and have them presented to the instrument one by one. Such autosamplers for flame work usually operate on what is popularly called the 'handshake' principle. When the sample container is in position, sample take-up is initiated and, after a suitable delay for absorbance stability to be attained, the sampler signals the main instrument for the integration period and data handling steps to be carried out. It should be possible to obtain duplicate or more integrated readings for each sample. At the end of the integration, the main instrument signals the autosampler to proceed to the next sample when a similar sequence takes place.

Autosamplers for electrothermal atomizers operate somewhat differently. The non-continuous nature of sample presentation for electrothermal atomization requires a wash-through for the sampling capillary between samples in order to minimize cross-contamination. The sample is then taken up and transferred into the atomizer usually by pipette, but in one example[493] by spraying into the warmed tube from a nebulizer. A signal from the sampler to the furnace controller then starts the heating programme. At an appropriate point in the heating programme, the furnace controller initiates the measurement and readout signal (as previously selected on the main instrument). During the cool-down period the controller also instructs the autosampler to change to the next sample, and to carry out the wash-through.

Autosamplers for both flame and electrothermal atomization should be capable of making the calibration procedure completely automatic, including the method of standard additions. The appropriate series of blanks and standards are loaded and the data handling system of the spectrometer programmed accordingly as mentioned in the next section. Automatic recalibration should also be possible at selected intervals during a long sample run.

With this type of accessory and printer output, the reading sequences, when the operator would otherwise have to sit for long periods in front of his instrument, are carried out automatically.

It would seem that this type of device could be fairly readily combined with a diluter, thus enabling the steps described in this and the previous section to be carried out in one accessory. This may also be achieved by using existing automatic analytical systems (e.g. Pye Unicam AC1 Automatic Processing System, Technicon Auto-analyser) though these often provide more facilities than are required. A simple dilution device for atomic absorption analysis[160] consisted of a T-junction capillary tube, one arm being connected to the uptake capillary, the second to the sample and the third to the diluent (a stream splitter working in reverse). By varying the diameter and length of the tubes to the sample and diluent, different dilution ratios could be obtained.

Similar devices have been described more recently by Rubeska[653] and others[774] for the addition of buffer solutions to samples. Branched uptake capillary nebulizers have also been used for the simultaneous addition of standards, buffers and releasing agents to the analyte solution.[259,514]

Multi-element Analysis. Although instrumental systems have been proposed or devised for the determination of several elements simultaneously, the idea of multi-element analysis by atomic absorption has not found general application, largely because of the limits imposed by the absorption law itself. The concentration range over which an element can be determined with acceptable accuracy in a given solution is comparatively small. It is unlikely therefore that when two or more elements are to be determined simultaneously in the same sample solution their concentrations will all occur in the correct range. Unlike emission techniques, where the sensitivity of individual detectors can be varied, and atomic fluorescence, where sensitivity can be increased with source power, in atomic absorption there is little that can be done instrumentally other than investigating the possibility that a compromised set of conditions may allow all the elements to be measured.

Atomization conditions cause most difficulty and must be chosen in order to atomize all the required elements. It is likely that a high temperature will be necessary with the possibility that some elements may still be incompletely liberated from their compounds while others may be largely ionized. Full use must therefore be made of spectroscopic buffer systems. Sensitivities can be altered relatively only by the use of alternative absorption lines.

If all these difficulties can be satisfactorily solved for a given elemental system, it remains to design an instrument in which the resonance radiation of each element is passed through the flame and then made to fall on a photodetector, having been separated both from all other modulated radiation from the source and from the comparatively very large amount of unmodulated emission from the flame. The latter consideration makes it essential that the dispersing device (monochromator, filter or resonance detector, p. 107) comes between the flame and the photodetector. Unless the incident resonance radiation from all elements is made to pass along the same path through the flame, a separate dispersing device is required for each element.

Walsh (1967) suggested systems whereby coincidence of resonance beams could be achieved.[771] Apart from the simplest one using a multi-element lamp, these usually resolve into the use of two monochromators—one used in reverse to combine beams from individual sources, so that they may pass coincidentally through the flame, followed by the second monochromator used to separate the beams again after the absorption process. Mavrodineanu and Hughes[501] described instruments based on this principle, and also two optical arrangements in which the same grating is used for both functions. Further detectors may be provided if emission measurements are to be made simultaneously.

The use of resonance detectors for each element would much simplify the optics, but, although one commercial double channel instrument using these devices was made available the principle has not become popular.

Data Handling. Circuitry which linearizes the calibration relationship, calculating a concentration value for display or print-out, and the possibilities offered by the incorporation of microprocessor-based data handling have already been mentioned. One or other of these systems is obviously a necessity for an automatic system, as they provide print-out of the actual analytical result with no intermediate step required by an operator. Many atomic absorption instruments now on the market offer these facilities, together with interfacing for on line calculators or computers and autosamplers.

REQUIREMENTS IN FLAME EMISSION AND FLUORESCENCE

A flame emission spectrometer consists of an atom source, monochromator and detector, while atomic absorption and fluorescence spectrometers consist of the same parts with the addition of a resonance radiation source shining through the flame and placed respectively on or at right angles to the pre-monochromator optical axis.

It is not surprising therefore that users may wish to employ their equipment for two or all three of these techniques. It is also desirable because some elements show their highest sensitivity in emission, some in absorption and some in fluorescence.

Both 'emission' (by which is understood thermally excited emission) and fluorescence are essentially emission phenomena, the radiation being emitted in all directions. Sensitivity is therefore improved if the optical system is arranged to gather light over the widest possible angle. A monochromator with a large aperture, e.g. $f.4$, and possibly with light-gathering mirrors placed behind the flame is therefore used. Longer slits can also be used in emission and fluorescence than in absorption in order to allow more wanted energy to reach the detector. Because emission spectra are more complex than absorption spectra, a monochromator with higher resolving power is needed if elements are to be determined in mixtures by emission. Conversely, as fluorescence spectra are simplest of all, measurements in fluorescence can be made adequately with lower resolution.

Flame emission

The only requirement for flame emission measurements, other than a circular Meker-type of burner head, is that the emission from the flame must now be modulated at the same frequency as the amplifier. This is achieved with a vibrating or rotating chopper between the flame and the detector which is brought into action when the instrument is switched into the 'emission' mode. Light from the hollow cathode lamps used in absorption must of course be prevented from reaching the detector.

Particular developments engendered by atomic absorption have re-stimulated interest in flame emission spectroscopy after a dormant period. Chief of these are the use of the nitrous oxide acetylene flame and of various separated flames. There is thus growing a modern literature on flame emission spectroscopy which features elements that would not have been considered less than ten years ago, but which is beyond the scope of this book. However, useful flame emission lines for a number of elements are included in the data given in Chapter 9.

Atomic fluorescence

Requirements for fluorescence measurements are more exacting. Although the fluorescence intensity depends directly upon the number of free atoms present in the flame and is therefore subject to the same interference effects as atomic absorption, it suffers from one further effect—that of quenching. By 'quenching' is meant the reduction in quantum efficiency caused by the presence of other species in the flame which exert a de-exciting effect on the excited atom. Both nitrogen and hydrocarbons give rise to quenching, and therefore in fluorescence it may be necessary to consider combustion gases which do not contain either of these materials. Hydrocarbon gases may be replaced by hydrogen, and air or nitrous oxide by argon oxygen mixtures. Unfortunately hydrogen–oxygen–argon flames do not have the same reducing properties as flames which produce carbon-containing radicals and they are not as effective for the metals which form refractory oxides. The flame gases may therefore have to be chosen more carefully with these two conflicting requirements in mind.

Another quenching effect known as 'self-quenching' depends upon the relationship between the fluorescence intensity and the atomic absorption coefficient. As derived by Winefordner[823] the full relationship is somewhat complex, but it was shown that the intensity passes through a maximum value as the number of fluorescing atoms increases. This results in calibration curves which flatten and bend over towards the high concentration values. It is therefore necessary to work in the lower concentration ranges. The critical concentration varies of course from element to element, and even below this value a very wide concentration range, e.g. $1:10^4$ or more, can be handled.

An important condition for sensitivity in atomic fluorescence is that there should be a low level of scattered light, particularly of the same wavelength as the fluorescence. This is again largely achieved by modulating the primary source and the amplifier at the same frequency so that continuous emission from the flame is not recorded. Light scattered by small solid particles is not prevented from being recorded as this will also be of the same wavelength as the primary source. This condition is thus more difficult to fulfil when resonance fluorescence is being measured, but the problem does not arise if stepwise or direct line fluorescence (see p. 13) can be utilized, as the fluorescence is then of a different wavelength from the primary radiation and scatter, which is therefore rejected by the monochromator.

From a practical point of view there are two further potential advantages of fluorescence over absorption. Since the monochromator only 'sees' the actual fluorescence radiation, and direct radiation from the source is directed away from the monochromator, the primary source does not specifically have to produce sharp line spectra. Therefore more intense sources such as electrodeless or vapour discharge lamps can be used with a proportional increase in sensitivity, the fluorescence intensity being directly proportional to the intensity of the primary radiation.

Taking this one step further, it is clear that fluorescence may equally well be excited by a continuum source, as the free atoms will only absorb at their resonance wavelength. The energy per unit spectral bandwidth of continuum sources, however, even of the high pressure xenon arc, is much less than for a line source over the resonance absorption line width and so sensitivities are much lower. But a single source can be used for nearly all elements.

Continuum Source Excitation. Although Winefordner and Vickers[823] first suggested the continuum source in 1964, the range of elements determined with this type of source has not altered appreciably since further work was reported from the same laboratory two years later. Then, Veillon *et al.*[764] investigated 13 elements using a 150 W xenon arc lamp in conjunction with a phase-sensitive amplifier. Good detection limits were obtained for bismuth, copper, gold, lead, magnesium, silver, thallium and zinc in an oxyhydrogen flame and for lead, magnesium and thallium in an air hydrogen flame. Scattered radiation was a serious interference, and Cresser[161] stated that even when the fluorescence spectrum was scanned, scatter contributed considerably to the background noise. He also suggested that the fluorescence excited by a continuum, in the type of flame necessary to ensure adequate atomization of many elements, was lower in intensity than the simultaneous thermal emission. Cresser and West[163] investigated the fluorescence of thirteen elements in a premixed air acetylene flame excited by a 500 W xenon arc. They found that the interferences were similar to those encountered in emission, though molecular band emission did not interfere. Detection limits for many of the elements including copper, silver, indium, thallium, cadmium, calcium, magnesium, chromium, manganese, cobalt and nickel were better in emission.

Hollow Cathode Lamp Excitation. Modern hollow cathodes often give sufficient intensity to excite fluorescence, though very good limits of detection cannot be expected. Nevertheless, West and Williams[790,791] determined silver and magnesium in an air propane flame with detection limits of 0.0009 and 0.0008 ppm respectively.

Electrodeless Discharge. Microwave excited electrodeless discharge lamps can be made for many elements. Where comparisons can be made, it is clear that this very intense source of radiation leads to an appreciable improvement in the detection limits. Electrodeless discharge lamps, both microwave and radio-frequency, are now available commercially. Under normal conditions the microwave type give the highest intensities but should be thermostatically controlled to achieve the desired degree of stability.[120]

Instruments for Atomic Fluorescence. It is perfectly possible to modify some commercial atomic absorption instruments so that they become capable of making atomic fluorescence measurements. This was particularly easy with some earlier 'open' instruments constructed on the optical bar principle. In the conversion of one type of enclosed instrument[80] the flame on a Meker burner is moved forward out of the optical axis and a series of mirrors arranged as shown in Fig. 39. Two mirrors are positioned to deflect and pass the excitation beam through the sample a second time. A third mirror collects the fluorescence radiation from the flame and passes it along the original optical axis into the monochromator.

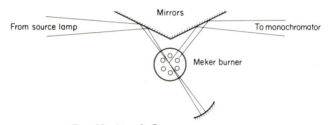

FIG. 39. Atomic fluorescence accessory.

Although electrodeless discharge lamps were used for cadmium and for zinc, the detection limits reported were no better than for atomic absorption, presumably because of the small solid angle over which the fluorescence can be directed to the monochromator.

A less sophisticated approach was made by Hubbard and Michel[332] who simply placed the source lamp in front of a single beam atomic absorption spectrophotometer at the same level as the optical axis, and close to and pointing into the flame. In two studies Winefordner and Staab[822] showed that a focusing lens for the excitation beam is not particularly helpful as the same result can be obtained by placing the lamp near the flame.

It is, however, necessary to prevent stray incident radiation from reflections near the cavity and other parts of the sample compartment of the instrument from passing towards the entrance slit. For this reason, Hubbard enclosed his electrodeless discharge lamp completely in a metal casing. In modern commercial electrodeless discharge lamps the tube and cavity are also completely encased, making them particularly suitable for this kind of application.

In further work on atomic fluorescence the same authors[333] converted another conventional flame absorption/emission instrument, incorporating a lock-in amplifier and preamplifier and high frequency modulation of the electrodeless discharge lamps. Sensitivity was increased by doubling the efficiency of the incident radiation with a concave mirror and using a system of condensing mirrors to increase the angle of collection of fluorescent radiation which was improved again later.[206] The device was used to study effects of different atomizing flames on the determination of tin in steel.[334]

DEVICES BASED UPON ATOMIC FLUORESCENCE

Resonance radiation detector

Principle. A resonance detector (sometimes called a resonance monochromator) is a device for detecting the resonance radiation of an element, based upon the fluorescence emitted when the radiation is passed through it.

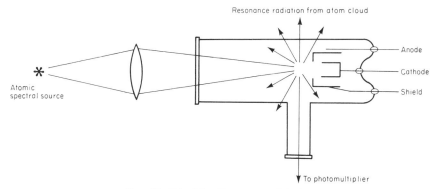

FIG. 40. Principle of resonance detector.

Resonance detectors were originally based on the design of hollow cathode lamps. A cloud of atoms of the elements to be measured was produced by sputtering from a conventional hollow cathode. Light from a second hollow cathode or other source passed through the cloud, exciting the atoms to emit fluorescence (Fig. 40). The fluorescence was measured with a photomultiplier. If a sample was aspirated into a flame between the source and the resonance detector (Fig. 41), any of the element present in the sample would cause absorption of the radiation falling on the detector and there would be a proportional loss of fluorescence intensity.

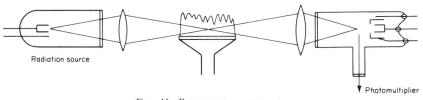

FIG. 41. Resonance spectrometer.

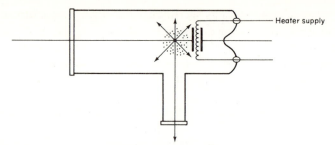

Fɪɢ. 42. Thermal detector.

The most important characteristic of a resonance detector is that it cannot be put out of adjustment by changes in temperature or mechanical shock. The effective resolution of a resonance detector is about 0.001 nm but the optical aperture is wide as compared with a conventional monochromator. Thus the amount of radiation from the flame that can produce noise on the output signal is far less than with a dispersion monochromator.

A disadvantage of the device is that it has a limited life in the same way as a hollow cathode lamp. Also, the conventional type of cathode may need careful shielding to prevent direct passage of light to the detector. These disadvantages are overcome for a very limited number of elements by the thermal type of resonance detector (Fig. 42). Here the atom cloud is produced thermally, the lamp requiring only a few watts to produce the optimum vapour pressure for volatile elements such as calcium, magnesium, copper, zinc and silver.

Uses of the Resonance Detector. Resonance detectors have been used to isolate lines used in the determination of lithium, sodium, potassium, magnesium, calcium, copper, zinc, silver, nickel, lead[771] and other elements. The use of resonance detectors simplifies the optics of multi-element systems, particularly if two elements can be atomized in each detector. Figure 43 shows one arrangement whereby four elements could be measured at the same time.

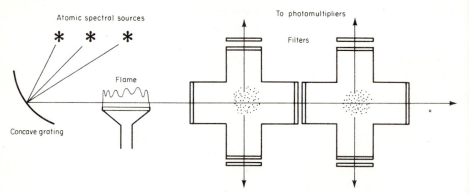

Fɪɢ. 43. Multi-element system with resonance detectors.

The wavelength stability of these detectors has led to their being suggested for portable analysers. A single detector containing calcium and magnesium could be furnished with photodetectors placed behind filters to select the wavelengths respectively of these two elements. A dual element source for calcium and magnesium would complete a very simple device for determining these two elements.

A device for determining copper, zinc, silver, nickel, cobalt and lead locally in ores overcomes relative sensitivity differences by taking different path lengths through a large circular burner. Sputtering-type detectors are used for all elements except zinc and lead, for which thermal detectors have proved superior. Analytical accuracies of about 5 % are possible.[772]

Simultaneous multi-element fluorescence systems

The simultaneous determination of a number of elements is well known in atomic emission, where arc or spark excitation can be applied easily to solid specimens but less readily to liquids. The limitations of atomic absorption in multi-element systems have been mentioned, but it appears that fluorescence is uniquely suited to simultaneous multi-element analysis of liquid samples. Some of the reasons for this are given.

Atomic fluorescence spectra are particularly free from spectral interference effects, even with transition and heavy elements.

The atomization and excitation functions are separated (indeed there may be a separate excitation source for each element) so good sensitivities can be attained for each element. In fact the sensitivity of each element is adjusted as required by altering the power of its resonance source.

There are no optical problems in focusing the source radiation into the flame, as, contrary to the requirements in atomic absorption, it can pass in any direction *except* directly into the monochromator.

Further, the concentration ranges of the several elements being determined in one solution are not restricted, either absolutely (provided they are not too high) or relatively, as they would be in atomic absorption by the absorption laws.

A device in which these advantages have been utilized was described by Mitchell and Johansson.[523] This is shown diagrammatically in Fig. 44.

Four element channels were provided, but in principle these could be increased. Light from the four hollow cathode lamps is focused into the same part of the flame and light from the flame is directed by means of a focusing mirror through a filter wheel on to a photomultiplier. A synchronous motor continuously rotates the filter wheel and an iron pulse-generating wheel at 1 Hz, producing a trigger pulse as each filter comes between the flame and photomultiplier. The wheel holds four interference filters selected to isolate the fluorescence radiation of the four chosen elements.

The trigger pulses are used for gating a 1 kHz modulated supply to the source lamp and the amplifier. The four channels then feed four integrators, the final

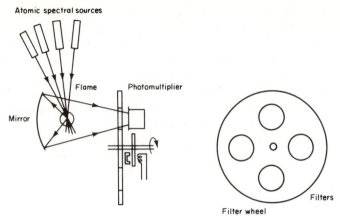

FIG. 44. Simultaneous multi-element fluorescent analysis.

voltages of which are printed out via a digital voltmeter. The gating pulse sequence is arranged so that the photomultiplier 'sees' the flame emission background in each filter bandpass just before the lamp is switched on, and this is corrected for.

The determination of copper, iron, magnesium and silver in an air hydrogen flame was described, this flame emitting the least molecular background continuum.

TABLE 4. Some detection limits in simultaneous multi-element fluorescence

Element	nm	Detection limit ppm	Flame
Aluminium	396.1	0.3	N_2O—C_2H_2
Silver	328.0	0.07	Air—C_2H_2
Calcium	422.7	0.003	Air—C_2H_2
Cadmium	228.8	0.03	Air—C_2H_2
Cobalt	240.7	0.06	Air—C_2H_2
Chromium	357.6	0.02	Air—C_2H_2
Copper	324.8	0.04	Air—C_2H_2
Iron	248.3	0.03	Air—C_2H_2
Magnesium	285.2	0.005	Air—C_2H_2
Manganese	279.5	0.003	Air—C_2H_2
Molybdenum	312.6	0.2	Air—C_2H_2
		1.0	N_2O—C_2H_2
		0.1	Argon-separated N_2O—C_2H_2
Nickel	232.0	0.08	Air—C_2H_2
Lead	405.8	0.07	Argon-separated N_2O—C_2H_2
Antimony	217.6	0.1	Air—C_2H_2
		(in MIBK)	
Selenium	204.0	1.5	Air—C_2H_2
Zinc	213.9	0.07	Air—C_2H_2

A further development of this instrument was described[193,521] with six channels, a nebulizer burner system for air hydrogen, air acetylene and nitrous oxide acetylene. There were both AC and DC amplification channels so that either fluorescence or emission could be measured. The detection limits for a signal to noise ratio of 2 are quoted in Table 4.

The six channel system was subsequently used by West et al.[368] to measure cobalt, chromium, copper, iron, manganese and zinc after preconcentration by extraction using sodium diethyl dithiocarbamate. The organic solution was atomized in a nitrogen-separated air acetylene flame.

Chapter 6

Analytical Techniques in Flame Atomic Absorption

Atomic absorption spectrometry is virtually a universal method for the determination of the majority of metallic elements and metalloids in both trace and major concentrations. The form of the original sample is not important provided that it can be brought into either an aqueous or a non-aqueous solution. This situation has been brought about by considerable improvements in instrumentation and also, perhaps partly as a result of this, a better understanding among analysts of the types of interference effect that may modify the expected response of a given element.

Atomic absorption methods combine the specificity of other atomic spectral methods with the adaptability of wet methods. High specificity means that elements can be determined in the presence of each other. Separations, which are necessary with almost all other forms of wet analysis, are reduced to a minimum and often avoided altogether, making a typical atomic absorption analytical procedure attractively simple. This fact, combined with the ease of handling a modern atomic absorption spectrometer, makes it possible for routine analyses to be carried out quickly and economically by relatively junior laboratory staff.

Usually, separations are required for only one of two reasons, to remove a major cause of interference or to concentrate the elements to be determined should they be present in amounts less than their detection limit. While separation procedures must therefore be quantitative for the elements concerned, they do not necessarily have to be specific as it is possible to determine a number of elements together in one solution. This concept leads to the separation of groups of elements rather than individuals, and indeed to a general philosophy of chemical preparation of samples for atomic absorption in which as many elements as possible are brought together for determination in the final analysis solution. This should always be the aim in method development.

OPERATION OF THE INSTRUMENT

Best results can only be obtained when the measuring instrument itself is maintained in good order and operated correctly. All manufacturers give specific instructions, and usually training, on these matters. However, certain points may be emphasized without in any way duplicating what is laid down in instruction manuals.

The instrument should always be placed under a suitable fume extraction hood. This is essential if a nitrous oxide flame or a halogen-containing organic solvent is to be employed or if toxic elements are being vaporized, in order to remove both toxic vapours and the 3 kW of heat generated by this flame. It has been shown[453] for example, that after aspiration of trichloracetic acid, burner gases contain chloroform, chlorine and phosgene.

The fume hood should extend over an area no less than about 30 cm square, and the open end should be between 15 cm and 30 cm above the chimney of the instrument. It is now recommended that its capacity should be such that it will remove 3 m^3 ($\sim$100 cu. ft) min^{-1}.

Optical alignment

To achieve best performance and sensitivity, the hollow cathode or other source should be positioned and its image focused on the monochromator entrance slit, the monochromator and pre-slit optics having been pre-aligned by the manufacturer. With this condition fulfilled, it is wise to check that the burner slot is aligned with the optical path, otherwise a considerable loss of sensitivity may result. Most instruments allow a lateral adjustment of the burner mount for this purpose. The adjustment is best carried out as follows: set up the instrument with no flame and any convenient source line to read 100% transmission (or zero absorbance), then position a rectangular piece of metal, the 'jig', so that one vertical edge is accurately halfway over the burner slot at one end of the burner. Note the reading. Now place the jig exactly halfway over the slot at the other end and again note the reading. From the two readings decide which way to turn the burner to make the readings given with the jig at each end of the burner slot equal. When they are equal, and with the jig still in position at either end of the burner, adjust the burner position laterally until the absorbance reading is 0.30 (transmission 50%). Subsequently, when new lamps are to be aligned it should only be necessary to position them so that the maximum output signal is obtained. This is usually done with an adjustment on each individual lamp holder.

Critical adjustment of the background corrector is described in Chapter 7.

Adjustment and care of nebulizer–atomizer systems

The number of free atoms actually produced in the flame 'cell' and therefore the *sensitivity* of the method as already defined depends directly upon the proper functioning of the sample handling system.

Nebulizers may be pre-set or adjustable. If the former, they will have been set in the factory to give the best nebulization efficiency. If adjustable, they can be set by the user to give either the greatest absorbance signal (which should correspond to best nebulization efficiency) or a required uptake rate. The first of these is obviously desirable in the majority of cases and as seen (p. 34), it is most likely to occur at a flow rate of 4–4.5 ml min^{-1}. Manual adjustment of the nebulizer to achieve maximum sensitivity can be done in one of two ways:

(i) With the flame running and the instrument already recording the absorbance of one of the 'easier' elements, the nebulizer is adjusted until the maximum absorbance signal is recorded.

(ii) Without the flame running but with the air supply connected, a vacuum gauge or simple mercury manometer should be attached to the sample uptake capillary. The capillary is first screwed back until a positive pressure is indicated. (In this condition the nebulizer would 'blow' instead of taking up the sample.) The capillary is then screwed forward until the gauge indicates the first vacuum maximum. It should in fact show only one maximum unless the capillary is not situated concentrically in the annulus.

Different flow rates may be selected for the following purposes:

A slow flow rate, for use with very small samples, to enable the recording system to reach an equilibrium value before the sample has been used up. The signal would be lower than in proportion to the new flow rate as the nebulization efficiency would also be lower. This could probably be accommodated unless traces near the detection limit were being determined.

A higher flow rate to give a higher absorption signal. This is a more doubtful proposition, as not only would the increase be less than proportional to the flow rate but some large droplets may well reach the flame, increasing noise and lowering the flame temperature with consequent loss of stability and reproducibility.

The nebulizer is likely to require little attention provided that it is not attacked or corroded by sample solutions. Modern nebulizers are usually lined with or made from inert materials where the sample comes into contact. They may, however, become blocked with solid material suspended in the sample and if not cleaned carefully some damage may result. Solutions filtered through paper, especially 'non-hardened' papers, are a frequent source of blockage as they contain suspended fibres. Millipore-type filters or centrifugation provide the best means of phase separation and should be employed where possible. Blockage of the nebulizer capillary itself can be avoided by placing a very short section of similar capillary tube in the sampling end of the uptake tube. Blockage then occurs at this point. The uptake tube is removed from the nebulizer, blown out and then replaced. Blockage of the nebulizer tube itself can usually be removed by passing a suitable length of wire in the sampling end of the capillary and pushing it through to the nose-piece. Wire introduced through the nose-piece may either damage or distort this end of the capillary with consequent alteration of the nebulizer characteristics. Alternatively the nebulizer can be removed from the spray chamber, and, while the air pressure is maintained, a rubber pad is held over the nose-piece. The air blows back through the capillary and should remove the blockage.

The spray chamber should be kept clean at all times. When purely inorganic solutions are being nebulized this is usually no problem, as the nebulization of pure solvent between and after the samples prevents evaporation of solutions with deposition of solute on the walls and on the flow spoilers.

Biochemical samples, particularly dilutions of blood, serum, proteins etc., tend to cause deposits to build up in the spray chamber in spite of interval washes. This can cause memory effects and lack of precision in the readings. Spray chambers ought therefore to be cleaned carefully after runs of such samples, and certainly at the end of each working day. It is necessary to wash through well or even replace completely the liquid trap between aspirated solutions which may cause precipitation or reaction, and a wise precaution, in any case, to dismantle and clean the spray chamber and flow spoilers at least once a week. During this weekly clean, too, the burner stem and any associated tubing should be dismantled and cleaned. Although in theory these should never become wetted by the sample mist, in practice some deposition does occur and if left can start spots of corrosion.

Burner shells should be cleaned weekly, though the jaws may require attention at more frequent intervals, possibly after each run of samples if these contain more than one or two per cent of dissolved solids. Some burner heads can be dismantled for cleaning, and often washing with water is sufficient. Organic or carbon deposits can be removed with a plastic domestic dish scourer and polished if necessary with '000' grade emery paper. Neither acids nor chemical polishes should be used as these start microscopic etch pits which develop into corrosion. The use of metal tools, e.g. screwdrivers, scrapers etc., or even razor blades should not be contemplated as their continued use will soon cause a rough surface and alter the burner characteristics. Even '000' emery paper should be used very infrequently to avoid widening the slot.

The slots of burners can usually be adequately cleaned with a piece of stiff paper (an unwanted visiting card is admirable) if it is inconvenient or impossible to dismantle the burner.

The top surfaces of burners are best polished with very fine emery paper, while deposits formed in the holes of Meker burners can be removed with soft wire.

Effects of organic solvents

Most organic solvents can be treated exactly as water except that problems sometimes arise when non-miscible aqueous and non-aqueous liquids are brought into contact in the drain tube and syphon. Precipitates or thick emulsions which prevent the steady flow of the waste liquid to the drain may be formed. If this is likely to occur the drain tubes should be emptied and refilled with the solvent to be used in the next run of analyses.

In nebulization systems with plastic tubes, spray chambers etc., it must be established that the organic solvent does not soften or dissolve the plastic. Hardened polythene, propylene and polytetrafluoroethylene seem to be impervious to all likely organic solvents, but PVC is soon softened by methyl isobutyl ketone.

It is important, too, to ensure that the critical C/O ratio is not exceeded, and for this reason it may be preferable to use a non-hydrocarbon flame, e.g. the nitrous oxide hydrogen flame, if solvents which cause very smoky flames have to be used.

Lighting and maintaining the flames

Some details have already been given about commonly used flames (pp. 28 *et seq.*). When lighting flames it is a good general rule to turn on the oxidant first and to turn it off last. Acetylene burning without premixed oxidant produces smoke and particles of soot and this condition is much to be avoided, even with an extraction hood. The oxidant should first be turned up approximately to the operating level, then the fuel gas turned on and the flame lit. The flows are then adjusted to their correct levels, and pure solvent is nebulized until the analysis run begins.

If a flame tends to lift off the burner head either one or both of the gas flow rates are too high. It is always better to have the flow rates too high than too low before lighting, however, as in the latter case there is more chance of a blow-back. This applies particularly to gas mixtures with a high burning velocity. The same principle applies when the flame is extinguished. The gases should *never* be turned down together otherwise the critical flame propagation velocity will be reached and, with the burner plates hot, a blow-back is likely to occur. The operating flow rates should always be such that if one gas is turned off or fails, the velocity of the remaining gas through the burner slot is still greater than the flame propagation velocity and the flame does not burn back through the burner slot.

Instrumental settings and the minimization of noise

Among the instrumental settings to be decided upon in any particular analysis are lamp current, photomultiplier gain, bandpass and integration time (or, in older instruments, amplifier gain and damping, and recorder expansion and damping). These are all chosen in conjunction to provide the most noise-free and responsive signal.

Ideally the signal as recorded on a chart recorder should appear noise-free and sharp as in Fig. 45 (a). Noise from various sources produces less distinct peak values (b). Sluggish response (damping) gives rounded off traces (c).

Noise on the signal originates in various ways throughout an atomic absorp-
ion system and it is worthwhile to consider these individually to see how they
each affect the final signal.

Flame Noise. This arises from refractive index variations in the region between
the hot parts of the flame and the cold surrounding atmosphere and from small
variations in effective path length of the flame cell. The flame is a highly dynamic
system which cannot be contained within closely defined dimensional limits, and
the convection currents induced around it cause such variations. Flame noise
always exists. It can only be minimized by good design of the burner ('laminar-
flow' burners are clearly the best), the flame compartment and chimney, and by
preventing draughts from windows and doors from reaching the flame. A fume
extraction system with too fierce a draught can cause turbulence effects within the
flame compartment.

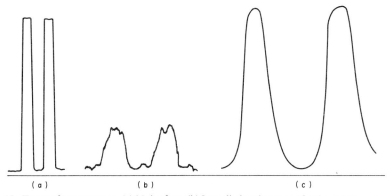

FIG. 45. Types of output trace. (a) Noise free. (b) Less distinct because of noise. (c) Overdamped.

Nebulizer ('*Concentration*') *Noise.* This is caused by fluctuations in the amount
of the element being measured which actually reaches the flame. Nebulization is
a quantized process and the droplets are produced in a slightly irregular stream.
The irregularities are smoothed out during the passage of the mist through the
spray chamber. There is thus a minimum effective size of the spray chamber and,
as indicated earlier, one which is too small, though giving somewhat higher
sensitivity values, will almost certainly produce higher noise.

Both flame noise and nebulizer noise result in short term fluctuations of the
absorption occurring in the flame, and hence, by whatever means the absorption
signal is recorded or amplified, noise from these sources will remain a constant
proportion of it.

Source Lamp Noise. This consists of long term ('drift') and short term noise.
The lamp discharge is a statistical mean of all the micro-scale sputtering discharges which can sometimes even be observed visually as a slight 'flicker'. There
is a minimum lamp current below which this effect is so great as to make the
discharge unstable. The lamp current must therefore be sufficiently high to

minimize this effect, but not so high that broadening of the resonance lines affects the sensitivity.

The intensity of the radiation from the lamp is the subject of all measurements, and any superimposed noise is therefore also amplified and recorded in the final output signal. The only exception to this is lamp noise of a frequency less than the chopping frequency in double beam spectrometers.

Noise Generated in Photomultipliers. This may be divided into dark noise and shot noise. As the dark current increases with cathode temperature so does the dark noise. While photomultipliers in commercial spectrometers hardly require special cooling facilities, in a well designed instrument they should not be subject to undue heating by other parts of the electronic system or by conduction of heat from the flame compartment. Shot noise is caused by irregularities in the electron showers produced by individual photons at the cathode and electrons at subsequent dynodes. The greatest contribution is thus at the cathode and there may be a small detectable effect at the first dynode. At later dynodes the effect is averaged out and it is here that the dynode voltage may be varied to produce a higher amplification factor for small incident light signals. As the shot noise is proportional to the square root of the incident intensity, photomultipliers should not be operated close to their threshold intensity values. When operated under ideal conditions, this detector can provide a very high, virtually noise-free, gain factor.

Amplifier Noise. In well designed amplifiers with reliable components this is entirely Johnson noise, which is so small that it is negligible compared with other sources of noise. This applies to both the gain and scale expansion facilities which, though similar, are separate in instruments reading directly in absorbance. Where a chart recorder is used the only further source of noise is usually the point of contact with the potentiometer or slide-wire. This is readily distinguished by its very irregular character.

In a correctly functioning instrument, the greatest source of noise is often the flame and this is usually made clear by comparing the character of the signal with and without the flame and sample nebulization.

Among the electronic parts of the instrument, the radiation generator is usually the next greatest source of noise and the lamp current should therefore be carefully optimized as suggested above. The object of the detector and amplifier is to provide a signal sufficiently large to drive the read-out device. In modern electronic design, the amplifier is usually run continuously on full gain because it is possible to control the photomultiplier gain without loss of stability. In this way it is possible to accommodate a large difference in intensities provided by source lamps of different elements.

Effects of Slit Width. In the foregoing observations it has been assumed that the width of the monochromator slits has been set to the best band pass for a particular analysis. The best value is again a compromise. The resolution of the monochromator is improved up to a limiting value (which depends on the remainder of the monochromator optics) as the slits are made narrower. On the other hand more light energy passes through the monochromator and into the detector as the slits are widened. The best value therefore depends on the following:

(i) The wavelength, if a prism monochromator is in use. The dispersion (in mm nm^{-1}) at 220 nm can be about 10 times as great as at 400 nm and twenty times as great as at 500 nm. Hence the slits can be considerably widened at lower wavelengths without loss of resolution but with an increase in energy falling on the detector.

(ii) The nearness to the resonance line of other lines emitted by the source. If non-absorbing lines originating in the spectrum of the element being determined or the lamp fill-gas are included in the spectral bandwidth of the monochromator these will fall on the detector, and however great the absorption of the resonance line the recorded signal will never fall below their intensity value. In general, therefore, elements with complex emission spectra, e.g. the transition and heavy metals, will require narrower slits than those with simple spectra, such as the alkalis and alkaline earths.

(iii) The intensity of the resonance line. The slit width can be increased if insufficient gain is available from the electrical system to operate the instrument satisfactorily. It may be regarded as a source of noiseless gain available at the cost of resolution and sensitivity. It is noiseless in the sense that it does not introduce a further noise factor. Existing noise is increased only in proportion to the signal.

With a low intensity source, therefore, better stability could be achieved by increasing the slit width, so that a signal of acceptable intensity falls on the photomultiplier. Simply to increase the gain of the photomultiplier operating at a low light level would amplify the higher shot and dark noise proportion of the output. Conversely the sensitivity (working graph slope) is best at the narrowest usable slit width.

(iv) Background emission from flame. The level reaching the photodetector is proportional to the square of the slit width and, as has already been noted, increases shot noise and the likelihood of saturating the detector.

(v) In the last resort, the matching of atomic spectral and continuum lamp intensities in instruments fitted with a background corrector, remembering that continuum lamp energy throughput varies with the square of the slit width, while line intensity is directly proportional.

Integration and Electrical Damping. This reduces the noise shown by a digital display, a meter or on a recorder trace. These readout devices themselves provide some inertia in the system but further variable damping should be provided. The degree of damping is usually defined by the time constant of the noise-filtering circuits. Typically this may be 0.2–2 s with a meter and 1–4 s with a recorder. The time required to make a reading is proportional to this time. If the time constant is too long, then at best, very rounded off traces (see Fig. 45c) or irregularly drifting traces or readings may be obtained. If it is too short an excessive noise amplitude will be apparent. However in cases of severe noise it is often easier to estimate a mean visually when the noise is reduced to say $\pm 3\%$ than it is to rely on the settling-down of an over-damped meter or recorder.

The noise effects which have been described would be observable only on instruments giving continuous readout. In the integrating or averaging systems employed to facilitate digital display or printout, the effects, though nevertheless present, may not be so apparent. The noise is manifest in variations in the averaged readings observed with the shortest integration time. This may be 0.2 or 0.4 s and thus gives virtually as much information as a recorder though in a form which may not be quite as readily appreciated. Conditions are adjusted as before until this variation has been minimized. When actual analyses are being performed the integration time is increased to the working value which may be 4 or 10 s or thereabouts.

CALIBRATION

Standard working curves

The relationship between the absorption indicated by the instrument and the concentration of the element which produces it is established in the calibration procedure. The theoretically linear relationship between the amount C of the absorbing species in the light path and the absorbance A, known as Beer's Law, is followed within certain limits in atomic absorption spectrometry. After an almost linear portion the calibration curve usually bends towards the concentration axis to a greater or lesser extent.

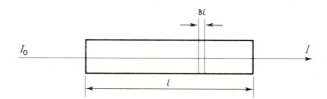

FIG. 46. Derivation of Beer's Law.

This relationship is easily derived as follows: if, in Fig. 46, the incident radiation, of intensity I_0, enters a flame or non-flame cell of length l, emerging after attenuation by absorption with intensity I, then, in some increment of the cell of length δl, the decrease in intensity $-\delta I$ is proportional to the incident intensity I_0, to δl and to the number of absorbing atoms which is proportional to their concentration C in the sample being atomized, i.e.

$$- \delta I = k I_0 C \delta l$$

Integrating this over the whole length of the cell:

$$-\int_{I_0}^{I} \frac{dI}{I} = kC \int_0^l dl$$

i.e.,

$$\log_{10} \frac{I_0}{I} = k'Cl$$

The value $\log I_0/I$ is defined as the absorbance A; thus

$$A = kCl$$

Now, if T is the transmission of the flame, then

$$A = \log \frac{I_0}{I} = \log \frac{1}{T} \tag{1}$$

The output of the detector, being proportional to the energy falling on it, is proportional to T, and hence the conversion from T to A must be made, by calculation, by a non-linear scale on the instrument meter or electronically in instruments provided with 'linear absorbance readout'.

In instruments giving linear transmission, the meter deflection $d \propto I$, so

$$A = \log \frac{d_0}{d} \tag{2}$$

The value of d with no absorbing species would be set to 100, hence the absorbance is 2 minus log of percentage transmission. It is sometimes inconvenient to set zero absorption to 100 on the meter or recorder, particularly when 'blank' absorptions may have to be observed or corrected for, so provided that the instrument is adjusted so that zero light gives zero deflection the zero absorption value can be set with the gain control to any convenient deflection d_0 and absorbance calculated from equation (2).

Instruments giving linear absorbance readout are automatically adjusted for zero light on a transmission scale, as zero energy corresponds to ∞A. Zero absorption would then normally be adjusted by amplifier gain controls to $0 A$, though as absorbances are directly subtractive it could be set if desired to any other convenient value on the scale, e.g. the background absorbance.

Non-linearity of Calibration Curves. The calibration curve of concentration against absorbance consists of an almost linear portion near the origin, followed by a section which bends to a greater or lesser extent towards the concentration axis. By convention, concentration (the independent variable) is the abscissa. The amount of bend and the noticeable point of onset of bend depend on the amount of unabsorbable radiation reaching the detector and on the presence of less sensitive or non-absorbing lines within the monochromator bandpass. Three possible calibration curves are shown in Fig. 47: (a) the ideal (almost impossible) case where all light reaching the detector is absorbed by the element being

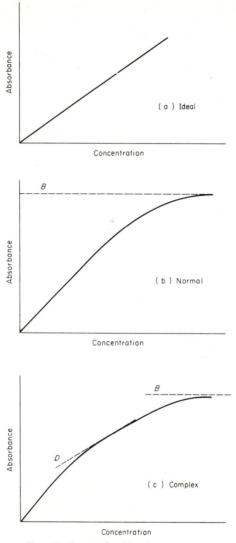

FIG. 47. Types of calibration curve.

determined to the same extent; (b) the normal curve, where B represents the transmission value of unabsorbed light, and the curve is asymptotic to this value. Unabsorbable light may be due to non-absorbing lines from either the source cathode material or the fill-gas which pass within the spectral bandwidth of the monochromator. It also occurs if the ratio of the half-widths of the emission line w_e and the absorption line w_a approaches or becomes greater than unity. Rubeska and Svoboda[657] showed that the calibration graph can only be linear if w_e/w_a is less than 1/5. When $1/5 < w_e/w_a < 1/1$ the graph is slightly curved, but if $w_e/w_a > 1/1$ the initial slope (which is a measure of the analytical sensitivity) starts to be decreased. Stray light of other wavelengths, caused by bad mono-chromator design or light of the correct wavelength which has not passed through the area of maximum atom concentration in the flame, also has the same effect. In Fig. 47(c) a further, more complex case is shown in which a line of lower absorbance occurs within the bandpass and the calibration curve first becomes asymptotic to D, its sensitivity, and then again to B, the unabsorbed light value.

A minor cause of curvature is the decrease in degree of dissociation with increasing concentration. This results in a lower proportion of free atoms being available at higher concentrations at a constant atomization temperature.

Many elements give linear calibrations up to about $0.5 A$ provided the instrument is set up and operated correctly and most of the remainder give only slight curvature to $1 A$. A notable common exception is nickel, which, because of the occurrence of the main resonance line at 232.0 nm in a densely populated spectrum region, invariably shows curvature which depends on the bandpass and available resolution.

Excessive curvature can also result if a nitrous oxide burner is used for air acetylene without readjustment of observation height or gas flow rates.

Occasionally, a calibration graph showing distinct curvature away from the concentration axis may be encountered. Typical examples are barium and europium in a nitrous oxide acetylene flame (see Fig. 48), or gold, sodium and

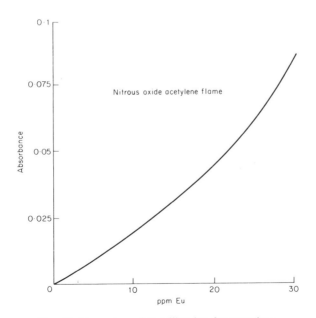

FIG. 48. Upward-curving calibration for europium.

potassium in the air acetylene flame. The cause is ionization. At very low concentrations a higher proportion of the element is ionized. This is also clear from the ionization curves for 10^{-6} and 10^{-8} atmospheres partial pressure in Figs. 5 and 6. The effect is noticeable at temperatures above the optimum for a given element. Correction, as will be seen when interferences are discussed, is by addition of an ionization buffer.

Procedure for Graphical Calibration. This entails drawing the calibration graph, a line or curve of best fit through a series of points obtained by measuring the absorbance of a series of standard solutions. The range of concentrations used is chosen to cover that in the samples to be analysed. The analysis result is then obtained by reading off the concentration corresponding to the absorbance given by the sample solution.

Procedure for Instruments with Curvature Control and Concentration Read-out. The procedure depends on the instrument in use, and the manufacturers' directions should be followed. However, this kind of instrument has three principal controls or groups of controls for use in calibration. These are the autozero control, the curvature controls and the scale expansion (or amplifier gain or 'concentration') control. The last two may be selected by the absorbance/concentration selection switch.

The calibration procedure usually calls for the use of a blank (zero standard) and two or perhaps three higher standards covering the required concentration range.

After running the blank, the zero absorbance is established with the autozero control. Readings are then obtained for the other standards. From the relationships between the concentrations and absorbance readings given by these standards, the degree of curvature is computed and entered in by means of the curvature control. It is in this step where procedures differ most from instrument to instrument. In the Pye Unicam system, for example, the two standards required must have concentrations such that $S_2 = 2S_1$, i.e. the top standard is twice the concentration of the lower one.

A circular nomogram is provided with the instrument by means of which the amount of curvature to be entered is quickly derived from readings given by these two standards. This system is based on the relationships of the stray light equation

$$A = \log_{10} \frac{(1-B)}{(T-B)} \qquad (3)$$

where A is the *true* absorbance (as distinct from the *observed* absorbance of equations (1) and (2) on p. 121) and B is the fractional transmission value of the stray or unadsorbed light. This equation is derived in the section on sources of error on p. 158.

In other systems at present available, entry of the correction coefficient may be by use of a formula deriving from the two standard readings or by means of an iterative procedure whereby the curvature control is adjusted using the two standards until linearity is achieved.

With curvature corrected in this way, the relationship between concentration and the instrumental output signal has been made linear, and application of the scale expansion control now converts instrument response directly to concentration.

Procedure for Instruments with Microprocessor-assisted Data Handling. Again, the manufacturers' or suppliers' instructions should be followed. There are advantages of a good programme applied to this type of system, other than the obvious ones of speed and avoidance of personal errors in computation of results. One should not be restricted to a particular curve correction system. The stray light equation is usually the best for flame atomic absorption, but in cases of complex curvature a polynomial may be more appropriate. There should not have to be any particular relationship between standard concentrations, though clearly the function coefficients are better established if the standard concentrations are spaced at approximately equal intervals. A good programme should also be capable of issuing certain warnings, e.g. of exceptionally high or reverse curvature; or that a particular sample is either just outside or considerably outside the calibrated concentration and curvature-corrected range. It should also have the ability of calibration and computation from the averages of a chosen number of standard and sample readings and to recalibrate on being given a new absorbance value for the top standard in order to eliminate the effects of sensitivity drift. It should also be able to correct minor errors in the automatically adjusted zero.

Method of standard additions

Sometimes it is not possible to overcome interference effects by matching standards with the samples, perhaps because the full composition of the samples is not known. Provided that sufficient sample solution is available, the analysis can then be done using the method of standard additions. It must be realized at the outset, however, that this method corrects only for effects which modify the slope of the calibration curve, i.e. the elemental sensitivity, and not those interferences which affect the absorbance at zero concentration.

The sample solution is divided into a number of aliquots; at least three are necessary. To all but one of these are added known increasing amounts of the metal to be determined and the samples then made up to the same concentration. The solutions are nebulized and the absorbance readings plotted against the added concentration of the metal (Fig. 49). The graph obtained is the heavy line and the absorbance of the unmodified sample is due to the element concentration sought (plus background absorbance and scatter). Ignoring the latter, the *true* calibration curve would be parallel to the one obtained, but passing through the origin. The required concentration may thus be read off at C_2. In fact if the graph drawn is produced backwards, its intercept C_1 on the C axis is, by similar triangles, equal to C_2, and this is where it is most conveniently read off.

The accuracy of an extrapolation method such as this is never as good as an interpolation method but sometimes it is the only one possible. The value found may well include background and scatter, and this must be corrected by one of the methods described on pp. 82 and 138. If possible it should also be checked that the element being measured gives the same response in the sample as in the additions, particularly if it is likely to be complexed or protein-bound.

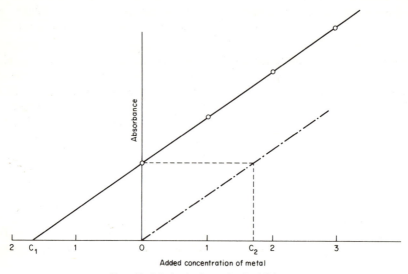

FIG. 49. Method of standard additions.

When an instrument with automatic zero internal calibration facilities and bipolar digits is in use this procedure can be simplified.[253] The sample solution is first aspirated and the display set to zero. The sample with the highest addition is then aspirated and concentration set to that of the addition. Linearity can be checked with a lower addition. Finally the blank solution is aspirated. The negative value then given on the display is then the actual value for the original sample. Reasoning for this will be obvious from Fig. 49.

Microprocessor-assisted instruments should be expected to have a standard additions programme included.

Internal standardization

The concept of measuring the ratio of the signals from the analysis line and an internal standard line is well known in emission spectroscopy, where the internal standard is either the major element (as iron in steel) or an added element of known concentration. In atomic absorption the method can be applied only when a double- or multi-channel instrument is in use, i.e. one which can measure simultaneously at more than one wavelength. Such instruments, though not common, are commercially available.

Internal standardization in atomic absorption was used by Butler and Strasheim[124] in an attempt to correct, at least partially, for variations in nebulizer and flame performance. Such variations are not corrected for by double beam spectrometers. When determining silver and copper in gold using a converted spectrograph, Butler and Strasheim found that the use of a non-resonant gold line as internal standard gave better results than a resonant gold line and that to plot the *ratio* of the intensities of absorbed copper or silver and non-resonant gold lines gave higher precision than plotting simply copper or silver absorbance.

Feldman[226] has described internal standardization techniques with a double channel spectrometer using absorption and emission techniques separately and in combination. When absorption is used in both channels, light from two hollow cathode sources is made to pass through the same flame. Light of the analysis and internal standard wavelengths is separated by means of two monochromators. The stability of the ratio of output signals was always superior to the analysis signal alone, and reproducibility of analyses as expressed by coefficients of variation was claimed to be better, usually by a factor of 2–3.

Some advantages of internal standardization were also mentioned by Kahn.[375] The response of iron falls off with increasing sodium chloride concentration, but the iron/manganese ratio remains constant. Manganese is thus a useful internal standard element for the determination of iron in presence of varying salt concentrations. Similarly, the lead/copper ratio was found to be constant in presence of sodium chloride.

A useful application could be to the rapid determination of several elements in ferrous alloys. Iron is a good internal standard element for copper, manganese and molybdenum. In calibration, therefore, element/iron absorbance ratios are plotted against element concentration. Because in this case the internal standard element is already present in the sample, both weighing and dilution steps need only be done quite approximately (even by visual estimation) without loss of accuracy.

When the internal element is selected, the major criterion is the relative flame sensitivity which should be matched as far as possible for all parameters affecting the elements' absorbance in the flame. An unmatched internal standard element will correct simply for dilution, flow and nebulization changes. Useful analyte–internal standard pairs recommended[226] are Ca/Sr; Al/Cr; Fe/Au or Mn; Mn/Cd; Cu/Cd, Zn or Mn; Cd/Mn; Zn/Mn or Cd; Pb/Zn; Si/V or Cr; V/Cr; Ni/Cd; Cr/Mn; Mg/Cd; Co/Cd; N or K/Li; Au/Mn; Mo/Sn.

Preparation of standard and reagent solutions

Standard Solutions. The general technique is to prepare relatively concentrated stock solutions for each element from which working standards are prepared by serial dilution. A stock should be a solution of a simple salt of one metal.

Reagents used in preparation of stocks need not be excessively pure as they are to be used at considerable dilutions. Analytical reagent grade is quite sufficient.

For the sake of convenience it is worthwhile to weigh out the exact quantity of the salt to give a stock solution of 1000 ppm of the metal ion. The pure metals themselves are used where possible, particularly if their salts are not stable enough on the shelf to be considered as primary solution standards. The metal should be in the form of foil, sheet, rod, wire or ribbon, and the surface should be de-oxidized either by dipping in acid or by abrasion. Powdered metals and sponges should be avoided as they may contain significant quantities of oxide.

Suitable reagents and solvents are given in the element tables in Chapter 9. Most of the stock solutions prepared as directed may be stored in polythene bottles for six months or more (gold is a particular exception). Very dilute standards, 1 ppm and less should not be used for more than one or two days.

As an alternative to preparing stock solutions oneself, these may be obtained commercially from good reagent suppliers including Hopkin & Williams Ltd., BDH (both UK), Merck (Germany), Arrow, Fisher and Spex (all USA).

Reagent Solutions are those which are added to standards and samples to overcome interferences, or simply to the standards in order to match the samples. They are invariably used in concentrated form, and therefore they must be of a high degree of purity, particularly with respect to the element being determined.

Lanthanum chloride is used extensively as a spectroscopic buffer as will be seen when interferences are discussed. Ordinary grade lanthanum salts are usually heavily contaminated with calcium and magnesium, the very elements which would benefit most by the use of lanthanum, and the amount of these elements added in the buffer may well be more than those being determined. Special atomic absorption grade lanthanum chloride and lanthanum oxide, which is especially low in calcium and magnesium, e.g. 0.2 and 0.1 ppm, is now available from reagent manufacturers. Additionally, a number of the alkali metal salts used as ionization buffers, strontium salts and EDTA (releasing agents), mineral acids and other dissolution reagents and extraction solvents are also available in 'atomic absorption quality'.

When the composition of a standard has to be matched with that of a sample, the matrix element to be added must be in a high state of purity, as there is no dilution of its own impurity elements compared with the sample. Only the purest forms available, e.g. 'Specpure' (Johnson Matthey), may then be good enough. Although 'blank' corrections can be made, the accuracy of the method falls off severely as the 'blank' value approaches the concentration being measured.

Non-aqueous Standards. When samples are dissolved in organic liquids, calibration can often be effected by dissolving suitable organo-metallic compounds in the same solvent. The choice of solvent may well be in the hands of the analyst, particularly if its purpose is simply that of dilution, and, in this case, methyl isobutyl ketone or white spirit are often preferred, as their physical properties approach those of water.

The solvent has a profound effect on both the physical and chemical properties of the flame and thus influences the precision and sensitivity of the analysis. Many common organic solvents are unsuitable for spraying into the flames produced by the usual nebulizer/spray chamber system as they are incompletely combusted and produce smoky yellow flames. The most suitable solvents are C_6 or C_7 aliphatic esters or ketones and C_{10} alkanes. Aromatic compounds and halides are usually unsatisfactory for the reason given, and the simpler solvents (methanol, ethanol, acetone, diethyl ether, lower alkanes etc.) because of their tendency to vaporize in the nebulizer causing an erratic response.

A range of organo-metallic compounds are now available for the preparation of standards. These are listed for individual elements in the Element Tables in Chapter 9. It should be noted that many of these compounds require a solubilizing agent before they can be made up into stock solutions. Some cyclohexane-butyrates, for example, may need to be dissolved first in 2-ethyl hexanoic acid and xylene before diluting with the principal solvent. A table of recommended solubilizing agents is included on p. 237 (Table 15).

SAMPLE PREPARATION

For the purposes of a general résumé of methods of preparation of materials for analysis by atomic absorption, samples may be divided into those which are already in a solution or liquid state, solid samples which contain a high proportion of organic matter and entirely inorganic solids such as metals, rocks etc.

Liquid samples

Aqueous Solutions. These can sometimes be nebulized without any prior treatment. Among these are domestic, natural and boiler waters, wines, beers and perhaps urine. A prior knowledge of the approximate concentration of the elements being determined is useful as it enables the operator to decide whether or not some dilution is desirable. If the element is in high concentration, e.g. more than 200 times the reciprocal sensitivity value, an appropriate means of reducing sensitivity should be applied. Removal of impact bead, burner rotation or use of alternative absorption lines are preferred (see p. 153), but if these are not practicable the solution should be diluted to bring it within the best range for measurement, i.e. 20–200 times the reciprocal sensitivity. If the concentration is very low, scale expansion or some chemical pretreatment may be required.

It may also be necessary to add a spectroscopic buffer to aqueous samples in order to suppress an interference effect. Such buffers are used in concentrated form in order to avoid undue dilution of samples where trace elements are to be determined.

Non-aqueous Solutions. These can also sometimes be run directly but this depends on whether or not the viscosity is similar to that of water, for which most nebulizers are designed. White spirit and methyl isobutyl ketone fulfil this condition and hence are recommended diluents for many organic liquids. Mineral oils and petroleum spirit, paints and drying oils, vegetable oils and organic liquids are some types of material that require minimal pretreatment. Standards can be made up in the pure basic solvent.

Solid samples

Organic. Many types of samples consisting of trace or minor elements in a largely organic matrix are best brought into aqueous solution after wet or dry oxidation of the organic material. Such samples include plant material, fertilizers, soils, food and feeding stuffs, biochemical specimens, etc. The dry oxidation procedure offers the advantage that the residue may be taken up in almost any desired acid medium but it cannot be used if volatile elements such as mercury, arsenic, lead, antimony and molybdenum are to be determined. Both oxidation procedures enable the trace elements to be concentrated and standards are made up in the same acid medium as the sample itself finally appears.

Organic chemicals and pharmaceutical products may also have to be treated by the above procedure but it is possible also that they may dissolve either in water or an organic solvent, in which case they can probably be presented to the instrument in this form.

A further successful procedure for organic-based solid (or liquid) samples is acid extraction. The sample usually has to be shaken or stood overnight in contact with hydrochloric or nitric acid. Quantitative extraction has been proved, e.g. for nickel in fats and a number of trace elements in plant material.

Inorganic. Recognized methods for the dissolution of metals, slags, ores, minerals, rocks, cements and other inorganic materials and products can be used, the analyst attempting as far as possible to bring all the elements to be determined together in one solution.

Where final solutions contain more than about 0.5% of dissolved material, the standards should also contain the major constituents, even where no chemical interference is expected, in order to match the viscosity and surface tension and avoid 'matrix' effects.

INTERFERENCE EFFECTS

Although atomic absorption spectra are very simple, there are a few known examples of actual spectral interference in atomic absorption, other than monochromator bandpass effects and in atomic fluorescence. These are discussed below.

Interferences which actually influence the proportion of atoms in the flame available to absorb the resonance radiation arise largely from chemical effects which originate either in the flame itself or in the sample solution. Such interferences are caused by the formation of stable compounds or by ionization.

In addition, minor interferences are caused by various physical phenomena including incomplete volatilization of the solid particles of samples formed in the flame, variations in the physical properties of the solutions (matrix effects) and scatter and background absorption.

Any effect which influences the number of free atoms in the flame, be it chemical or physical, affects the results obtained in emission, absorption or fluorescence equally. Absorption measurements are free from those interferences which arise from interactions between excited atoms, e.g., the transfer of energy by collisions of the second kind which affect emission and the similar quenching effects which have been noted in fluorescence.

Spectral interference

Comparatively few examples of actual spectral interference have been reported, either in atomic absorption or atomic fluorescence. This implies that the possibility that a resonance line emitted from the source lamp may overlap with an absorption line of another element in the atomizer is very small. Spectral 'interferences' can take several forms:[460]

(i) more than one absorbing line in the spectral bandpass;
(ii) non-absorbed line emitted by excitation source;
(iii) spectral overlap in atom source;
(iv) continuum or broad band absorption and scatter.

Case (i) above has already been mentioned as it has the type of effect illustrated in Fig. 47(c) on calibration curves. Case (ii) would normally be referred to as a band-pass effect. Although the line profiles of the resonance and offending non-absorbed lines would be well separated by a higher resolution monochromator, they are sufficiently close to pass through the exit slit of the atomic absorption monochromator, giving rise to unabsorbable radiation, leading to the more conventional type of calibration curvature described by the stray light equation. The effects of case (iv) are discussed in detail in the sections on Background Correction. Only case (iii) is discussed here.

Spectral 'overlap' suggests that the interfering line is sufficiently close to the wavelength of the analysis absorption line actually to absorb some of the radiation from the emission line. If the analyst is unaware of the overlap, he may obtain an erroneously high reading for the wanted element if the element giving the interfering line is also present in his sample.

The problem is only overcome by removing the interfering element from the sample, or better, if possible, by using another absorption line for his analysis. If the analyst is aware of the overlap, he may use it to advantage as it may enable him, under other circumstances, to employ the emission line of the first element as an alternative and less sensitive line for the second.

A number of cases of spectral overlap were quoted from previous reports, and others discovered, by Norris and West.[555] These are summarized in Table 5. It will be noted that a number of resonance lines (marked *) are quoted in the 'source' column. As most of these would be in normal use in atomic absorption, the presence of the element in the 'analyte' column could cause a spectral interference. In these cases, the roles of sought element and interference could also be reversed. Otherwise, the lines given in the 'source' column can be used as alternative lines for the element given in the 'analyte' column. The atomic absorption sensitivity reported in the literature is included as a guide.

It would seem from the above that atomic emission and even atomic absorption profiles are not always as narrow as is usually suggested, the interference

TABLE 5. Known cases of spectral overlap affecting atomic absorption

Source line[a]	Emission nm	Analyte	Absorption nm	Separation nm	Sensitivity ppm	Reference
Aluminium	308.215	Vanadium	308.211	0.004	800	224
Antimony	217.023	Lead	216.999	0.024	250	688
Antimony	217.919	Copper	217.894	0.025	100	555
Antimony	231.147*	Nickel	231.095	0.052	35	555
Antimony	323.252	Lithium	323.261	0.009	200	555
Arsenic	228.812	Cadmium	228.802	0.010	45	261
Copper	324.754*	Europium	324.753	0.001	75	224
Gallium	403.298*	Manganese	403.307	0.009	15	50
Germanium	422.657	Calcium	422.673	0.016	6	737
Iodine	206.163	Bismuth	206.170	0.007	10	171
Iron	271.903*	Platinum	271.904	0.001	40	224
Iron	279.470	Manganese	279.482	0.012	0.04	242
Iron	285.213	Magnesium	285.213	<0.001	10.0	242
Iron	287.417*	Gallium	287.424	0.007	250	555
Iron	324.728	Copper	324.754	0.026	0.8	242
Iron	327.445	Copper	327.396	0.049	1.1	242
Iron	338.241	Silver	338.289	0.048	150	555
Iron	352.424	Nickel	352.454	0.030	0.1	242
Iron	396.114	Aluminium	396.153	0.039	50	555
Iron	460.765	Strontium	460.733	0.032	20	555
Lead	241.173	Cobalt	241.162	0.011	15	555
Lead	247.638	Palladium	247.643	0.005	3.5	555
Manganese	403.307*	Gallium	403.298	0.009	25	50
Mercury	253.652*	Cobalt	253.649	0.003	100	481
Mercury	285.242	Magnesium	285.213	0.029	200	555
Mercury	359.348	Chromium	359.349	0.001	250	555
Neon	359.352	Chromium	359.349	0.003	0.1	553
Silicon	250.690*	Vanadium	250.690	<0.001	65	224
Zinc	213.856*	Iron	213.859	0.003	200	565

[a] Source lamps hollow cathode except neon, iodine, arsenic and mercury (electrodeless discharge).

* Indicates resonance line from source.

being experienced at line separations even greater than 0.04 nm. The arsenic–cadmium, iodine–bismuth and neon–chromium overlaps have also been used to stimulate fluorescence in the second element of each pair. Other overlaps have been reported in atomic fluorescence with mercury and cadmium sources.[565]

Chemical interferences

It has been the practice among atomic absorption analysts to differentiate between 'enhancing' and 'depressive' interferences. In all cases however the presence of substances which produce an improved sensitivity for a given element actually do so by reducing an existing depressive interference. It would appear, therefore, to be more logical to look upon any effect which causes an element to give less than its maximum possible absorption as a depressive interference. The maximum possible absorbance would be given when an element is 100% atomized, but this condition, though readily defined, is unlikely to be attained in practice unless there is considerable separation of the atomization and ionization curves (see p. 18).

Stable Compound Formation is the best known of the interferences. It is responsible for many of the depressive effects reported over the years in both emission and absorption flame photometry. It arises because compounds or radicals containing the element being measured are not broken down into individual atoms at the temperature of the flame being used. Stable compounds may even be formed in the flame. This condition exists below the atomization curves of Figs. 5 and 6.

Typical examples are the lowering of alkaline earth metal absorbances in the presence of aluminate, silicate, phosphate and some other oxy-anions, the low sensitivity of metals which form very stable oxides, such as aluminium, vanadium, boron etc. or even carbides, and the depression of calcium in presence of protein. All these occur in the air acetylene flame.

The effect of oxyacids on alkaline earth metals is normally appreciable only when the anion is in a chemical excess over the metal. This suggests that an equilibrium is being set up between the species M—O, or M—Cl if the materials are dissolved in chloride medium, and M—O—X. In the presence of excess —O—X, the equilibrium of the first part of equation (1) will tend to move to the left preventing formation of free M. High absorbance readings are obtained at higher temperatures, and also with hydrocarbon fuel gases as compared with hydrogen. There would therefore appear to be a further mass action effect which, according to the conditions in the flame and concentrations of reacting species in the flame plasma, favours the production of either the free atoms or the stable associated M—O species. Increased temperature results in the formation of free atoms of the wanted metal M according to the second stage of reaction (1).

$$M-O-X \underset{\substack{\text{excess}\\\text{oxyacid}}}{\overset{\text{heat}}{\rightleftharpoons}} M-O \underset{\substack{\text{low flame}\\\text{temp}}}{\overset{\text{heat}}{\rightleftharpoons}} M+O \overset{C}{\rightleftharpoons} M+CO \qquad (1)$$

Lower temperatures and/or excess of the oxyacid thus favour the persistence of the stable oxy-salt. Because the concentrations of oxyacid radical and carbon in the flame are usually present in a large excess compared with M, the metal being atomized and measured, the reaction equilibrium is usually independent of the concentration of M.

The action of a releasing agent is to influence the chemical equilibrium in the desired direction:

$$M—O—X + R \xrightleftharpoons{\text{excess R}} R—O—X + M \qquad (2)$$

If R is a metal which forms a similarly stable compound with the oxyacid, and is present in excess, then mass action dictates that reaction (2) must proceed to the right, producing a higher proportion of free atoms of M.

Good releasing agents are therefore those metals which themselves form stable oxysalts, and for this reason strontium and lanthanum have been most used.

From the second and third stages of reaction (1) it is clear why metals forming refractory oxides give low sensitivity at low flame temperatures. An increase in temperature, such as is achieved when nitrous oxide is used as oxidant instead of air, will itself tend to improve the situation. It has already been seen (p. 31) that the presence of free carbon or carbon-containing radicals up to the critical C/O ratio removes the excess of oxygen as carbon monoxide pushing the equilibrium further to the right. In this way carbon itself behaves as a releasing agent. The practical results, that hydrocarbon based flames give better sensitivity for refractory metals, might thus be predicted.

Some of the so called refractory metals also tend to form stable carbides, and thus show worse sensitivity in hotter richer hydrocarbon flames. Carbide formation by vanadium and molybdenum in a nitrous oxide acetylene flame can be minimized by the presence of excess aluminium.[471] This is probably another example of releasing action in the flame.

The effect of metal–oxygen bond energy on the sensitivity, in the nitrous oxide acetylene flame, of a number of metals showing high bond energy was investigated by Sastri et al.[662,663] Solutions of simple salts and oxysalts were compared with solutions containing metallocenes and fluor-complexes where the metal is not oxygen-bonded. In the latter cases the sensitivity of the metal was always higher. An irregular correlation was then shown to exist between the increase in sensitivity and the metal–oxygen bond energy. A further type of releasing action is thus the prevention of the formation of metal oxygen bonds in solution by suitable complexation. The absorbances of titanium and zirconium are improved for example in the presence of hydrofluoric acid. The general applicability of the silicate dissolution method in which the principal anion formed in the final solution is fluoroborate[610] is also ascribed to the prevention of M—O species in the flame and the more ready breakdown of M—B—F species at high temperatures.

Another example of a stable compound formed in solution is the calcium–protein complex which, present in blood serum, appears to persist in the air acetylene flame. Cooke and Price[155] showed that the calcium absorbance is not a direct function of the dilution factor, even when lanthanum chloride is added as the releasing agent. When EDTA is used instead, the sample can be used at any dilution and the correct concentration of calcium calculated.

This is explained as the calcium EDTA complex is stronger than the calcium–protein complex in solution, but the latter is more readily dissociated in the flame. Like hydrofluoric acid in the previous example, EDTA behaves as a pre-flame releasing agent.

Ionization. Both the emission and absorbance of low concentrations of an alkali metal can be increased in a hot flame by adding a second alkali metal. This used to be called an enhancing interference but is in fact an example of the decrease in the degree of ionization in the presence of another easily ionized metal. When the second metal is present as an impurity, perhaps in unknown or varying amounts, the degree of suppression of ionization also varies and the interference effect results.

As the concentration of the wanted metal increases, when the second metal is not present, the proportion of ionized atoms decreases. This, as mentioned on p. 123 where calibration curvature was discussed, results in curvature of the working graph away from the concentration axis. The different degrees of ionization of potassium at two different concentrations (partial pressures) are shown in Fig. 5.

Both of these effects are eliminated by the addition of an ionization buffer. This is a second, easily ionizable, metal added in excess.

$$Li \overset{heat}{\rightleftharpoons} Li^+ + e \tag{3}$$

$$K^+ + e \rightleftharpoons K \tag{4}$$

Lithium, when present in excess, can provide sufficient electrons to prevent the ionization of potassium, although the higher alkali metals, with still lower ionization potentials, are even more efficient.

If the absorbance of an ionizable metal in the nitrous oxide flame is plotted against concentration of added ionization buffer, a plateau is usually reached, e.g. at approximately 1000 ppm of potassium when added to calcium, strontium or barium. This indicates that the ionization has been reduced to virtually zero.

The tendency to use hotter flames, to increase the sensitivity of the refractory metals and to widen the scope of atomic absorption analysis, has increased the number of ionization problems. Calcium (see Fig. 6) is almost 50% ionized at the temperature of the nitrous oxide acetylene flame, and both barium and strontium, though showing an increase in sensitivity of over twice when nitrous oxide acetylene is substituted for air acetylene, give *further* sensitivity improvement factors of about four times when potassium is present in excess as an ionization buffer.

Many metals, including aluminium[57] and silicon,[606] are ionized to an appreciable extent at these same flame temperatures. Some degrees of ionization

TABLE 6

Element	Ionization potential (eV)	% Ionization in N_2O/C_2H_2
Be	9.3	0
Mg	7.6	6
Al	6.0	10
Yb	6.2	20
Ca	6.1	43
Sr	5.7	84
Ba	8.3	88

in the nitrous oxide acetylene flame are given in Table 6. The use of an ionization buffer is thus nearly always recommended with these very hot flames.

It is interesting to note that the ionization potential of lanthanum (5.61) is very comparable with that of lithium (5.39) and lanthanum therefore acts as an ionization buffer (in addition to its duties as a releasing agent) for many metals including calcium, magnesium, silicon, aluminium, etc., in the hot flames. This is partly the explanation of the versatility of lanthanum salts as spectroscopic buffers. In a complex matrix it may well be difficult to decide which particular function it is fulfilling.

Physical causes of interference

Incomplete Volatilization. This implies that, at the temperature of the flame, the droplets produced by the nebulizer have given rise to solid particles which, because of their high vaporization temperature, their speed through the flame, or both, are not completely converted to a vapour. The degree of atomization is therefore lower than would be expected, given the other chemical conditions in the flame. This type of interference effect is usually experienced under reducing conditions and is then caused by the formation of metal/metal solutions of high boiling point.

Baker and Garton[72] pointed out that when volatilization of particles in the flame is incomplete, there was likely to be a noticeable departure from linearity of the absorbance/concentration relationship. A feature of this type of interference is greater curvature of calibration graphs.

The depression of the chromium and molybdenum response by high concentrations of iron in the air acetylene flame is explained by incomplete volatilization.[645] The depression has been noted previously[78,178,534] and has been reduced in the presence of ammonium chloride and aluminium chloride.

The depression is most significant in a fuel-rich flame, is less in a lean flame, and is not experienced at all in the hotter nitrous oxide acetylene flame.[732] The degree of depression increases gradually as the iron concentration is increased, levelling off only when a large excess of iron is present. This does not support the theory of the formation of a definite compound, e.g. a spinel. Furthermore, in the reverse situation, the iron absorption is little depressed by a large excess of chromium.

That this is a metal/metal effect, not an example of mixed oxide formation, is also confirmed by the severity of this interference effect in different acid matrices, which decreases in the order hydrochloric acid > sulfuric acid > phosphoric acid. This exactly parallels the ease with which iron is able to atomize from these media. If the presence of oxides were the controlling factor, the above order would be expected to be reversed.

Aspiration of chromium/iron solutions will cause relatively large solid particles, which after reduction by the flame gases consist of chromium (boiling point 2480 °C) in a matrix of iron (boiling point 3000 °C). These are not completely vaporized and the atomization efficiency of the chromium is low. In the reverse situation—iron in a matrix of chromium—the chromium vaporizes at a lower temperature and the iron is atomized.

A similar case is that of rhodium, the response of which in an air acetylene flame is much influenced by the presence of different acids and other metals. Addition of sodium sulfate increases the absorption above that given by a solution containing only rhodium[362,378] (as rhodium sulfate or ammonium chlororhodite) and appears to overcome interferences. These interference effects also are not observed in a nitrous oxide acetylene flame.

As rhodium salts are readily reduced to rhodium metal, it must be assumed that the small solid particles obtained after evaporation of the solute contain rhodium largely as the uncombined metal, the boiling point of which is over 2500 °C. There is thus incomplete vaporization and atomization. The presence of sodium sulfate prevents the formation of metallic clotlets, and as this evaporates, the rhodium is released in atomic form.

A similar explanation can be given for the releasing action of ammonium chloride, aluminium chloride and alkali sulfates on chromium in presence of iron. Instead of a solid solution of very high vaporization temperature being formed, the wanted metals are brought into solid solution in small particles of these substances. In fact, ammonium chloride and aluminium chloride would not melt, but would sublime. Sodium sulfate would also probably form sodium oxide, which also sublimes. It is likely therefore that their releasing action is a result of the rapid liberation of the metal either as an atomic or molecular vapour, both of which would make for high atomization efficiency and absorption response.

Studies on a number of cation interferences in the nitrous oxide acetylene flame, which are relevant in this context, have been made by Marks and Welcher.[487] The conclusion is reached that, though flame conditions are by no means always adequately defined in the literature, the probable explanation of many interference effects is the difference in volatility of the analyte when it is accompanied by other metal species. Most 'concomitants' actually increase the volatility of an analyte, and the majority of the interferences reported are positive. The volatility of the matrix as described by the boiling point and heat of vaporization is not always the most important consideration as the evaporation rate of small salt particles in a flame is a complex function of drop size, diffusion coefficient of the evaporating species, surface tension and heat conductivity.

Matrix Effects. These influence the number of atoms actually entering the radiation beam rather than their effectiveness once there. They usually arise from differences in the physical properties of samples and the so-called matching standards. A difference in acid concentration, for example, can cause a difference in viscosity or surface tension. The viscosity affects the rate of uptake of the sample by a given nebulizer and the surface tension affects the size distribution of the droplets formed and hence the nebulizer efficiency.

Some workers recommend that matching of standards and samples is necessary in respect of components in concentrations greater than 1% of the final solution. While this may well be sufficient for the solute, a closer match may well be necessary for the solvent, particularly where neutral sample solutions, e.g. natural waters, are to be compared with synthetic standards made up in acid medium. It is good practice to follow the same rule for both sample and standard, making both up to a stated acid concentration (0.1% hydrochloric acid, 1% nitric acid, etc.) and for accurate work to ensure that this is maintained to within 10%.

The solvent vapour pressure also directly affects the droplet size distribution both at the nebulizer and during transport to the burner. Mixtures of solvents, particularly when low boiling point solvents are involved, must therefore be carefully controlled, as must the temperature of the solvent, particularly if samples are nebulized at temperatures near their boiling point.

Purely physical effects like those described would be expected to affect all elements in exactly the same proportion. This is often not the case, however, and some other effect is presumed also to be present. This is most probably a further result of different particle size distributions, the larger particles being incompletely vaporized. Stupar and Dawson[710] showed that some interference effects, particularly those of aluminium and silicon on magnesium, previously thought to be entirely chemical in nature, could be drastically reduced when an ultrasonic nebulizer providing a more uniform smaller droplet size was employed. The type of ultrasonic nebulizer that they employed (flow rate less than 0.1 ml min^{-1}) may not be suitable for general analytical work, but the conclusion that good nebulizer performance characteristics and efficient spray chamber operation are essential to good analytical sensitivity and precision can still be drawn.

Background Absorption and Scatter. While these increase the actual readings obtained, the sensitivity is not improved but a spurious absorbance is added to the true value. Scatter is analogous to turbidity in molecular spectrometry and is the result of the presence of small particles in the resonance beam. Such particles may remain because of the flame's inability to vaporize a high dissolved solids content of the sample solution, or may be due to the formation of particles, e.g. of carbon, in the flame itself. It is well established that droplet sizes passing into the flame in a direct injection burner are much larger than with a spray-chamber system and thus greater scattering background is experienced with the former.

The causes of spurious absorption signals, their effects and instrumental methods of correction have been critically reviewed.[31]

The magnitude of this effect varies considerably with the wavelength at which measurements are being taken. For particles having a diameter of less than one tenth of the wavelength of the incident radiation, the amount of scatter according to the Rayleigh theory is proportional to λ^{-4} so that towards the lower wavelengths a very sharp increase in the effects of scatter is experienced. Light scattering therefore affects particularly those elements that absorb at the lower wavelengths, especially arsenic, selenium, zinc, cadmium and lead.

Spurious atomic absorption readings are also caused by the absorption of light by radicals and molecular species. Such species are always present in flames, resulting from components of the prepared sample solution or from the combustion reactions in the flame itself. A well known example of molecular absorption is the CaOH radical band in the 548–560 nm region, which has a maximum approximately corresponding to the barium resonance line at 553.6 nm. Molecular dissociation bands, unlike the scattering continua, may have comparatively steep sides. In the case of molecular electronic spectra, the fine structure may give rise to sharp discontinuities, of which the cyanogen bands formed in the nitrous oxide acetylene flame are an example.

Kirkbright has demonstrated that sodium iodide, bromide and chloride produce strongly banded absorption spectra in the 190–390 nm region[16] in a cool flame. As would be expected, the effect of the background absorption of these molecules increases as the melting points, boiling points and dissociation energies increase from chloride through bromide to iodide.

Interference from molecular absorption spectra is reduced in hotter flames, particularly the nitrous oxide acetylene flame, as the species become dissociated at higher temperatures.

The effects of background absorption and scatter are manifest in two ways. Firstly, a spurious increase in the absorption signal is produced. This may lead the unwary analyst into reporting results considerably higher than the true ones. Secondly, since this spurious absorption assumes the noise characteristics of the flame, an extra noise component is included in the measured signal, which close to the detection limit of the element being measured may be greater in amplitude than the noise on the true absorption signal, or even greater than the true absorption signal itself.

The presence of background absorption and scatter is easily detected. All calibration curves should pass through the origin. If a particular one does not, either there is a reagent blank present in the standards or there is background. The background is evaluated in terms of its equivalent absorbance by several different kinds of method.

Two Line Method. This requires the measurement of absorbance, firstly at the resonance line wavelength of the element to be determined. This will give the total absorbance value of both atomic absorption and the unwanted background $A_{(S+B)} = A_{(S)} + A_{(B)}$. A second measurement is then made at a closely situated gas line or other non-absorbing line. This gives the absorbance value of the background which is then subtracted from the first reading giving the wanted absorbance value. This method can be used with simple single beam instruments but has two disadvantages. The hollow cathode lamp non-absorbing line may not be sufficiently close to the resonance line for the assumption that the background value is the same at both wavelengths to be valid. There should in any case be not more than 20 nm (or 5 nm at wavelengths below 220 nm) separation between the two lines used. Further, two such absorption measurements are necessarily made at different times and the population of non-atomic species may not be constant. This is more likely to apply to electrothermal atomization than to the flame where a sample solution can be aspirated twice under acceptably reproducible conditions.

This method can readily be applied in 'simultaneous' form on a double channel atomic absorption spectrophotometer. The choice of non-absorbing line would be made easier because the light source in the second channel could be chosen especially to give spectral proximity to the resonance line. The disadvantage of time difference would not apply.

Continuum Method. The continuous spectrum of a deuterium arc or deuterium hollow cathode lamp (and tungsten halogen lamp in the visible region) can be utilized to measure the continuous background absorption or scatter at the same monochromator wavelength setting as the resonance line. The attenuation by atomic absorption of light from the continuum lamp is insignificant compared with the total amount of continuum light passing into the monochromator because less than 1% of the normal bandpass is occupied by the width of the absorption line. The atomic absorption plus background reading is therefore made first within the atomic spectral lamp and the background measurement made subsequently with the continuum lamp. As in the two-line method the difference between the two absorbance values obtained is the true atomic absorption.

This method can also be used with single beam atomic absorption instruments but retains the disadvantage of the two measurements being made at different times. This disadvantage would not be present if the continuum lamp were used in the second channel of a double channel instrument.

Simultaneous Background Correction. It is much more convenient in flame work, and essential with electrothermal atomization, to make the total absorbance and 'background-only' measurements simultaneously. This is readily done on double channel instruments as already mentioned, but can also be done on modified double beam and single beam instruments. Many commercial instruments now offer this facility within their design.

A rotating chopper or static beam splitter (usually present in double beam instruments) is positioned so that the light beam from the atomic spectral lamp and that from the continuum lamp are directed through the flame or atom cell. The mechanics of this system are illustrated in Fig. 35 and described in more detail on p. 82. The 'double beam electronics' system then records or displays the required atomic absorbance value.

In many double beam spectrophotometers used in this way, the real reference beam is inoperative and the advantages of double beam operation are lost. However, this criticism does not apply where both atomic and continuum sources operate in double beam mode using four electronic channels.

The noise component associated with high levels of scattering is normally much more difficult to deal with. Its effect is, however, very much reduced with simultaneous background correction, provided that an adequately fast chopping frequency, e.g. not less than 40 Hz, is employed. Circumstances in which a large scatter signal is generated are clearly best avoided wherever possible. If it is caused by the presence of large concentrations of dissolved solids in the sample solutions, it may be necessary to separate the wanted components before atomic absorption measurement. It is good practice too, to avoid sample preparation procedures whereby high concentrations of dissolved solids are introduced. Fusions with alkaline salts can, for example, often be replaced by dissolution with hydrofluoric-based acid mixtures. If the scatter originates from the flame itself, the flame conditions may have to be modified. A highly reducing flame, though an efficient atomizer, may cause too much scatter and a compromise would have to be sought between sensitivity and noise in order to achieve the best detection limit. When the conditions permit a more transparent flame could be used, e.g. the hydrogen argon flame is satisfactory for arsenic and selenium.

Testing the Efficiency of Background Correction. A simple method of testing a simultaneous background corrector is to insert a gauze or neutral filter into the sample beam of the atomic absorption spectrophotometer and note the corrected absorbance reading. If the instrument is previously set up to read zero, then, because the gauze has no atomic absorption, it should remain on zero with the gauze in place. The gauze must be inserted at a point on the optical path which does not correspond to a focused image, otherwise, if the holes in the gauze are comparable in size with the focus of the monochromator entrance slit, a small change in the position of the gauze could produce a large change in absorbance.

Under certain circumstances this test may indicate good correction even when the system is not working satisfactorily. Even if the two beams are out of alignment for example, they would still be subject to the same 'absorbance'. The situation that obtains with most atomizers is more severe. The flames in normal use are very narrow, and in the extreme case, if the beams are not aligned, one may pass along the flame but the other may miss it partly or even completely. The effects of even slight beam misalignment in electrothermal atomizers will be shown to be very serious. The best test of correction efficiency is therefore one which simulates real sampling conditions.[31]

A common example of a matrix component which causes background

absorption, particularly at low wavelengths is sodium chloride. Its effect on the
determination of arsenic with flame atomic absorption provides the basis of a
realistic test.

> Three solutions should be prepared, using the purest available grade of sodium chloride,
> e.g. 'Puratronic': (a) 20 ppm arsenic; (b) 5% w/v sodium chloride; (c) 5% w/v sodium
> chloride plus 20 ppm arsenic. The instrument should be set up as follows: wavelength
> 193.7 nm; bandpass 0.4 nm; arsenic lamp (preferably electrodeless discharge); observa-
> tion height 10 mm; air acetylene flame (stoichiometric). With the background correction
> system not operating, set the reading to zero on deionized water, then aspirate the above
> three solutions in the order given and note the readings. Repeat the procedure with the
> background corrector in operation.

Under the above conditions, the background signal from the sodium chloride
should be almost equal to the arsenic absorbance signal, itself equal to just over
0.100 absorbance units. With the background correction system operating
properly, solution (b) above should give a signal no greater than 0.002, and the
difference between the absorbances given by solutions (a) and (c) should likewise be less than 0.002.

TRACE ANALYSIS

Sensitivity and detection limit

The high sensitivity of many metals in atomic absorption suggests that this is a
good technique for determining traces of metals; indeed it has been regarded by
many as primarily a trace technique. In this situation, the absorbance range
giving best quantitative accuracy cannot be used and the instrumentation is
being operated at, or near to, its maximum sensitivity.

The *reciprocal sensitivity* of an element, originally termed the 'sensitivity' by
atomic absorption spectroscopists, is defined as the concentration (usually
quoted in parts per million) which will absorb 1% of the incident resonance
radiation of that element, i.e. it gives an absorbance of 0.0044. In most analytical
techniques sensitivity is understood to be the rate of change of measurement
with concentration, i.e. the slope of the working curve, dx/dc. Reciprocal sensitivity is also sometimes called the 'characteristic concentration'. It will be
noticed that 'one per cent absorption' is something that happens in the flame,
and does not depend upon the instrument or measuring circuits employed,
though a wrong reading may well be given if these are not efficient in any way.

In trace analysis one is more concerned with the limit of detection. This has
been discussed at great length by many authors, but the most serviceable definition of the detection limit in atomic absorption spectrometry is the concentration
corresponding to twice the standard deviation of a series of not less than ten
readings taken close to the blank level. The determination of the detection limit
is then the statistical evaluation of the smallest quantity detectable with a 95%
certainty. As it is not defined as a direct function of the instrumental noise level

it can be used in respect of all types of instruments including those with integrating read-out.

Detection limit is directly related to reciprocal sensitivity as defined above (for if the slope of the curve is doubled the concentration corresponding to a given absorbance is halved) but is lower, i.e. better, by a factor which depends on the stability and reproducibility of the read-out system.

In order to avoid the effects of small variations in slope near the zero point of the calibration curve, the following procedure is often adopted for deriving detection limits of elements in atomic absorption under experimental conditions being investigated:

Prepare solutions of two concentrations such that the lower is approximately equal to the anticipated detection limit and the other is about twice this value. Different sets of volumetric glassware can be used for these solutions to minimize errors caused by contamination. Readings for each standard, alternating with the blank solution, i.e. blank; low standard; blank; higher standard, are taken, the sequence being repeated twenty times. From the data obtained, calculate the detection limit as follows:

(i) Take the average of the two blank readings made before and after each standard reading. Subtract this average from the standard reading they apply to.
(ii) Calculate mean and standard deviation for twenty corrected low standard readings and also for twenty higher standard readings.
(iii) If the ratio of means is different within experimental error from ratio of prepared concentrations, the test should be rejected and repeated.
(iv) If the ratios in (iii) agree then

$$\text{det. lim} = \frac{2 \times \text{Standard deviation} \times \text{concentration}}{\text{Mean absorbance}}$$

The accepted detection limit is the average of the results calculated for the two standards independently.

Atomization efficiency

In order to improve the detection limit in a given analysis, it is first desirable to see if the sensitivity can be improved. This is aimed at producing the greatest possible absorbance in the flame for a given quantity of the absorbing element. When instruments employing nebulizer–spray chamber systems are being used the following points should be checked:

(i) Nebulizer take-up rate and efficiency, remembering that the efficiency of a pneumatic nebulizer is only about 10% (20% at most when one of the other devices described in Chapter 3, p. 35, is employed). If the operator is able to adjust the nebulizer himself, this is best done while the equipment is operating so that he may be sure of adjusting to give the highest absorbance. If sample volume is no problem a little greater absorbance might be obtained by using a nebulizer with an increased uptake rate, though the increase in absorbance will not be proportional and the sample or concentration noise may be increased. If the volume of sample is strictly limited, the use of one of the total sample devices referred to on p. 52 *et seq.* or an ultrasonic nebulizer, to obtain higher efficiency at lower sample uptake rates, might be considered.

(ii) Spray chamber cleanliness and efficiency. The spray chamber should be clean to ensure best response time and reproducibility. It is also worthwhile to compare sensitivity and detection limit with and without flow spoilers. As suggested on p. 39 these should be optimized for a given nebulizer and some adjustment may be desirable.

(iii) That the best flame conditions for a particular element are in use. Some guide to the best lines and flame types is given in Chapter 9.

Optical efficiency

Best analytical sensitivity is obtained only when the source produces a sufficiently narrow resonance line and the monochromator isolates that line from all other radiation.

The quality of the resonance line from the source should be checked to ensure that it is not self-absorbed in the source lamp itself and that it is not likely to be interfered with by other lines emitted by the source. As hollow cathode sources age, the intensity of the wanted line decreases and that of unwanted lines may well increase. When the situation cannot be restored by decreasing monochromator slit width and increasing amplifier gain, the source lamp must be replaced.

The condition and type of the source is most important in trace analysis, for both affect the width of the emitted resonance line. Rubeska and Svoboda[657] showed that if the ratio of half-widths of emitted resonance line to absorption line is greater than one, not only is the calibration graph appreciably curved, but the reciprocal sensitivity also is affected. The geometry of modern sources, hollow cathodes in particular, is arranged so that line broadening is minimized; so, for this and the reason given in the next paragraph, it is not usually possible to improve the situation by running the lamp at a lower current.

Under the conditions of trace analysis, where absorbances are low and light levels at the photomultiplier are high, the main limitation is most likely to be shot noise. The light level L given by the lamp is a function of the lamp current i:

$$L \propto i^n$$

where $1 < n < 2$. Shot noise $\propto L^{\frac{1}{2}} \propto i^{n/2}$ so that if lamp current is halved the corresponding shot noise is reduced only by the factor 0.7–0.5. To improve the signal to noise ratio and hence the detection limit, this must be accompanied by an improvement of at least 30–50% in reciprocal sensitivity. As the likelihood of such an improvement is remote, it may well be better actually to *increase* the lamp current in order to improve the detection limit.

The level of unabsorbed light reaching the detector is reduced when the mono-chromator and source optics are matched to each other and optimized to the flame. Different elements may require different conditions in this respect. Calcium, for example, is atomized maximally within a quite narrow region of the flame. Sometimes, therefore, an optical field aperture is provided whereby the monochromator only receives a narrow pencil of light. The burner height is then adjusted until the maximum absorbance is obtained. The sensitivity of calcium and some other elements can thus be quite materially improved. A field aperture restricts the amount of energy falling on the detector, however, and the signal to noise ratio at the read-out stage and hence the detection limit may not be improved proportionally.

The unabsorbed light level also depends on the proximity to the resonance line of other lines emitted by the source. Their effect depends upon the resolution of the monochromator, and the slits can be closed until the limit set by resolution is reached. This is yet another adjustment which improves reciprocal sensitivity at the expense of the amount of energy falling on the detector. A compromise giving the best detection limit must therefore be reached.

Noise and detection limit

The detection limit is usually a concentration of the element in solution, which we have defined on p. 142, and is thus comparable between instruments and between types of analysis. If one can improve the limit of detection of an element in simple aqueous solution there is a good chance that it can be improved by similar measures in the presence of other substances. Another definition of limit of detection, in terms of the smallest *mass* of an element that can be detected, is often used in connection with 'total sample' techniques particularly when a solid sample is used.

Although the detection limit is not now defined directly in terms of the noise level on the output signal, the latter affects it strongly. If the readout is in the form of a recorder trace, the position at which the analyst chooses to draw the 'average' value is less readily determined at high noise levels (see Fig. 45). Instrumental averaging or integration devices, though faithfully recording the information with which they are fed cannot produce the true result correspond-ing to an infinite integration time over a short period of time, and hence for a series of such readings a reproducibility which is some function of the noise level will result.

In the section on the operation of the instrument (p. 116) the sources of noise were discussed and possible means for reducing them suggested. Electronic ('Johnson') noise, lamp noise and photomultiplier dark noise are characteristic of a particular piece of equipment and may not be readily influenced by the analyst using it. These should be low compared with photomultiplier shot noise, which as shown in the previous section is likely to be an important source of noise in trace analysis. The effects of shot noise are increased where signals resulting from reduction of energy from the source lamp require greater amplification factors.

There is thus an optimum combination of lamp current, photomultiplier EHT, slit width and amplifier gain which gives the least noisy readout signal. This is ascertained in the absence of the flame.

In measurements near the detection limit, flame and nebulizer or concentration noise do not usually contribute a large proportion of the total noise signal, unless there is background absorbance or scatter. Thus flame conditions should be chosen to give the maximum sensitivity, unless considerable background absorbance or scatter are thereby introduced, e.g. for elements like tin or molybdenum, which require a very rich luminous flame if being determined with air acetylene.

Scale expansion

As the reciprocal sensitivity of a given element (1 % absorption) is usually by no means the lowest concentration that can be determined, scale expansion should be used to measure such concentrations with a higher degree of confidence. A scale expansion facility of up to $50 \times$ is often provided in commercial instruments though factors of more than $20 \times$ are not particularly useful. In digital instruments with curvature correction and concentration readout the latter is often achieved either by scale compression or by scale expansion. It should be remembered that when curvature correction has been applied, a given amount of scale expansion has a much greater effect on noise at the higher end of the calibration because of the now-hidden curvature.

The use of scale expansion in trace analysis thus permits greater accuracy of reading of the scale where small scale deflections are involved. Both the noise and the drift associated with the output signal are also expanded by the same factor. The highest scale expansion factors are useful only in circumstances where an inherently stable flame and source can be employed. Elements which are atomized in air acetylene, whose resonance lines do not occur at low wavelengths and for which reliably stable hollow cathode sources are available, therefore permit the highest scale expansion factors. Copper and cobalt are good examples. It should be possible with good equipment to scale-expand copper readings corresponding to concentrations of 0.1 ppm by a factor of 20 or 40, and thus obtain a detection limit of 0.002 ppm or better. Cobalt gives a detection limit of 0.01 ppm in a similar way. On the other hand, such high expansion factors are of no avail for arsenic and selenium, where the detection limit is much affected by the prevailing flame noise.

Use of method of standard additions

While standard curves would be employed wherever possible, the problem of determining a trace element in an otherwise largely unknown matrix often occurs. Sufficient accuracy for trace levels is usually obtained with the method of standard additions (p. 125) as under these circumstances the calibration graph is most likely to be linear. The method is much facilitated by the use of scale expansion, though it is necessary to check and correct any background absorbance interfering at the resonance line being measured.

Chemical separations and concentration

The sensitivity of atomic absorption is such that many metals can be determined in trace quantities in the presence of major constituents. Nevertheless it is sometimes necessary to include a chemical separation in the preparation of the sample for atomic absorption analysis. The two main reasons for taking such a step are:

(a) when the concentration of the element to be determined is below the detection limit after normal preparation;
(b) when a separation is necessary, either from an excessive concentration of other dissolved solids (which cannot be handled by the nebulizer), or from an overwhelming chemical interference effect.

Many separation methods have been used in atomic absorption analysis. These fall into a few well-defined categories largely because of the general principle of separating, where possible, groups of elements rather than individuals, as suggested on p. 112.

Evaporation of the Solvent. This can be applied where no major constituents are present, e.g. in the analysis of water from various sources or the analysis of organic solvents. Although this method is used for concentration factors of 10–100, there is theoretically no limit to the concentration possible. The metal to be determined must not, however, be present in a volatile compound.

Deproteinization of Biochemical Samples. When trace elements are to be determined in whole blood or serum, dilution to decrease viscosity and total solids results in concentrations below the detection limit. A simple method of removing the protein is to add 1 ml of 10% trichloroacetic acid to 1 ml of the serum sample, shake and centrifuge. The supernatant liquid is then aspirated direct.

Most metal–protein bonds are broken by trichloroacetic acid and the metals themselves, including copper, zinc and iron, transfer to the supernatant.

Removal of Organic Matrix by Wet- or Dry-ashing. Dry-ashing procedures are preferred where possible, as the residue can then be taken up in the simplest possible acid solvent. Many organic matrices are completely oxidized at 550–600 °C.

When volatile metals and metalloids are to be determined, wet oxidation must be used instead. Wet oxidation procedures are fully described in other publications[365] and, for some specific applications, later in this book.

Most analysts are aware of the dangers of perchloric acid in wet oxidation procedures and it is felt by many that these are much outweighed by the advantages this acid has to offer in atomic absorption. Chief of these advantages is that it causes interference effects with very few metals. It also allows other acid radicals to be removed easily by fuming and thus facilitates the preparation of exactly matching simple aqueous standards.

Separation and Concentration by Co-precipitation. This often allows the exchange of a complex matrix by a much simpler one. Provided the co-precipitation system is carefully chosen, a concentration step can be achieved simultaneously. The determination of traces of strontium is facilitated by co-precipitation with added (or existing) calcium as oxalate. Zirconium and other 'group 3' metals are separated by co-precipitation on ferric hydroxide.

Removal of Matrix Metals in Metallurgical or Inorganic Analysis. This depends on the chemical properties of the matrix metal. Electrolysis may be used for copper and also for ferrous alloys, where iron, nickel and chromium may be removed together, facilitating the determination of trace elements. Iron is also removed specifically by solvent extraction with isobutyl acetate from strong hydrochloric acid solution.

The hydride separation method for arsenic and similar elements may be subject to chemical interferences in solution at the hydride generation stage. Copper, for example, present in large concentrations inhibits hydride formation by these elements. Also, if copper is removed by electrolysis these tend to be lost by hydride formation in the electrolysis cell. The problem has been overcome for bismuth[85] and for arsenic, selenium, tin and tellurium[86] by the co-precipitation of the analyte metals on lanthanum hydroxide. The copper is filtered off and, on redissolution, the lanthanum has no interfering effect.

Solvent Extraction of Trace Metals. This is probably the most widely used separation technique, as it can be reduced to its simplest form. It is possible, and often desirable, in atomic absorption to extract more than one element at one operation. Specificity resides in the measuring technique. The choice of chelating or complexing reagent is therefore not limited as in colorimetry to one which gives a strong colour for the metal being determined, and complicated methods involving extractions and back-extractions in order to improve specificity are avoided.

For this reason, complexing reagents deemed unsuitable in colorimetry for their non-specificity can be used to advantage. Among these are dithizone, and the various thiocarbamate derivatives. One of the latter, ammonium pyrrolidine dithiocarbamate (APDC), first investigated by Malissa and Schöffmann (1955)[477] was reintroduced for determining traces of lead, copper and zinc by Willis[806,808] and Allan.[48,49]

The claim that this reagent can be used to complex some 30 elements has been confirmed[778] and most of these can be extracted into organic solvents. The pH range for formation and extraction of complexes is given in Table 7.

A number of advantages result from the extraction of APDC–metal complexes into a suitable organic phase. The metal may be concentrated by a factor of one hundred or more. Wanted metals can be separated from high concentrations of other solutes which cause difficulties in nebulization and atomization. The

TABLE 7. pH ranges for formation and extraction of APDC–metal complexes

Periodic group	4a	5a	6a	7a	8		1b	2b	3b	4b	5b	6b
Element	V	Cr	Mn	Fe	Co	Ni	Cu	Zu	Ga		As	Se
pH for formation	2–6	2–9	2–12	0–14	1–14	1–14	0–14	1–14	2–10		0–6	2–10
pH for extraction	4–6	3–7	4–6	1–10	1–10	1–10	0–14	1–10	3–8		0–4	3–6
Element	Nb	Mo		Ru	Rh	Pd	Ag	Cd	In	Sn	Sb	Te
pH for formation	2–4	2–6		1–14	1–14	1–14	0–14	0–14	2–10	3–7	2–9	2–6
pH for extraction	2–4	1–3		1–10	1–12	1–10	0–14	0–11	2–9	4–6	2–5	3–5
Element		W	Re	Os	Ir	Pt	Au	Hg	Ti	Pb	Bi	
pH for formation		1–3	1–14	1–14	1–14	1–14	0–14	0–14	1–14	0–14	0–14	
pH for extraction		1–3	1–10	1–10	1–14	1–10	0–14	0–10	2–12	0–8	1–10	
Element	Th	U										
pH for formation	4–8	2–5										
pH for extraction	4–6	3–4										

atomic absorption signal for nearly all metals is enhanced by a further factor of 3–5 when aspirated in an organic solvent instead of aqueous solution.

APDC complexes are soluble in a number of ketones. Methyl isobutyl ketone which is a recommended solvent for atomic absorption allows a concentration factor of ten times, while *n*-amyl methyl ketone may be used if concentrations of up to fifty times are required. Concentrations greater than 50 times are possible using chloroform. APDC has a very high partition coefficient when extracted into chloroform, and chloroform has very low solubility in the aqueous phase. The chloroform solution, though perfectly suitable for use in electrothermal atomizers, is not very suitable for aspiration into a flame, and a procedure modified by wet-ashing of the chloroform extract and redissolution in 50/50 aqueous acetone mixture is recommended.

The two procedures given by Watson[778] are as follows:

(i) Extraction using methyl isobutyl ketone
APDC solution: dissolve 1 g of APDC in water. Dilute to 100 ml and filter before use.
To 50 ml of sample solution add 5 ml of APDC solution, and adjust to the required pH with acetic acid or caustic soda solution. The pH should be 5 except for arsenic, molybdenum, thallium and tungsten (pH 3). For manganese, raise to pH 12, mix, stand for 2 min then adjust to pH 5. With chromium and molybdenum, heat to 80 °C

for 5 min before proceeding. Transfer the solution to a 100 ml separating funnel, extract the complex (which may have precipitated) into 4 ml of methyl isobutyl ketone by vigorously shaking for 30 s, then stand for 2 min. Transfer the aqueous phase to another separating funnel and repeat the extraction with 1 ml of methyl isobutyl ketone. Discard the aqueous phase (which should now be colourless), combine the extracts in the first funnel, mix and filter through a cotton wool plug into a small beaker. This extract is aspirated in the atomic absorption spectrometer.

Standards should be prepared for all elements by the same procedure, starting with mixed aqueous standards.

If a sample volume greater than 50 ml is required to increase the concentration factor, substitute n-amyl ketone which has a lower solubility in water than methyl isobutyl ketone. Watson showed that at least the following elements are quantitatively extracted with this procedure: bismuth, cadmium, cobalt, copper, iron, lead, manganese, mercury, nickel, tin and zinc, unless the limiting solubility of any complex is approached, e.g. 800 μg for mercury and 40 μg for nickel.

(ii) *Extraction of smaller quantities using chloroform*
Purified APDC solution: grind 5 g of APDC with 50 ml of acetone, collect the solid in a porosity 3 sintered glass crucible, wash with 20 ml of acetone, dry in air and dissolve 1 g of the dried APDC in 100 ml of water.

Chloroform, sulfuric acid 98% (18 M), nitric acid 70% (16 M), perchloric acid 60% (10 M) should all be of high purity.

To a volume of solution (up to 2 l) contained in a separating funnel add 2 ml of purified APDC solution and adjust the pH as in the first method. Allow to stand for 2 min, then extract the complex by vigorous shaking for 1 min with 10 ml of chloroform. Allow the phases to separate and transfer the chloroform extract into a 100 ml Pyrex conical beaker. To the aqueous phase add a further 2 ml of APDC solution and repeat the extraction. Combine the extracts. If electrothermal atomization is to be used, make these up to a convenient volume, e.g. 25 ml, and inject suitable aliquots into the atomizer.

If flame atomization is to be used, the solvent may be changed as follows: add 5 ml of sulfuric acid and heat to dense white fumes. Cool, add 0.5 ml of nitric acid and again heat to dense white fumes. Repeat until a clear brown solution is obtained, then add perchloric acid dropwise until a clear or pale yellow liquid is obtained. Evaporate to about 1 ml, add 4 ml of water, cool and transfer to a 10 ml volumetric flask. Wash the beaker with two 2 ml portions of acetone, add the washings to the volumetric flask then make up to the mark with acetone. This solution is aspirated in the flame.

Mercury, tin and bismuth are volatile under the conditions given.

Other extraction systems have been used, e.g. dithizone forms extractable complexes with about twenty metals, and the diethyl dithiocarbamates are almost as numerous as the pyrrolidine dithiocarbamates. These reagents complex many of the same metals as APDC however, and have little or no advantage.

8-Hydroxyquinoline ('oxine') is a useful chelating agent, as a number of metals form oxinates which do not react with APDC. In particular, aluminium, calcium, strontium and magnesium form extractable complexes, though not all under the same conditions, which at least allow a good concentration factor from dilute aqueous solutions.

For further information on the possibilities offered by various solvent extraction systems the reader is referred to the texts on this subject.[160a,532,703]

Also used for concentration–separation steps in sample preparation for atomic
absorption are certain ion association extraction procedures which are usually
capable of extracting larger actual amounts of metal ions than the chelation
procedures described above. They have the further advantage of not requiring
the expensive chelating reagents. A well known example is the extraction of
Fe(III) from strong hydrochloric or hydrobromic acid solution into diethyl
ether. This kind of separation is perhaps most useful for the removal of an interfering bulk matrix, leaving the analyte element in its original concentration in the
aqueous solution. Its possibilities may also be assessed from the specialist
literature.[532]

Ion Exchange. Though invariably slower than solvent extraction, ion exchange
techniques have been used to separate certain groups of metals from an undesirable matrix. Perhaps the most useful separation of this type is of trace heavy
metals from higher concentrations of alkali metals as described by Biechler.[99]
Especially for the analysis of industrial effluents, the method entails passage of
the sample at pH 5.2 through 50–100 mesh Dowex A.1. resin in a 10×1 cm
column. Copper, lead, zinc, cadmium and nickel are retained with 1 litre
samples, but complete retention of iron requires a sample of less than 500 ml.
These metals are eluted with 25 ml of 8.0 M nitric acid and the eluate made up to
50 ml. The reference solution should be 4.0 M nitric acid.

Phosphate may be separated from solutions containing calcium and strontium
on a de-acidite FF (acetate form) column like the one described by David[176]
though the need for this particular separation is probably lessened in the light
of hotter flames and knowledge of releasing processes.

Two chelating resins, Chelex 100 and Permutit S1005, were investigated by
Riley and Taylor[633] and were found to retain a number of elements, including the
rare earths. Bismuth, cadmium, cerium, cobalt, copper, indium, lead, manganese,
nickel, scandium, thorium, yttrium and zinc were completely recovered by
elution with mineral acid (2 M) and molybdenum, tungsten and vanadium by
elution with ammonia (4 M). Rhenium and silver were 90% recovered.

There are a very large number of examples in the literature in which ion
exchange has been used to separate an interfering matrix, either by retaining the
analyte elements as in the examples above or by retaining the matrix element. In
the latter case, relatively large amounts of the ion exchanger may be required and
no actual concentration of the analyte is achieved in the process.

Again the reader is directed to the specialist literature [160a,346] for more information on the kinds of separation that may be applicable to atomic absorption
methods.

DETERMINATION OF MAJOR COMPONENTS

The high sensitivity of many elements should not lead to the assumption that atomic absorption spectrometry is not to be considered for the determination of higher or major concentrations.

Provided that stable and reproducible instrumental performance can be established and that as much analytical care is taken as would be expected in volumetric or gravimetric assays, there is no reason why coefficients of variation of 0.5% and better should not be achieved. It is believed that many analyses which are difficult and time-consuming by classical methods, and not sufficiently reliable by newer physical methods can be tackled by atomic absorption with advantages in speed and convenience, and with sufficient accuracy for all except perhaps the most stringent assay requirements. Examples of such analyses occur in complex oxide materials, slags, cements, glasses, rocks, ores etc., some of the more complex alloys, plating solutions and also wherever metal-containing additives have to be controlled in an essentially organic matrix.

The achievement of high accuracy in atomic absorption determinations is both an analyst's and an instrumental problem.

Sample preparation and calibration

The accuracy of an analytical result cannot be better than the accuracy of the methods used to prepare the sample itself. Random errors are statistically additive and they must therefore be minimized throughout the whole procedure. The weighing of the original sample and the dissolution process must be carried out with the same care and refined technique employed in volumetric or gravimetric assay. Subsequent dilutions must be performed in high quality volumetric ware.

The accuracy of calibration standards is a *sine qua non* in this type of analysis, but the question of interferences must also be carefully examined so that standards and samples can be appropriately matched. Even minor interferences are now important. This problem may be circumvented if chemically analysed standard samples of identical type are available. These may be obtained in the UK from the Bureau of Analysed Samples and in the US from the National Bureau of Standards. Many types of alloy, rocks, ores etc. are included in their catalogues. Either the calibration can be established using several such standards, if available, or the calibration curve shape can be established with synthetic standards and the overall slope with one reference standard near the top of the calibration range. A programme for this was mentioned in the section on microprocessor-assisted calibration. These standards may also be used, as in the author's laboratory, to check that all interferences have been allowed for in the making-up of independent synthetic standards, thus improving confidence in the final results.

Releasing agents and ionization buffers must, of course, be added in equal quantities to both samples and standards. The effect of all the other major constituents of the sample on the one being determined may have to be examined. The materials from which standards are prepared may also have to be chosen carefully, for example, if an alkali salt such as potassium dichromate is used to prepare high concentration standards, the ionization buffering effect of the potassium, present only in the standards, may cause low results to be returned for the samples. Such effects can usually be foreseen and, if not preventable, can often be overcome by adding a small excess of the offending constituent to both samples and standards.

Because most of the best-known resonance lines of the elements are extremely sensitive, either the preparation of the analysis solution must incorporate considerable further dilution steps, or other means must be sought to decrease the sensitivity so that the measured absorbance falls within the instrumental range of greatest accuracy. Large dilution factors are not ideal as they incur manipulative errors and an unnecessary increase in the time spent on the analysis.

Decreasing analytical sensitivity

The alternative is to decrease the absorption signal in the instrument itself. This must be achieved without loss of stability and reproducibility which therefore precludes modifications to the nebulizer or its associated gas flows. Flame conditions and observation height, too, should only be adjusted to give the best stability and freedom from noise.

Sensitivity can be decreased by choosing another absorbing line of lower absorptivity. Most elements have sufficient resonance lines within the effective instrumental wavelength span to allow a wide range of concentrations to be determined. The use of more than one line per element, of course, is normal practice in emission spectroscopy. Useful alternative lines are quoted, together with an estimate of their relative sensitivities in the element tables in Chapter 9. It should be noted, however, that the most sensitive line is usually also the strongest line and therefore higher amplifier gain factors are nearly always required for the alternative lines. A practical point, too, is to ensure that the alternative line is in a part of the spectrum transmitted by the particular source lamp in use. The alternative line for magnesium at 202.6 nm for example is not transmitted by a lamp with a 'UV glass' window.

Sensitivity is also effectively decreased by rotating the long path length burners through an angle about their vertical axis, so that the resonance beam passes through only a narrow section of the flame. The rate at which the absorbance decreases with the angle of rotation (see Fig. 50) is obviously very high at small angles of rotation and thus it is difficult to set the burner angle accurately to ensure a decrease of sensitivity by a given factor. It is probably best to use the burner only fully angled, i.e. at 90° from its intended position, when about one eighth of the normal sensitivity is obtained. Rotation of the burner is clearly preferable to the use of an alternative resonance line if the latter requires gain factors to be increased to the point where noise on the output signal becomes noticeably greater.

When a circular Meker head is provided for emission measurements, this will give an effective path length of about one fifth that of the 10 cm air acetylene burner, with a sensitivity decrease in approximately the same ratio. Note, however, that the air acetylene emission burner cannot be used safely with nitrous oxide acetylene, nor will an air propane emission burner support an air butane flame.

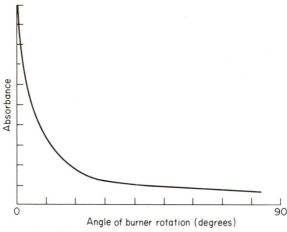

Fig. 50. Effect of burner rotation on sensitivity.

Instrumental performance and reading accuracy

If an analytical accuracy of 0.2 % is to be achieved, the stability and reproducibility of both zero absorbance base-line and measured absorbance signal must also be of this order.

Observable drifts in the base-line or resonance line absorptivity cannot therefore be allowed during the period of the actual analysis, i.e. when samples and standards are being aspirated. While a double-beam spectrometer corrects for variations in source lamp intensity, it may not reveal absorptivity drifts due to lamp warm-up or alteration in flame conditions, and so, whatever instrument is in use, it is wise to check for drifts originating in any part of the system by repeating one of the high calibration standards at frequent intervals, perhaps after every four or five samples, and certainly at the end of the analysis.

Good reproducibility requires, in addition to general stability, an output signal with the least possible associated noise. The origins and suppression of noise were discussed (p. 116 *et seq.*) and as lack of sensitivity is rarely a problem in this type of analysis, the instrument conditions can be chosen with low noise as the primary objective.

The part of the instrument read-out scale in which the greatest reading accuracy is obtained depends on numerous factors but, as pointed out by Weir and Kofluk,[781] in the most precise work absorbance readings must be limited to a relatively narrow range. They demonstrated that coefficients of variation usually have minimum values for absorbance between 0.5 and 0.8 (see Fig. 51) when the instrument is set up to give best precision.

The optimum absorbance at which maximum photometric accuracy can be obtained may be derived from consideration of the signal to noise and other instrumental characteristics.

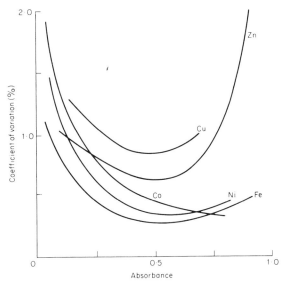

FIG. 51. Coefficients of variation at different absorbance levels reported by Weir and Kofluk (Ref. 781).

In the following, ΔA and ΔT are random errors associated with a single reading in the absorbance and transmission readout modes respectively.

By definition
$$A = -\log_{10} T = -\log_e T . \log_{10} e \qquad (1)$$

so
$$dA = -(\log_{10} e) \frac{dT}{T}$$

or, for finite increments
$$\Delta A = -(\log_{10} e) \frac{\Delta T}{T} \qquad (2)$$

Below are listed the sources of random error, together with their associated error function, i.e. the way in which the incurred error is a function of the readout signal:

1st Category $\Delta T \propto T$ (i) source lamp fluctuation
                                         (ii) gain drift
                                         (iii) optical stability

    (iv) flame scattering effects

    (v) readability of linear absorbance readout scale

2nd Category ΔT is constant   (vi) electronics (Johnson) noise

    (vii) readability of linear transmission readout scale

    (viii) flame emission noise breakthrough

3rd Category $\Delta T \propto T^{\frac{1}{2}}$ (ix) photomultiplier shot-noise

4th Category $\Delta A \propto A$ (x) variations in analyte concentration, i.e. nebulizer noise

    (xi) variations in absorptivity coefficient, e.g. atomizer noise

    (xii) variations in cell path length, i.e. flame 'flicker'.

The first three categories are common to all types of absorption spectrometry and were summarized by Ford.[240] The fourth type, characterized by Roos[641] as $\Delta T \propto T \log_e T$, arises essentially from the dynamic nature of the flame absorption cell.

The reason for the particular error function ascribed to each error source will in most cases be obvious. In (i)–(iv), variations directly affect the amount of energy falling on the detector. In (v), ΔA is constant, so, from equation (2) $\Delta T/T$ is constant, so $\Delta T \propto T$. In (vi), (vii) and (viii), ΔT is clearly constant for a given instrument setting. Johnson noise will be greater, however, at high gain settings (low source intensity) but is much less likely to be a precision limiting factor than shot noise. Shot noise itself (ix) results from the quantized nature of electromagnetic radiation and the relationship $\Delta T \propto T^{\frac{1}{2}}$ is proved by statistical theory.

If analytical precision S be defined as $C/\Delta C$, then, when Beer's law is true

$$S = \frac{C}{\Delta C} = \frac{A}{\Delta A} \tag{3}$$

Combining this with equations (1) and (2)

$$S = \frac{A}{\Delta A} = \frac{T \log_e T}{\Delta T} \tag{4}$$

For category 1 error functions, where $\Delta T \propto T$, from equation (2) ΔA is constant, thus $S \propto A$. Hence for all errors in this category the precision should increase directly as the absorbance increases. There are of course practical limits to this which are dictated by the presence of stray light and other deviations from Beer's law.

In category 2, where ΔT is constant, from equation (4)

$$S = \frac{T \log_e T}{k}$$

So
$$k \frac{dS}{dT} = 1 + \log_e T = 0 \text{ for maximum } S$$

therefore
$$\log_e T = -1$$

and from (1)
$$A = \log_{10} e = 0.43$$

This is the well known derivation of this value for absorption spectrometers where the factor limiting accuracy is the readability of the T scale. In atomic absorption this error function is more likely to apply only in the case of serious flame emission noise breakthrough, when precision would, in any case, be relatively low.

In category 3, substituting $\Delta T = k'T^{\frac{1}{2}}$ in equation (4)

$$S = \frac{T \log_e T}{k'T^{\frac{1}{2}}} = \frac{T^{\frac{1}{2}} \log_e T}{k'}$$

Now,
$$k' \frac{dS}{dT} = T^{-\frac{1}{2}}(1 + \tfrac{1}{2} \log_e T) = 0 \text{ for maximum } S$$

Hence $\log T = -2$, and from equation (1) $A = 0.86$.

This optimum absorbance value is usually quoted for spectrometers with photo-multiplier detectors, where reading accuracy is limited by shot noise. In cases where highly stable instrumental and flame conditions can be achieved, this would be the limiting factor in atomic absorption spectrometry too.

In category 4, which is probably most serious in atomic absorption, $\Delta A \propto A$, so $S = A/\Delta A = $ constant [and it follows from equation (4) that $\Delta T \propto T \log_e T$] thus the precision is independent of the absorbance, the transmission and concentration.

These results are easy to interpret for instruments with transmission readout. Category 1 fluctuations are proportional to the scale reading and thus, within the limits of calibration linearity, precision becomes greater toward low transmission values. In category 2, best precision is around T values of 37% and in category 3 around 14%. In category 4 precision is independent of T.

Deviations from Beer's Law. The above is purely an idealized approach to the error problem, for, not only do errors from all four categories operate simultaneously, but deviations from Beer's Law considerably modify these results. The effect of such deviations can be calculated assuming that they are largely the result of unabsorbed (and unabsorbable) light reaching the detector.

If s is unabsorbable and l is absorbable then

$$(s+l)T = s + l.10^{-abC}$$

where T is observed transmittance, a absorptivity, b pathlength and C concentration.

i.e.,
$$sT + lT = s + l.10^{-abC}$$

so
$$10^{-abC} = \frac{s - sT - lT}{l}$$

hence,
$$abC = \log_{10}\frac{(l+s)-s}{T(l+s)-s}$$

$$= \log_{10}\frac{(1-B)}{(T-B)} \tag{5}$$

where $B = \dfrac{s}{l+s}$ i.e. the proportion of stray light.

Equation (5) is the 'stray light equation' referred to on pp. 92 and 124.

Differentiating,
$$ab.dC = -\log_{10} e.\frac{dT}{(T-B)}$$

so
$$S = \frac{C}{\Delta C} = \frac{C}{dC} = -\frac{(T-B)}{\Delta T}.\log_e\frac{(1-B)}{(T-B)} \tag{6}$$

The four error functions can now be substituted in this equation. By judicious choice of the value of the constant k for each case, the relationship between precision and absorbance values for various values of B, i.e. for different degrees of variation from Beer's Law, can be calculated.

For category 1 errors, $\Delta T = kT$ so equation (6) becomes:

$$S = \frac{(T-B)}{kT}.\log_e\frac{(1-B)}{(T-B)}$$

If, by way of example, the source of the worst error is the absorbance scale, which should be readable with an accuracy of $0.001A$, $\Delta T \approx 0.002T$. Figure 52(a) shows the values of S plotted against A for the values of B: 0, 5, 10, 20, 30, 40 and 50%. As already derived, when $B = 0$, i.e. Beer's Law is obeyed, the relationship is linear, and precision increases with the absorbance reading. However, when stray light is present, the precision is seen to pass through a maximum. At 5% stray light the maximum precision $S_{(max)}$ is at absorbance 0.9, and for higher values of B, the maximum precision occurs at lower values of A.

Note that the slopes of the curves at any point are directly proportional to $1/k$, so if k is greater than 0.002 the values on the ordinate may be reduced in proportion. For a serious flame scatter interference, k could be 0.01 or greater. Thus, while for a given value of B the absorbance at which maximum precision occurs may be expected to remain the same, the precision itself is less, i.e. the error is greater, in proportion.

It would be hoped that in a good modern atomic absorption spectrometer, the other sources of category 1 error (viz. source lamp fluctuation, gain drift and optical instability) would be small compared with the examples given.

Category 2 errors where $\Delta T = k$ give the basic equation

$$S \doteq -\frac{(T-B)}{k}.\log_e\frac{(1-B)}{(T-B)} \tag{7}$$

Considering the readability of the T scale, ΔT should not be greater than 0.1% of T, $k = 0.001$.

Relationships are plotted in Fig. 52(b) assuming this value. As already calculated, $S_{(max)}$ for $B = 0$ occurs at 0.43 but as B increases above zero, $S_{(max)}$ occurs at progressively lower absorbance values. The maximum precision values are themselves relatively low and this suggests a real disadvantage of instruments which read out linearly in transmission, particularly when higher concentrations are being determined.

In linear absorbance instruments the worst source of error in this category would be emission noise breakthrough. This has been experienced by several workers when attempting to determine, for example, barium in presence of high concentrations of calcium. The value of k may then rise to between 0.01 and 0.1. As the slopes of the curves are proportional to $1/k$, with such values very low precisions are indicated, particularly at low absorbance values and when there is deviation from Beer's Law.

Category 3 errors, due to shot noise, where $\Delta T = kT^{\frac{1}{2}}$ produce the basic equation

$$S = -\frac{(T-B)}{kT^{\frac{1}{2}}}.\log_e \frac{(1-B)}{(T-B)} \qquad (8)$$

With a modern photomultiplier in use under average conditions in an atomic absorption spectrometer, shot noise should be such that k is not greater than 0.001. Figure 52(c), in which this value has been used, confirms that, when $B = 0$, $S_{(max)}$ occurs when absorbance is 0.87. Again, as deviations from Beer's Law increase, $S_{(max)}$ occurs at lower absorbances, e.g. for 5% of stray light the maximum precision is at about 0.65 absorbance units.

A somewhat different form of relationship is shown by the category 4 error function. The basic equation becomes

$$S = -\frac{(T-B)}{kT\log_e T}.\log_e \frac{(1-B)}{(T-B)} \qquad (9)$$

If total atomization noise produces a possible reading error of 1%, i.e. $\Delta A = 0.01A$, then from equation (4), $\Delta T = 0.01\,T\log_e T$. The precision/absorbance relationships, using the value 0.01 for k, are plotted in Fig. 52(d).

As deviations from Beer's Law are introduced, precision falls off markedly with increased absorbance. The comparatively low precision values are also a feature of this error category, which is probably the greatest source of inaccuracy in the conventional atomic absorption technique.

These relationships may be used to deduce actual precisions in particular practical cases but it must be realized that they apply to a single instantaneous reading. Where the effect of the source of error is to introduce a noise component on the output signal this can be reduced by taking averaged readings or by integrating, both over a selected period of time.

Nevertheless, computed precision values are useful in two particular ways:

(i) They enable an operator to compare directly the effects of different sources of error upon the final result; for example, it is often assumed that photomultiplier shot noise is the main limitation on precision when the flame is stable and transparent. The foregoing results show that, in the case of an instrument with transmission readout, this may well not be true.

(ii) They enable the effects of different degrees of variation from Beer's Law to be calculated.

The value of 10% for B is not unrealistic or unusual when a resonance line occurs close to or in a group of non-absorbing lines. This is the case where the calibra-

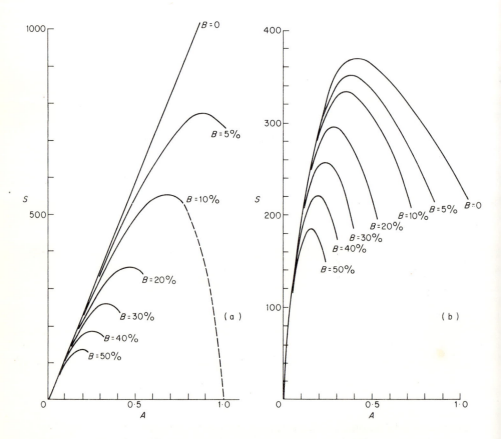

FIG. 52. (a) Plot of precision against absorbance for category 1 error function, $\Delta T = 0.002\ T$. (b) Plot of precision against absorbance for category 2 error function, $\Delta T = k = 0.001$.

tion graph is asymptotic to the horizontal at absorbance 1.0. The precision values are the inverse of the errors, which are statistically additive. So, for example:

$$\frac{1}{S^2} = \frac{1}{S_1{}^2} + \frac{1}{S_2{}^2} + \frac{1}{S_3{}^2} + \frac{1}{S_4{}^2}$$

where S is the overall precision and S_1, S_2, S_3 and S_4 refer to the precision values relative to the various error functions. If just the errors due to absorbance scale reading, shot noise and atomization noise are added in this way for $B = 0$ and $B = 10\%$, at a typical absorbance value of 0.50 and using the error constants from which the graphs in Fig. 52(a), (c) and (d) were drawn, we obtain the result:

$$\begin{aligned} \text{for } B &= 0 & S &= 97.4 \\ \text{for } B &= 10\% & S &= 82.8 \end{aligned}$$

and these are the expected precisions of a single reading in these two cases (these would be popularly expressed as errors of '1 in 97.4' and '1 in 82.8' respectively).

Conclusions from Treatment of Error Functions. The values of the constant k chosen for use in the calculations which gave the graphs in Fig. 52 are typical of

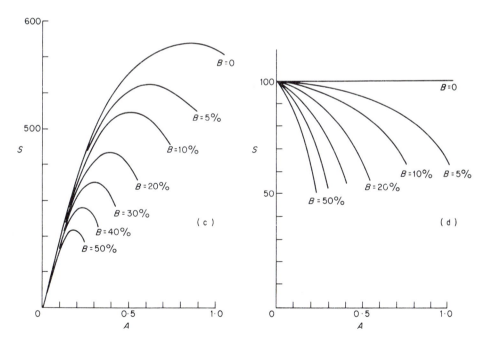

FIG. 52. (c) Plot of precision against absorbance for category 3 error function, $\Delta T = 0.001\, T^{1/2}$. (d) Plot of precision against absorbance for category 4 error function, $\Delta A = 0.01\, A$ i.e. $\Delta T = 0.01\, T \log_e T$.

those encountered in traditional atomic absorption techniques. The different error functions contribute different magnitudes of error (roughly in the descending order: categories 4, 2, 1 and 3) though these are directly affected by the chosen values for the constant k.

The importance of stable flame conditions is thus again emphasized, as improvements here will generally contribute the greatest gain in repeatability of results and therefore in overall precision.

All error sources operate simultaneously to a greater or lesser extent. It would therefore appear that, for low deviations from Beer's Law, taking the relevant graphs into consideration, the maximum precision is most likely to occur between the absorbance values 0.5 and 0.8. This observation agrees with the practical data given by Weir and Kofluk[781] which has already been mentioned.

METHOD DEVELOPMENT

Preliminary indications

The development of methods in atomic absorption is essentially no different from other types of analytical chemistry, and there can be no hard and fast rule, for the procedures adopted will depend upon the particular analysis in hand, and on the pre-knowledge of the analyst himself.

In general, the analyst should first acquaint himself with the reported behaviour of the elements to be determined under the commonly accepted standard conditions. These are summarized in Chapter 9.

If the sample is a dilute aqueous or non-aqueous solution of the elements in question, then a comparison is made between the reported sensitivity and concentration expected to be present in the sample. If the sample is solid material, it must first be assumed that it will be diluted by a factor of $100 \times$ on being taken into solution.

For best quantitative accuracy the concentration in the prepared solution should, as shown in the previous section, be between 20 and 200 times the reciprocal sensitivity. To achieve this, total dilution factors of up to $1:10^4$ are regarded as permissible. If higher dilutions should be required, the sensitivity also should be reduced (see p. 153).

If the concentration lies between $1 \times$ and $20 \times$ the reciprocal sensitivity, the situation can either be accepted as a 'trace' analysis, when a result of the correct order but with a lower precision can be reported, or steps may be taken to increase the concentration in order to maintain precision.

For concentrations below the reciprocal sensitivity, the procedure depends upon the noise level and attainable detection limit, but a concentration step is usually made unless a purely semiquantitative indication is all that is required.

Calibration

With a series of standard solutions covering each concentration range expected, the flame and instrumental conditions should be adjusted to give best sensitivity or best stability as required. The effects of the addition of other major constituents are then checked. Interference effects are then countered by use of the appropriate type of spectroscopic buffer and/or by modification of the flame conditions. It may be necessary, for example, to employ a hotter flame and an ionization buffer, or satisfactory accuracy may be achieved simply by careful matching of the standards to the samples. The latter expedient is usually satisfactory only as long as the concentrations of the major constituents remain virtually constant and provided they do not exert too great a degree of depression.

The calibration and general procedure is then checked against standardized samples, if available, of the same type as those to be analysed. If chemically analysed standard samples are not obtainable, it is necessary to establish the degree of recovery of accurately known amounts of the wanted elements added to the sample solutions. Provided the method development and analytical procedure have been carefully carried out, recoveries should be within $\pm 2\%$ (relative) of the amounts added. This would normally be sufficient to establish the validity of a method where the sample and standard solutions actually being measured contain mixtures of inorganic ions. In the presence of organic material such as protein, the metal already existing in the sample may be bound differently from that added subsequently and give rise to a different response in the flame. In such cases the results obtained with the organic matter present must always be compared with those given after the organic material has been destroyed by dry- or wet-ashing.

Precision

An indication of the accuracy expected from the developed method is always of value and is essential when different methods are to be compared. Errors may be random or systematic. Systematic errors almost invariably indicate a difference in chemical constitution between prepared sample and standard solutions which should have been taken into account during the method development investigations. In principle, therefore, it is always possible for systematic errors to be eliminated. Random errors cause poor repeatability between results obtained in one laboratory and poor reproducibility between those obtained in different laboratories.

A measure of the precision is therefore obtained with a reproducibility test in which the standard deviation or coefficient of variation is calculated from at least ten and preferably twenty replicate results. To check the instrumental reproducibility, the test is performed on one prepared solution, a blank or pure solvent being aspirated between successive readings. Finally the test should be carried out on the same number of individually prepared solutions from the same original bulk sample, in order to show up errors originating from the sample itself (heterogeneity) or the preparation procedure.

In a good quantitative procedure, coefficients of variation of $\pm 1\%$ and better are obtainable, so that the precision can be improved by replication. If n replicate samples are analysed, the precision is $1/n^{\frac{1}{2}}$ times the coefficient of variation. This will also be the accuracy of the method in the absence of systematic errors.

Chapter 7

Analytical Techniques in Electrothermal Atomization

The establishment of electrothermal atomization as a standard method in analytical laboratory practice has taken much longer than flame methods, even though they were suggested very soon after the introduction of analytical atomic absorption itself.

By the present time, electrothermal methods must have been attempted for nearly every possible elemental analysis but with widely varying degrees of success.

There are three principal areas where electrothermal atomization offers advantages over the flame. The first of these is where detection limits are sought which are below those given by the flame. As a very general rule, the reciprocal sensitivities given for the majority of elements by electrothermal atomization are about one thousand times better than by the flame. Detection limits are often about the same as reciprocal sensitivities and hence are improved by a smaller factor. Secondly, electrothermal methods are to be considered when the amount of primary sample is strictly limited, e.g. serum from a child, forensic samples, inclusions in metals and minerals, etc. The normal volume of solution taken for this method is 50 μl or less, as compared with the 0.5 ml necessary to obtain one integrated reading with the flame. Thirdly, electrothermal atomization can offer the possibility of analysing solid samples without dissolution. This statement, though true, must be interpreted with some caution, because solid sampling presents a number of problems which are obviated in solution methods. Principal among these is the difficulty of achieving accurate calibration unless solid standards are available. For this reason electrothermal atomization of solutions will be discussed in detail first, and solid samples will be dealt with towards the end of the chapter.

The most successful applications of electrothermal atomization so far reported are to samples with very volatile matrices, or to samples where the volatility of the matrix is very different from that of the element to be measured. This includes most of the field of water analysis, biochemical and other organic samples which are readily pyrolysed or ashed, leaving the element to be measured

165

in the mineral residue and some metals and refractories where the trace elements being measured are completely volatilized before the matrix material.

Precisions obtained with replicate firings of the same sample or standard may be as good as 1 %, or better with an autosampler. Analytical accuracy depends on a number of influencing factors and may not be as good as has come to be expected with the flame. Very great accuracy is, of course, not often called for at the trace and ultratrace levels at which electrothermal methods excel.

Some of the essential differences between flame and electrothermal techniques contribute to the greater possibility of inaccuracy in the latter. Among these are the greater difficulty of weighing or measuring the microvolumes of sample employed, the very much greater effect, relatively speaking, of sources of contamination, and the greater uncertainty in the measurement of the fast transient absorbance signals rather than integration of a virtually steady signal. In addition to these instrumental or human sources of error, chemical interference effects in furnace atomization are mostly quite different from those in flames.

It is only after appreciation that such difficulties exist that the analyst will be able to refine his technique sufficiently to obtain useful results.

OPERATION OF THE INSTRUMENT

In this context, the instrument consists of an atomic absorption spectrophotometer in which the flame and its attendant gas supplies and spray chamber system have been replaced by the electrothermal atomizer.

The spectrophotometer may be a conventional 'flame' instrument though clearly better results will be obtained with an instrument whose optics and electronics, as described in earlier chapters, have been designed with the special demands of electrothermal atomization in mind. The electrothermal atomizer will be controlled by a special unit to provide the versatility of heating programmes required by different types of sample.

It is again strongly recommended that the spectrophotometer and furnace head be placed beneath a fume extractor. Unless they constitute the sample matrix, actual amounts of toxic elements vaporized into the laboratory atmosphere are likely to be very small indeed but the ashing of samples with organic matrices can cause working conditions to become very unpleasant. To be effective a higher rate of extraction is necessary than with flame work, as the flame is no longer present to assist with its high rate of convection. An effective forced draught also helps to prevent deposition of contaminating dust particles in the vicinity of the furnace.

Optical alignment

Alignment is achieved in much the same way as for the flame. The position of the atomic spectral source is first adjusted for maximum energy throughput, the furnace itself not being in place. When the furnace is installed, vertical, rotational and lateral adjustments are made until the position is reached which causes minimum attenuation of the optical beam. In a well designed system such attenuation should amount to only about 10% or less.

The alignment procedure may be somewhat modified when the simultaneous background corrector is to be used. It is of prime importance that the atomic spectral beam and the continuum beam should follow exactly the same light path through the atomizer. The manufacturer's operating instructions should be the best guide to achieving this. The best way of checking the efficiency of the background correcting system is described later in this chapter.

Instrument settings

The instrument is set up for best noise-free conditions compatible with good sensitivity. 'Flame' and 'nebulizer' noise are now absent and only optical and electronic sources of noise remain. Noise in the output signal, whatever its origin, can seriously distort the absorbance peak maximum, and can therefore affect accuracy if peak height readings are taken. The effect of noise on peak area measurements is less important.

Lamp current should be as high as permissible to ensure the least proportion of undesired radiation. It may, however, be dictated by the need to match the intensity of the continuum source of the background corrector. In well designed instruments there are also means for adjusting the intensity of the continuum source.

The furnace tube position should then be checked for maximum energy throughput.

Slits should in general be narrow in order to minimize the effects both of continuous emission from the furnace and of continuous absorption or scatter background relative to the atomic line signal. It will be remembered that the energy transmitted by a monochromator varies directly with the slit width for a line source, but with the square of the slit width for continuum sources. The effects of continuum emission, which may tend to overload the photomultiplier and cause non-linearity of response is likely to be worse at higher wavelengths, while the effects of scatter become worse towards the low wavelengths.

Photomultiplier gains are usually preset or adjusted to be compatible with source intensity. It is worth remembering, however, that background attenuation in electrothermal methods can reach values even higher than 1 absorbance unit, and therefore the whole system should be capable of recording these low intensities with best possible freedom from noise.

The read-out system

Any time constant or response lag in the measuring system will clearly have the effect of distorting the shape of the true overall response function. The higher or sharper the peak, the greater will be the distortion and the greater will be the relative error at the peak value itself. This effect can contribute markedly to the curvature of calibration curves in electrothermal atomization methods.

The main sources of electronic response lag are in the decoder and logarithmic conversion circuits, though the limiting time constant may be in a chart recorder if this is used.

Thus, because the overall response function may have half widths of less than one second in cases of fast atomization, the recorder should have a full scale deflection capability of not more than half a second. Time constants of the spectrophotometer measuring system, particularly if this has been designed for electrothermal work, are likely to be less than that of the recorder, which is why peak measuring circuitry is the best way to ensure measuring precision and ultimate accuracy.

Peak Height or Peak Area. The question of which measurement is to be preferred has been much discussed.[45,713] Much early work was done with peak heights, though a number of factors contribute to preventing the recording or measurement of the true absorbance peak. In addition to the instrumental time discussed above, these include slow atomization caused by use of too low an atomization temperature, and too high a dead volume in the furnace tube. Slow atomization has the effect of lowering the peak absorbance and increasing the half width, as fewer than the maximum number of atoms contribute to the peak absorbance. In large tubes there may be a volume containing atoms capable of absorbing radiation but which is missed by the radiation beam. This simply results in lower peaks.

It has already been mentioned that from the theoretical viewpoint, integration, i.e. peak area, is preferable for measurements at relatively low temperatures or low atomization rates. Following on from this, however, variations in sample matrix composition can cause differences in atomization rate and consequent differences in peak height. Peak area measurements can solve this problem, but it is also avoided by close matching of samples and standards.

On the other hand, integration can give erroneous results in the case where multiple peaks due to background signals appear during the atomization phase.

An experienced operator is usually able to distinguish such spurious readings visually on a strip chart recording, but they are not readily differentiated by the integrating system.

A possible advantage of peak area measurements is that the linear calibration range can be extended, provided it is not already limited by deviations from Beer's law, by the deliberate use of comparatively low atomization temperatures in order to reduce analytical sensitivity. Peak heights are measured:

(i) as the height of the maximum of the recorded absorbance trace above the zero absorbance baseline;
(ii) electronically, using the digital peak-recording circuitry generally built into a modern atomic absorption spectrophotometer.

Peak areas are measured:

(i) by physically measuring the area beneath the recorded absorbance trace using conventional methods;
(ii) by using the integration facilities of the instrument readout system.

In view of the foregoing, it is clearly advantageous for the analyst to be able check the general shape of absorbance peaks either on a fast recorder or by oscilloscopic presentation, in order to be able to decide whether peak height measurements are likely to be satisfactory in a given analysis, or whether peak areas may be better.

SELECTION OF OPERATING CONDITIONS

The furnace controller enables the four or five basic steps—dry, ash, atomize, tube clean (and tube blank)—to be carried out in a sequential form, each stage governed by the particular phase of the controller programme.

Each of these stages has to be carefully optimized to obtain the best results for any particular analysis, therefore we will consider them in sequence.

Drying

The importance of correct drying conditions cannot be overemphasized. During this phase the solvent is evaporated from the sample solution injected into the graphite tube. This step has to be accomplished in a controlled manner such that there is a slow, even evaporation of the solvent from the matrix leaving behind the solid dried sample. The solvent must not be removed from the droplet by rapid boiling because the spitting and frothing actions which would then occur cause particles of the sample to be ejected. With the gas flow these particles can be carried out of the graphite tube and lost.

The best method of ensuring the correct drying time and temperature is by actually viewing a droplet's behaviour as the drying sequence is carried out. The droplet should slowly and evenly reduce in size and eventually disappear; at no time should any excessive movement of the droplet be observed. As an empirical rule, a setting of 105 °C is high enough for aqueous samples. The time necessary varies according to the sample size but generally a time in seconds up to $2\times$ the sample size in μl should be used, i.e. 20 μl of aqueous sample generally needs between 30 and 40 s at 105 °C to dry correctly. With solvents of different boiling points the temperature and time necessary will obviously be different from this example, but can easily be determined by the suggested method.

It is possible also at this stage to achieve a useful concentration effect by multiple injections and dryings. For example a sample may give an analyte concentration below the detection limit. By injecting 20 μl of sample, drying this, adding another 20 μl of sample, drying this and repeating as many times as necessary it is possible to achieve an analyte concentration above the detection limit. Obviously care has to be taken because the matrix is being concentrated likewise and this may cause problems later on, i.e. during ashing and/or atomization. When using this concentration method it is best to let the programme continue into the planned ashing sequence for several seconds to make sure that condensed solvent does not remain in the tube ends or baffles. It may be of advantage when using multiple dryings of a sample with a heavy matrix to allow the programme to complete the planned ashing stage and to stop just before atomize. In this manner the amount of matrix present when the programme is selected to go to completion is very much reduced. Care must be taken here to allow adequate cooling time after the end of the ashing sequence before the next injection is attempted or the micropipette tip may melt.

When samples dissolved in organic solvents are being dried care has to be taken because these substances have low surface tensions and easily wet graphite. As they become warm during the drying stage they tend to spread out over the tube length. It is possible when using large volumes that the liquid might even run out of the tube ends. Thus, it may not be possible to inject the same maximum volume of an organic solvent as for an aqueous solution and volumes should be restricted to half of this value. Furthermore organic solvents tend to soak into the graphite far more than aqueous samples. This means that the appearance time of the analyte on atomization (especially that which has soaked deepest into the graphite) can be markedly different from that achieved with aqueous samples.

This problem is much reduced by use of pyrolytically coated graphite tubes. The effect can also be reduced by injecting a similar volume of the pure organic solvent immediately before the injection of the sample volume and drying both together. The injection of the pure solvent tends to fill the graphite's porous surface and prevent the sample from soaking in when it is injected on top of the solvent. Alternatively, the sample can be slowly injected into the furnace tube which is already held at the required drying temperature. In this situation the solvent then evaporates from the sample immediately on contact with the warm graphite surface and does not penetrate the surface layers.

Ashing and atomization

When trace metals are to be determined in pure aqueous solutions the ashing phase has no significance and can safely be ignored. When trace metals are to be determined in varying amounts of matrix, however, the ashing stage is the most significant stage in the whole furnace programme. The whole success of the analysis depends on the correct selection of the ashing conditions. Proper thermal destruction of the matrix in a sample depends on the matrix itself and no general rules can safely be applied.

One general principle should be especially noted: the use of too high an ashing temperature or time results in the loss of significant quantities of the analyte before the atomization stage. This is particularly true of the more volatile elements such as mercury, arsenic, cadmium, lead, tellurium, etc. The best method to determine the limits for both ashing and atomization temperatures is the construction of ash/atomize curves for the elements and matrices involved.

Preparation of ash/atomize curves

The first stage is to prepare an aqueous solution containing only the required element at a reasonable concentration (perhaps sufficient to give approximately a 0.1 A signal). Should the sample contain the element as a particular compound, e.g. chloride or nitrate, then the aqueous solution should contain the element in that form. This is because certain compounds are volatile as molecular species at temperatures well below the atomization temperature of the element itself. Metals which do this are notably aluminium, iron, nickel, arsenic, cobalt, etc., and the halides are normally the most volatile. Most metal oxides or hydroxides are stable with regard to temperature and hence volatile species should be converted to either of these forms.

After preparation of the solution, the best drying temperature can be established. Then, omitting the ashing stage, an initial atomization time and low temperature (such as 5 s at a setting of 1400 °C) are set on the controller. The tube clean conditions are set at 5 s at nearly maximum power. A suitable volume, such as 20 μl of the aqueous standard solution, is then injected into the graphite tube and the complete furnace programme is carried out. The resulting peak height (if any) recorded during the atomization step is plotted on a graph of peak height versus temperature. A total of three injections is carried out at each set of conditions and the atomization temperature setting is increased before the next set of injections. As the power is increased the peak height also increases until the point is reached where the peak height does not increase (or the rate of increase becomes much smaller). This gives the lowest temperature for complete atomization.

For the sample where a continually increasing peak height for increasing temperature is obtained, the point at which the rate of increase becomes smaller than the initial rate is chosen. Then the selection of a lower temperature than maximum is normally advantageous because it results in increased tube lifetime, better precision of results and, in many cases, reduced emission interference. These results are illustrated graphically in Fig. 53(a) and (b) respectively.

When the best atomization temperature is selected, the correct time can be established by using a fast recorder speed at a known chart time and measuring the time taken for the peak to form and decay. Normally the time taken is that in which the recorder pen returns almost to baseline again, as shown in Fig. 54.

There are occasions, however, when a blank signal due to emission and other effects can be measured at the chosen atomize temperature and in this case careful selection of the atomization time can give the best peak height and minimum blank signal, as shown in Fig. 55.

Here time A includes all peak of height (b) but also a significant blank (a).

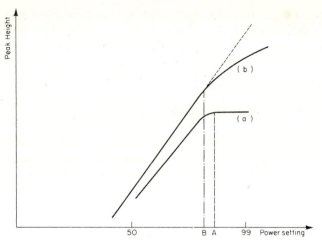

FIG. 53. Selection of atomization temperature.

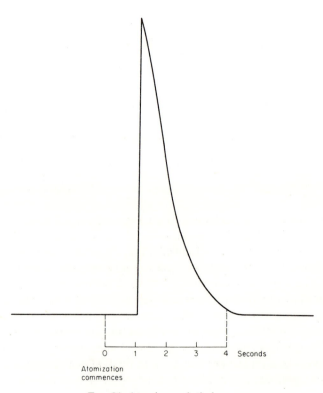

FIG. 54. Atomize peak timing.

By choosing time B the maximum peak height (b) is still obtained but the blank (c) is much smaller.

The selected conditions for atomization are then left untouched on the controller. To determine the correct ashing conditions a similar procedure is followed. Leaving the dry, atomize and tube clean conditions constant the ash conditions are changed progressively. A time of 30 s is chosen for ashing, this being a reasonable time to ash a typical matrix.

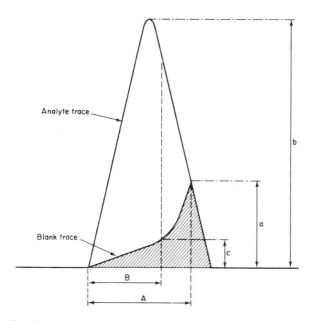

FIG. 55. Atomize peak timing in the presence of a blank signal.

The ash temperature settings are then increased slowly from a lower setting of 150 °C. Again triplicate peaks are obtained at each setting and a graph of peak height versus ash temperature is obtained. This is typically a mirror image of the atomize graph because the measured peak height remains essentially constant until the point is reached where analyte begins to be lost. At this point the peak height diminishes rapidly with increased temperature setting. The safe ashing temperature is taken as the point 100 °C approximately cooler than the turn-over point (see Fig. 56). We can now set the time and temperature for the ashing sequence. One must bear in mind that should a longer ashing time be necessary, e.g. 60 s instead of 30 s, the ash curve might be modified and analyte lost at a lower temperature. Conditions should therefore be re-checked with the longer ash time.

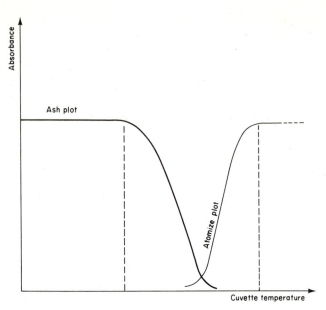

FIG. 56. Ash/atomize curve.

A successful ashing sequence should result in almost no smoke being produced during the atomization phase but this is not always possible. The smoke formed during atomization has therefore to be compensated for by using background correction or by careful choice of atomization conditions.

Tube clean and tube blank

The tube clean step is normally set several hundred degrees above the atomization temperature for about 5 s. This is usually sufficient to clean the tube effectively.

A matrix such as blood or serum, however, leaves a deposit behind in the tube which may resist removal by any means other than physically brushing it out with a pipe-cleaner or such item. This deposit eventually physically blocks the light beam and adversely affects the precision of measurements. Also, with some elements, carbide (or nitride if nitrogen purge gas is used) formation occurs and gives a memory effect. In this case repeated high temperature firings might be required to bring the blank firing peak back to an acceptable level.

If a tube blank sequence is provided it may be used to check that no memory effect is present, i.e. the tube blank should give no appreciable peak at all except in conditions where some analyte remains after the atomization step. With a new tube, the residual peak may be due to impurities in the tube material itself. Even high purity graphite contains certain impurities as will be evident from the manufacturer's specification. The tube blank time and temperature are set the same as for atomization.

CALIBRATION

There are no differences in principle between calibration procedures for flame and non-flame methods. Calibration standards should therefore match the samples as nearly as possible with respect to major components. In the case of samples of unknown constitution, the method of standard additions can be employed. In practice, however, the procedures actually adopted result from a new possibility of greater flexibility of approach.

Direct calibration

In many cases, the ashing step destroys the sample matrix before atomization takes place. The analyte is therefore atomized from a simpler compound than the original sample or the sample solution. The calibration standards should therefore be formulated to produce that same compound on ashing. For example, when trace elements are to be determined in an aqueous or organic matrix, the latter may be 'mineralized' in the ashing step in presence of an oxidizing acid, leaving simply a residue of metal oxides. The standards need contain only the same acid as the solvent medium as it will leave the metals in the same form after ashing. Major metallic constituents, e.g. of metal or other inorganic samples, should be matched in the standards.

The calibration relationship may be expected to be linear in the range from 0 to 0.3 or 0.4 absorbance units. The sample should be diluted so that the element sought will respond in this range.

There are several possible procedures for calibration.

(i) Prepare a set of standards (usually three are sufficient) and a reagent blank covering the concentration range in which the sample is expected to lie. Inject the selected volume (usually 20 μl, but circumstances may dictate more or less) and take each standard through the chosen atomization programme two or three times to ensure good agreement. Measure the resulting peak heights or peak areas. The blank, containing solvent and all added reagents except the element being determined, is always necessary in order to correct for residual amounts of the element in the water or other solvent, the reagents and even the graphite tube. Construct a calibration graph. If this cuts the absorbance axis at the blank value, subsequent readings are automatically blank corrected. Then treat samples in the same way and read concentrations from the graph.

(ii) An alternative procedure is to prepare the reagent blank and just one standard. The points on the calibration graph are obtained by injection of different volumes, e.g. 5 μl, 10 μl and 20 μl. This procedure usually gives a distinctly more curved calibration because the degree to which the solutions spread to different temperature zones along the tube is different for different injected volumes.

(iii) A better way, and a variation of the one-standard procedure, is to obtain higher points on the calibration graph by multiple injections. The furnace is allowed to run only through the drying step between such injections, so that the second and third are added to a dry residue. The spreading of solution along the tube is thus much reduced.

Most workers agree that procedure (i) gives the best analytical accuracy of the three.

Calibration solutions

Stock solutions for electrothermal analyses can be prepared by the analyst himself. The materials and weights given in Chapter 9 for flame work are applicable, though wherever possible the standard should be made up in nitric acid. Alternatively, commercially prepared solutions may be used.

Working standards are prepared from the 1000 ppm stock solutions by diluting in steps using clean volumetric glassware. Glassware is suitable for dilution but not for storage. Although the use of a micropipette to prepare a working standard in one single step (e.g. 10 μl of 1000 ppm made up to 1 litre gives an 0.01 ppm working standard) is not conducive to the very highest accuracy, this may well be the best way to avoid contamination through excessive handling. The final working standard should then be acidified with an appropriate acid, i.e. the same acid or acid mixture used during preparation of the samples, normally avoiding hydrochloric acid for this purpose because it leads to the formation of volatile chlorides.

Method of standard additions

For more complex solutions and samples where it is not possible to remove the matrix during the ashing step it may be necessary to use the method of standard additions. Some workers advocate that this method should always be run initially to check for interference effects so that the best calibration procedure can be selected. If the standard additions graph and the direct calibration graph were parallel, freedom from interferences in the sample would be indicated, i.e. the element is in the same form in sample and standard immediately before atomization, or the two forms give the same absorption response.

The principle of the method of standard additions is described on p. 125. It is advisable that all readings be obtained in duplicate or triplicate. There are again variations of the procedure.

(i) Three or four aliquots of the sample solution are transferred to volumetric flasks and known, different amounts of a standard solution of the analyte element are added to each of these except one. The solutions in the flasks are then made up to volume. These solutions plus a reagent blank are then run in the electrothermal atomizer. Each reading should have the blank subtracted before plotting. The sample result is then itself blank-corrected.

(ii) The standard additions can be made in the furnace tube itself thus avoiding dilution of the sample and possible contamination from volumetric ware. The selected volume of sample is injected into the tube, dried *and ashed*, the programme being stopped before atomization. After cooling, a known amount of the analyte element is injected on top of the ashed sample and the complete atomization programme run through. This is repeated for two more different additions and, of course, for the sample with no addition. The standard addition graph is plotted in the normal way, absorbance peak versus amount of added analyte element, and the amount of the element in the sample computed. Note that, again, the different additions can be made either as fixed volumes

of different concentrations, different volumes of the same concentration, or multiple
additions of the same concentration. The drying step should be run between multiple
additions.

The validity of the method of standard additions depends on the forms of the
analyte element in the sample and in the added standard responding in the same
way during the atomization step. This may not always be so in practice. For
example, in determining lead in whole blood by this method it would have to be
proved (or assumed) that lead added as a lead nitrate solution is atomized to the
same extent as lead bound organically in the sample. This is most easily checked
by running a certified standard material through the procedure. If no such
reference standard exists, as with the blood example put forward above, the
sample should also be run after pre-treatment for removal of the matrix, e.g. wet
or dry ashing, and the results compared.

Two further limitations apply to the use of the method of standard additions.
Simultaneous background correction must be employed because of possibly
varying amounts of matrix material present in the tube during the several firings
needed to make one determination. Furthermore, all readings must be within
the linear portion of the calibration graph in order that meaningful results may
be obtained. It is, of course, perfectly permissible to use a linearization function,
either as provided in the atomic absorption spectrophotometer itself or as may
be devised with a programmable desk top calculator.

SAMPLE HANDLING AND PREPARATION

The increased sensitivity which is the main feature of electrothermal atomization
methods introduces a number of difficulties connected with the handling and
preparation of samples. We attempt to offer some practical guidance on the
avoidance of errors through contamination and on the choice and use of micropipettes. The analyst must appreciate, however, that we are dealing with a
technique of ultramicroanalysis, and any advice or experience that he can make
use of on that subject will be entirely relevant here. Some of the problems have
been defined by Mitchell.[524]

Avoidance of contamination

The importance of clean-air rooms, or at least clean areas and benches, though
emphasized by Fuller,[19] is a point which has often been omitted from the
earlier literature.

Atomic absorption with electrothermal atomization typically involves measurement of less than 1 ng of an element during each analytical sequence. At such
extremely low levels, contamination of the apparatus with detectable amounts of
common elements is a severe problem. Contamination may occur at any stage in
the procedure. It may arise from the reagents used in sample preparation, from

the vessels used during preparation, or from the laboratory atmosphere at any stage in the procedure even when the sample is actually situated within the graphite tube. Sodium, magnesium and zinc are often detected as contaminants after conventional laboratory washing procedures, and zinc particularly is often found in cleaning agents. Other elements which frequently figure as contaminants are iron, copper and potassium. If the laboratory deals with a particular type of sample, then the matrix of that sample is a potential contaminant.

The precautions recommended to avoid contamination are detailed below and are divided into two sections, those considered essential to enable electrothermal atomization to be carried out successfully, and those considered desirable.

Essential Precautions

1. All glass or plastic vessels to be used for electrothermal atomization work should be washed, rinsed and then soaked in 2% v/v nitric acid for at least 24 hours and then thoroughly rinsed in high purity deionized or double-distilled water.

2. A 'clean' bench area should be reserved for solution preparation for use with the furnace.

3. The volumetric and storage ware used for solutions for electrothermal atomization should be kept separate from apparatus used for conventional laboratory work.

4. Solutions of low concentration should be prepared immediately before use and after preparation should be transferred to a suitable plastic container for storage.

5. When solutions of the same concentration of an element have to be prepared regularly on a routine basis it is advisable to keep the same apparatus for the same solutions.

6. Efficient fume extraction.

7. A high purity water supply, either deionized having a minimum resistivity of 10 MOhms cm^{-1}, or double-distilled. It is preferable to produce water as it is required rather than to store it for later use. For this purpose a deionizing system is usually more convenient.

8. Micropipette tips may introduce contamination. If this is excessive it may be necessary to soak the tips in dilute nitric acid and then wash with high purity water before use. In any case, with each solution, the micropipette tip should be washed through twice with injections which are discarded into a beaker before it is used for the actual injection into the graphite tube.

Desirable Precautions

1. The complete electrothermal atomization system should, preferably, be in a room separate from the general laboratory, and well away from sample preparation procedures.

2. The room should be under a small positive pressure, supplied from a pump system with filtration for dust particles.

3. The room should not be used by personnel as an access room for other parts of the laboratory.

Plastic Ware. All solutions should be stored in plastic bottles, as glass vessels usually give greater contamination and also adsorption of the required element from the solution onto the glass surface. Polypropylene and polyethylene are the best general purpose storage vessels, but even these cause solution deterioration, noticeably with solutions having concentrations below the ppm level. This effect is activated by acidic solutions and soaking in water is preferable to dissolve any contaminating salt. If soaking in acid is necessary then acid concentration no greater than a few percent v/v should be used. Polystyrene vessels should not be used because they are particularly susceptible to adsorption effects. The best storage vessels are those made from PTFE or FEP material, which give very few problems from contamination or formation of bonding sites. These are very expensive, however, compared with polyethylene and polypropylene.

Micropipette Tips. Micropipette tips may give contamination from materials used in construction or from packing, but more usually contamination comes from handling in the laboratory. For this reason it is very good practice to put the tip on the micropipette from within the supplier's packet by handling the outside of the packet only. Also, when the pipette is put down between operations it is essential to rest it in such a way that the tip does not come into contact with any bench surfaces or other objects. Use of the same tip for many injections is acceptable providing one tip is used for only one solution as cross-contamination may otherwise result. It is good policy to dispose of the tip when changing solutions so that it is not left for later use, by which time it could be contaminated by contact or by atmospheric dust.

In some cases the dyes used for colouring tips have been found to contain contaminants. Pink tips have been found to contain cobalt, yellow tips may contain cadmium, and white tips may contain lead. This source of contamination is now less common and should not cause any significant problems if the tips are purchased from a reputable manufacturer.

As a check on contamination from this source it is essential, of course, to make at least two injections from each solution in order to check reproducibility.

Use of micropipettes

Unless the special type of autosampler can be used, micropipettes are an essential part of electrothermal atomization techniques, as they give one of the most reliable methods of introducing small volumes of liquid samples into the graphite atomizer.

The pipettes are all based on air displacement with the simple plunger and are provided with non-wettable plastic (usually polypropylene) disposable tips to contain the solution, preventing any contamination of the pipette itself. Most micropipettes have a double action plunger system, i.e. calibration and overshoot positions, which ensures that the sample is completely dispensed.

There are many manufacturers offering a complete range of volumes: in addition some have available a selection of pipettes with adjustable volumes.

Choice of Micropipette Tip. The tips used with the micropipette contribute greatly to the accuracy of the analysis and different types are better suited to certain uses and volumes. In general, the 'dead' air space between the sample and the plunger seal should be kept to a minimum since this can seriously affect both accuracy and precision, due to the air expanding in response to even very small changes in temperature, such as could be produced by a loose or a firm grip.

The dart-like tip shown in Fig. 57 is ideal for injections of smaller volumes, e.g. of 50 μl and less. This allows only a small 'dead' air space, and the tip reaches almost to the surface of the graphite tube to deposit the sample.

A more cylindrical tip (Fig. 58) is best suited for higher volumes.

FIG. 57. Dart-like micropipette tip recommended for aqueous solution injection.

FIG. 58. Tip best suited for large volumes (i.e. 100 μl).

Operating Technique. Correct operation of a micropipette is essential to enable volumes to be dispensed precisely. The plunger of the micropipette should be depressed slowly but firmly to the first stop position, and the tip just inserted in the solution to be tested. The plunger should be allowed to rise carefully over a period of 3–4 s. To inject the sample into the atomizer, the plunger is slowly (3–4 s) depressed fully to expel the sample completely. The pipette and tip should be completely removed from the tube before the plunger is gently released to avoid taking the sample back into the tip.

Accuracy and Precision. Accuracy is usually within 1 % of the stated volume, and most manufacturers will specify each pipette to be accurate within this limit. It is however important to maintain the instrument in good order according to instructions.

Since atomic absorption is a relative technique, relying on standard solutions for calibration, the absolute accuracy of the micropipette is not critical providing the same one is used to dispense both known standards and unknown samples; the prime requirement is therefore the ability to reproduce the volume, i.e. good precision. Precision should be within 1 % RSD with volumes down to 20 μl, but will only be within 2 % RSD at 10 μl. Volumes below 10 μl should not be used if at all possible because of the loss of precision. These figures represent results of sampling very dilute aqueous solutions. As the total dissolved solid content of the solution rises, so the viscosity will increase, and as this becomes significant the precision will deteriorate. The technique of reverse pipetting[33] gives better precision but is not suited to injection in a graphite tube.

Pipetting precision of blood or serum is improved by taking up a volatile liquid, e.g. n-heptane, between the bottom and first stops of a double action micropipette. The sample is taken up in the calibrated part and both are then injected into the furnace. The heptane washes out the sample completely and is evaporated off at the drying stage.

Injection of organic solvents

Standard tips are made from polypropylene which has been treated to be non-wetting for aqueous solutions. When non-aqueous solvents such as ketones, alcohols and chlorinated hydrocarbons, etc. are to be handled, however, the precision or repeatability is very much poorer, anything from 10 % or worse. There are two main reasons for this. The non-wetting coating or finish is rapidly destroyed by organic solvents and, although some tips might work properly, it is likely that droplets of liquid will remain behind on the inner walls of the tip. One recommended procedure is to pre-wash the tips in the solvent to get uniform performance; this not only destroys the coating, however, but encourages droplet formation. Also, when standard tips are used for volumes between 1 μl and 20 μl, there is a comparatively large dead space above the liquid. This results in premature sample ejection, due to the build up of solvent vapour pressure in this space which forces the liquid out. Low boiling point solvents such as chloroform are particularly prone to this problem.

Choice of Solvent. The problem can be alleviated to some extent by careful choice of solvent. Chloroform could be replaced, for example, by 1,2-dichloroethane which has a much lower vapour pressure. Avoidance of solvents of low boiling point reduces the problem of expulsion of the sample from the pipette tip, but does not prevent the formation of droplets within the tip.

Choice of Tip Type and Material. The dead-space above the sample in the dart-like tip is much reduced by the use of the type of micropipette which uses fine capillary tips. The Oxford Ultramicro-sampler is an example. Use of this device completely overcomes the solvent expulsion problem. The principal disadvantage from the point of view of electrothermal atomization with a graphite furnace is that its maximum capacity is 5 μl. This may be too little for the sensitivity of some elements.

The tip material provided with this syringe is still prone to droplet formation, but this can be replaced by PTFE capillary tubing of the correct dimensions, e.g. Polypenco size TW 24, which appears to overcome the problem completely. Good precision should be attainable with tips made from this material, provided the tips are cut across at 90° to the tube axis. Chamfered tips give rise to a variable position of the meniscus with consequent loss of reproducibility.

Sample preparation

In many analyses, preparation will consist either of dilution of an already liquid sample or dissolution of a solid sample followed by dilution.

Contamination introduced by the diluents or solvents used can be significant. The need for high quality deionized water has already been emphasized. Normal analytical reagent grade solvents and reagents are generally not sufficiently pure and materials of a higher degree of purity, e.g. BDH 'Aristar', Merck 'Suprapur' grades or their equivalent, must be used. Sometimes solvents even in these grades give rise to a significant blank and this can only be reduced by further distillation.

Minimum quantities of acids are needed if dissolution is carried out in a PTFE pressure dissolution vessel such as that originally described by Bernas[97] and which is capable of operating at temperatures up to 180 °C. With organic matrices, however, only the special versions of this apparatus capable of withstanding very high pressures should be used. For completely inorganic dissolutions, there is also an all-PTFE vessel, which is limited to temperatures below 120 °C, but which completely removes the risk of contamination from the metal outer casing.

A vapour phase acid attack method has been described by Mitchell and Nash.[525] The acid is slowly evaporated in one part of the apparatus so that its vapours attack the sample which is in another part and not in contact with the liquid acid. Less pure acids may therefore be used. A simplified arrangement operating on the same principle was used by Woolley[832] for the decomposition of high purity glasses.

A so far little used technique in which organic matrices are oxidized at low temperatures in an atmosphere of electronically excited oxygen was introduced by Gleit and Holland some years ago.[282] This would avoid the need for using either a muffle furnace with its risk of loss of volatiles, or wet oxidation with its attendant risk of higher reagent blanks.

Chemical separations

The types of separation procedure described elsewhere in this book for the improvement of sensitivity and for matrix separation in flame atomic absorption analysis can, in principle, be employed for the same purposes before electro-thermal atomization.

In solvent extraction methods only reagents and solvents of the very highest purity can be used and it is most important that blank extractions should be run wherever appropriate. Electrothermal atomizers allow a greater degree of flexibility than the flame, because the solvent does not have to be flame-compat-ible. The whole range of organic solvents can thus be considered. In particular, the separations using APDC and 8-hydroxyquinoline may, under the correct conditions, be more successful using chloroform than methyl isobutyl ketone which is recommended for flame work. The chloroform version of this extraction is included on p. 150. Therefore the best solvent for a particular separation can be employed, the only limitation being imposed by the difficulties of handling organic solvents with micropipettes discussed on p. 181.

It has been confirmed[42,150,287] that elements in compounds in organic solvents usually give the same response in electrothermal atomization as the same element in an aqueous standard solution. This is because the solvent is removed at the drying stage and most organic complexes are converted to stable inorganic compounds at low ashing temperatures.

Ion exchange separations may also be carried out in preparation for electro-thermal atomization. An interesting variant on this method is where the resin itself, containing the bonded analyte element, is subjected to direct analysis in the solid phase. In one example[542] one litre of seawater was passed through 500 mg of chitosan (a natural chelating polymer). The resin was then homo-genized and 5 mg samples of this were analysed for vanadium. Response of vanadium from the resin and from aqueous standards was shown to be the same.

BACKGROUND ABSORPTION AND SCATTER

Background absorption and scattering effects are commonly encountered in electrothermal atomization.

Broad band absorption of radiation by molecules occurs because of molecular species formed or vaporized in the atomizer during the atomization step. Massmann[492] and others[35,164] discussed the absorption spectra of alkali metal halides as an interference in electrothermal methods, showing that the spectra may range steeply from comparatively high peaks to low troughs (see Fig. 59). It is clear, therefore, that not only must the correction be carried out simul-taneously, but it should also be done at the same wavelength, i.e. no variation of the two-line method is suitable, unless there is known to be no molecular absorption in a particular case being investigated. The effect is also less common at wavelengths above 350–400 nm. A particular kind of background absorbance which is difficult to detect and which is not corrected for by the usual methods is caused by the comparatively sharp fine structure of the vibrational component in molecular electronic spectra. Some examples of this are known: e.g. the cyanogen bands produced if nitrogen is used as the inert gas, and Massmann has also drawn attention to the SO and SO_3 bands produced during vaporization of sulfates; but fortunately they are few.

Light scattering gives a virtually flat absorbance continuum over the mono-chromator band pass. It is caused by the recondensation of sample matrix,

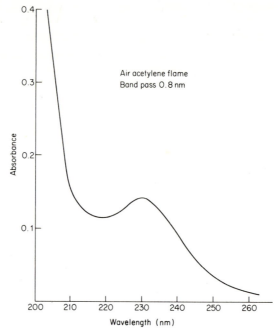

FIG. 59. Absorption and scatter signal from 10% w/v 'Puratronic' sodium chloride, measured using emission lines from various source lamps.

usually near the ends of the furnace tube, after vaporization during the atomiza-tion step, to form a smoke. It also appears to be caused by particles of graphite being ejected from the walls of the tube when the tube is suddenly subjected to a very high temperature. The light scattering effect, according to Rayleigh's law for particles whose diameter is less than the wavelength of the incident light, is proportional to the inverse of the fourth power of the wavelength and proportion-al to the square of the particle volume. It thus increases markedly with particle size, but more important, it can increase very rapidly towards lower wavelengths, e.g. it could be sixteen times worse at 200 nm than at 400 nm. This is yet another reason why correction should be done at the same wavelength.

Critical alignment of light sources

To achieve best coincidence of the atomic and continuum beams through the atomizer, which is particularly important in electrothermal analysis for reasons already discussed, the following procedure may be adopted with instruments which have an energy meter as well as absorbance readout:

(i) Set up the instrument and background corrector as for an analysis according to operating instructions, and, with no absorbing substance in the beam, set the absorbance display to zero.
(ii) Note the energy reading in the sample beam.
(iii) Slide the vertical edge of an opaque card horizontally across the right-hand end of the furnace tube until the energy meter indicates that about half of the radiation has been obstructed.
(iv) If the absorbance readout is outside the range ± 0.030 A remove the card and make a horizontal adjustment to the position of the hollow cathode lamp.
(v) Repeat (iii) and (iv) until absorbance display remains within the range ± 0.030 A.
(vi) Repeat (i) and (ii), then slide the horizontal edge of the card vertically upwards across the end of the furnace tube until again the energy meter indicates that about half of the radiation is obstructed.
(vii) If the absorbance readout is outside the range ± 0.030 A, remove the card and make a vertical adjustment to the position of the hollow cathode lamp.
(viii) Repeat (vi) and (vii) until absorbance display remains within the range ± 0.030 A.
(ix) Check that energy levels in both channels are satisfactory, adjusting gain if necessary. The two beams are now aligned for electrothermal measurements.

Testing the efficiency of background correction

The method of simultaneous background correction using a double beam or dual beam spectrophotometer has been described. The importance of optical coincidence of the atomic and continuum beams must be re-emphasized.

The method of testing the efficiency of a background correction system using a gauze inserted into the beam (Chapter 6) is even less satisfactory for electrothermal methods than for flame. The background absorbance may vary considerably over the diameter of the tube. Smoke from the sample matrix may be subject to turbulence from the inert gas flow and therefore the absorbance pattern is continually changing. The gauze test does not take account of this and so it follows that the only satisfactory test is one performed under real conditions, using a sample specially formulated for the purpose. Such a sample is 10% w/v sodium chloride ('Puratronic' grade or equivalent) and the test may typically be performed at the lead line 217.0 nm as follows:

(i) The instrument and background corrector are set up according to the instructions in the users' manual.
(ii) The electrothermal atomizer is set to give the following programme:

	Dry	Ash	Atomize	Clean
Time (s)	20	30	4	5
Temperature (°C)	110	300	2450	2750

(iii) Set $\times 0.5$ scale expansion, or recorder 0–2 A full scale.
(iv) *Without* background correction, inject 10 μl of the sodium chloride solution, run the atomizer programme and record absorbance during the atomize step. Repeat four times.

(v) Set × 1 scale expansion, or recorder 0–1 A full scale.

(vi) *With* background correction, inject 10 μl of the sodium chloride solution, run the atomizer programme and record absorbance during the atomize step. Repeat four times.

In this test, the peaks obtained without background correction should be greater than 1 A. Those obtained with correction should be about 0.01 A, as shown in Fig. 60. If the results of the test are much different from this, then results given in an actual analysis could be in error.

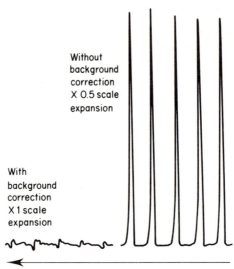

Without
background
correction
X 0.5 scale
expansion

With
background
correction
X 1 scale
expansion

FIG. 60. Results obtained for the Electrothermal Atomization Background Correction Test with the Pye Unicam SP2900 and SP9-01.

INTERFERENCE EFFECTS

Many interference effects in electrothermal atomization have been reported in the literature but not so many have been satisfactorily explained. It is becoming clear, however, that as in the early days of flame atomic absorption many of the interferences reported were associated with flame-type, so in electrothermal methods many of the interferences are atomizer-dependent; that is, they are effects of the temperatures employed and of the geometry and other characteristics of a particular type of atomizer.

It frequently happens that, in the presence of a particular matrix, the release of an atomic species is retarded. The peak height would then be less but it may well be that the integrated signal (peak area) is not changed. This is probably best considered as a temperature effect rather than a true chemical interference. If the total number of atoms actually produced is less, however, as in 'stable compound formation' in the flame, resulting in both the peak height and area measurements being decreased, then a real chemical interference may be said to have occurred.

Interference effects may thus be categorized as being physical or chemical in the mechanism by which they are produced.

Physical interferences

The physical effects of viscosity and surface tension have a less important part to play than in flame analysis. They may affect the reproducibility of micro-pipetting to some extent, but their main effect is in the degree to which the sample spreads in the graphite tube after it has been injected. Tubes designed to ensure the best isothermal heating and use of peak area measurements will minimize this effect.

In addition to the background absorption effects fully discussed earlier, the emission continuum from the furnace tube may distort the baseline and even affect the photomultiplier response, particularly at higher (visible) wavelengths. This effect can be detected if the atomizer is run through its normal programme at the appropriate wavelength without any sample.

Incomplete atomization or an ineffective subsequent tube clean step can result in an enhancement of the analyte response in further analyses. This problem is worst with elements that form very stable refractory oxides. It may be necessary to use very high atomization and tube clean temperatures, or longer times for these steps.

In the chapter on theoretical aspects of electrothermal atomization, the influences of the geometry of the furnace tube and of the rate of heating and other physical characteristics on the absorbance signal were discussed. Insofar as any of these factors prevents the maximum possible signal from being recorded it may justifiably be construed as a physical interference.

Chemical interferences

Chemical interferences in electrothermal atomization methods are probably best categorized as volatile compound formation, in which the analyte element is wholly or partly lost from the furnace tube before atomization temperature is reached, and stable compound formation, where atomization is either wholly or partly prevented or much retarded by the formation of complexes that remain unatomized at the normal atomization temperature.

Volatile Compound Formation. The analyte element may be lost during the ashing stage for two reasons: it may be present in the original sample in a compound which is appreciably volatile at the ashing temperature employed or it may be converted into such a form by the sample matrix or solvent (organic or inorganic). The use of the ash–atomize curves described on p. 171 will immediately indicate whether this can occur, and also shows the best ashing and atomization temperatures.

For this reason oxyacids are preferred to halogen acids as solvents for inorganic samples because halides are in general much more volatile than oxides. If, therefore, they are not lost at the ashing step, they may well vaporize during the early part of the atomization step and be expelled from the furnace tube before the atomization temperature is reached.

Some elements, particularly those of the arsenic and selenium groups, are themselves comparatively volatile and so are many of their compounds.

These problems can often be reduced or even overcome by a study of the chemistry of the analyte metal or the matrix in which it is present, with the object either of making the matrix more volatile than the element, or, what is the same in effect, converting the element to a more stable form. In either case, the matrix is removed before the analyte element.

The presence of high concentrations of alkali chlorides for example, not only causes high background absorbance problems, but also encourages formation of volatile chlorides of the analyte element. Such a matrix can be removed by the addition of an excess of ammonium nitrate to the sample solution.[210] The following conversion then takes place during the ashing step:

$$\text{NaCl} \quad + \quad \text{NH}_4\text{NO}_3 \quad \rightarrow \quad \text{NaNO}_3 \quad + \quad \text{NH}_4\text{Cl}$$

boiling point	decomposes at	decomposes at	sublimes at
1413 °C	210 °C	380 °C	335 °C

Thus all the resultant compounds are volatile below 500 °C. This is particularly valuable for sea water analysis, as most elements are not vaporized until higher temperatures than this are reached; even lead, for example, remains and is subsequently vaporized at about 750 °C.

There are several examples of ways in which volatile elements are stabilized so that they remain in the atomizer until after the matrix has been removed.

Selenium may be rendered involatile in the presence of nickel, molybdenum or copper.[210,316,347] The nickel or copper is added as an equal volume of a 2 mg ml^{-1} solution, on top of the sample in the atomizer which has been subjected only to the drying step. The sample can then be ashed safely at temperatures up to 1200 °C where selenium would normally be lost at about 500°. Arsenic and tellurium can also be similarly stabilized.

Cadmium is similarly stabilized up to 1000 °C by addition of 2% ammonium phosphate,[191] where it is normally volatilized at 500 °C.

Mercury is also stabilized up to 300 °C by nitric acid/ammonium sulfide solution.

It would appear that whenever an analyte presents a problem due to its high volatility, or indeed, where an analyte and matrix volatilize at inconveniently close temperatures, a study of a chemical textbook or handbook to determine whether any compounds of the analyte have high melting points is recommended. If such a compound exists, then the addition to the sample in the atomizer of the appropriate element or anion to produce it either in solution or during the ashing step could well effect a substantial reduction of the problem.

Stable Compound Effects. The response obtained for a particular element is likely to be reduced if it is present in the furnace tube as a compound which is not easily dissociated in the environment of the tube at atomization temperatures. Most oxides are in fact reduced under these circumstances, though some elements can form carbides at temperatures lower than those required for gaseous atom formation. This would have the effect of considerably slowing down the rate of atom production, or of preventing it altogether. Elements that form stable carbides which are not readily dissociated include barium, vanadium, tungsten, molybdenum and tantalum. The extent and ease of carbide formation is known to depend on the state of the surface of the graphite atomizer. It should therefore always be maintained in the same condition for reproducible response from such elements, and this is most easily achieved by using, or continuously reforming, a pyrolytic coating.

A suitable metal atomizer would also minimize this effect, the probability of a carbide being formed from an organic matrix being usually low.

Some workers[740] report that a few elements, in particular barium, molybdenum and titanium, are liable to form stable nitrides if nitrogen is used as the inert shield gas. It is to be expected in this case that the effect on samples and standards will be similar and therefore errors incurred are not usually great.

SOLID SAMPLES

The possibility that solid microsamples could be introduced directly into a graphite tube atomizer is an attractive one for several reasons. Without a dissolution step not only is the overall time taken for an analysis appreciably shorter but much intricate manipulation is avoided, and the chances of contamination from reagents and volumetric apparatus are much reduced. Relative detection limits should also be lower because the sample is not diluted. Much work is currently in progress with the object of achieving these advantages. It must be admitted, however, that the present status of solid sampling methods, with respect to general applicability and reliability, leaves much to be desired.

The weight of a microsample to be introduced into the atomizer must of necessity be low, e.g. in the 0.1–10 mg range. To weigh these with the desired degree of accuracy requires an accurate and robust microbalance. Suitable balances are made by both Cahn and Mettler.

Weighing the microsample is comparatively easy. Transferring it quantitatively to the atomizer without loss and without contamination is less easy.

To overcome this problem, tantalum and graphite boats have been designed to be positioned within graphite furnaces, and graphite boats or cups for use in furnaces of special configuration have also been introduced commercially. It is likely that this kind of sample handling is the best for solid microsamples, particularly for powders which must be weighed out from a primary sample. In the case of metals or other samples in the form of drillings or chips, the transfer difficulty is less acute because a piece of sample of suitable size can more readily be picked up and placed in the furnace by means of a specially made pair of tweezers or miniature tongs. Nevertheless, use of a graphite boat prevents build-up of matrix metal or carbides and thus helps to prolong the life of the atomizer tube.[548]

Fuller[260] has recently suggested the use of a thixotropic thickening agent for suspending powdered minerals and similar samples. This enables small amounts of such samples to be transferred easily and reproducibly to the atomizer.

Accurate calibration may not be possible unless reference standards of the same material as the samples are available. Such reference standards, though on the market for most metals and many types of refractory, are rarely calibrated for the trace element concentrations for which electrothermal atomization of solid samples is likely to be carried out. The procedure in which solid samples are calibrated using solution standards needs careful consideration before application. For metal samples it will almost certainly be unsuccessful because they are atomized directly as metal vapours while atomization of the standards is by way of reduction of the metal oxides.

As has already been noted, the ashing of some materials, particularly of those with a mainly organic matrix, usually results in the formation of a stable inorganic residue. Such a residue can be converted to the oxide form in presence of an oxyacid such as nitric acid. It is then most likely that an analyte metal in this residue will respond in the same way as a synthetic nitrate solution taken through the atomization programme. In such cases, calibration of solid microsamples using solution standards can be successful.

A method of forming secondary solid standards has been proposed by Gries and Norval[298] in which a suitable metal salt is introduced into molten urea. This is allowed to solidify and then finely ground. Standards made in this way are homogeneous, readily pyrolysed and stable at room temperature. Further, urea can be obtained in a pure state and when molten will dissolve many inorganic compounds. Such a procedure has attractions under some circumstances, but it may be wondered whether solid standards made up in this way are any more convenient to handle than solution standards, and whether the form of the analyte after the ashing step would not be the same were a solution standard to be used instead.

A further problem with solid microsamples lies in the possibility of heterogeneity of the bulk sample from which the analytical sample is taken. This difficulty would be associated with microsamples which are required to be representative of a larger sample such as powders, drillings or crushings of metal, etc. It can only be minimized by the usual processes of sampling from bulk by comminution, mixing, quartering, etc. carried out on the smaller scale involved. This problem clearly does not apply to microsamples which are samples in their own right, e.g. individual hairs, inclusions isolated from a piece of metal or rock, splinters of glass, etc. and it is probably for such microsamples that solid sampling techniques are, at the present time, of most value and interest.

Chapter 8

The Applications of Atomic Absorption Analysis

In view of the considerable literature that now exists on atomic absorption analysis, it would require a very large volume indeed to discuss in any detail all the known applications. Wherever metals are determined, and in whatever matrix, atomic absorption can be applied. The features of atomic absorption make it particularly suitable for the analysis of micro-samples. The author has discussed this aspect elsewhere.[600]

There is very little difference in the methods required to handle the materials within certain well defined groups. One such group includes waters from various sources, wines and beers. Another includes food products, feedingstuffs, plant materials and biological tissues, and a third contains various complex oxide mixtures such as slags, rocks, ores, cements, glass and ceramics. When prepared for analysis, the solutions contain much the same elements and therefore the interference problems are also similar. To differentiate between 'industrial' and 'geochemical' applications or even between 'industrial' and 'biochemical' is often pointless.

When seeking information on a particular analysis, therefore, a worker may need to look beyond the confines of his own field, as identical problems may have been solved elsewhere.

We attempt to divide the whole analytical scene broadly into waters, metallurgical, inorganic, organic, and biochemical sectors, though even here some overlapping must occur.

Unless special details are given, the reader should refer to the element tables in Chapter 9 for element sensitivities and generalized operating conditions.

THE ANALYSIS OF WATERS AND DILUTE AQUEOUS SOLUTIONS

Although waters might be thought to be one of the simplest possible types of sample for atomic absorption analysis, they can be dilute (and sometimes not so dilute) solutions of almost anything. Probably for this reason, more work has been published about the atomic absorption analysis of water than of any other single matrix. A monograph on the chemical analysis of water includes atomic absorption as a standard technique.[815] Published work resolves into three main areas: purified waters; a group containing ground waters, river water and industrial effluents; and seawater. The elemental sensitivities and detection limits published in the scientific and the manufacturers' literature usually relate to aqueous solutions. These are therefore the levels of performance expected in water analysis. Considerable use has also been made of the three main concentration techniques: evaporation, solvent extraction and ion-exchange.

Evaporation is a simple, though often lengthy and therefore inconvenient method of achieving concentration, particularly if improvement factors of 10 or more are required. It could also have the disadvantage that volatile compounds of some elements, particularly mercury and the arsenic group, could be partly lost.

Several authorities[220,399] agree that the most useful extraction system is APDC–methyl isobutyl ketone. Instability in APDC complexes may be overcome by evaporating the MIBK solution to dryness and redissolving in acetone + 0.1 M hydrochloric acid.[563]

The use of ion-exchange resins for concentrating certain groups of metals has already been mentioned on p. 151. One of the most useful of these continues to be Chelex 100[664,717] on which a range of metals may be concentrated at pH 6–8 and subsequently eluted for atomic absorption analysis with 1 M nitric acid.

Purified water

Techniques for determining zinc, lead, copper, nickel, calcium, magnesium, iron and manganese were described by Boettner and Grunder.[105] The APDC–MIBK method was used by Fishman and Midgett[234] for cobalt, nickel and lead, the extraction being carried out at pH 2.8. In an examination of some Indian water supplies, Soman et al.[696] used dilution and concentration by evaporation as appropriate to determine cobalt, chromium, copper, iron, lithium, manganese, rubidium, strontium and zinc.

The more stringent requirements of the steam water circuits of power stations were described by Wilson[814] who used the APDC–MIBK extraction as necessary, though direct analysis gave detection limits in $\mu g\, l^{-1}$ of: iron 30, copper 15, zinc 15, nickel 60, calcium 10, magnesium 1 and sodium 4. Some 'pure' waters can therefore be analysed without prior chemical treatment, but where levels below their detection limits are to be determined, one of the concentration procedures, particularly the APDC extraction, must be employed, as described on p. 148.

For most elements in water matrices other than seawater and other salines, it is generally possible to inject the sample acidified with nitric acid directly into the electrothermal atomizer. Special care must be observed to protect both sample and standards from the effects of contamination. Background correction may not be found to be necessary for very pure waters, and published sensitivities may be expected to be attained for most elements because interferences are negligible.

Chakrabarti and co-workers[476] made a study of the determination of a number of elements in pure water by electrothermal methods, showing that contamination, adsorption and ion-exchange reactions occur between the analyte ions and the surface of the sample container. Preservation of such samples was discussed subsequently.[716]

In the determination of very low levels of sodium (<0.1 μg kg^{-1}) in ultrapure water, particular problems arise in the production of reliable standard solutions. These may be overcome[270] by a continuous flow method of producing the standards. In this, water is maintained pure (sodium-free) by circulating in a closed circuit system through an ion exchange column. Required sodium levels are then obtained by diverting the pure water to an exit pipe via a T-piece at which it is combined in known ratio with a standard sodium solution by means of a metering pump.

Rivers and industrial effluents

Unhappily it is a sign of the times that these should be grouped together in the eyes of many analysts. The essential differences from purified water are the presence of suspended organic matter, and the increase in concentration of many, usually toxic, elements.

Many river authorities and water boards employ atomic absorption in water pollution control. The methods are usually similar to the procedure suggested by Price.[601]

Suspended matter in the sample is first separated, or, if it contains any of the elements to be determined, it must be homogenized and dissolved or extracted with nitric acid. In extreme cases, the solid matter is centrifuged off and treated separately as a sludge. Clear or clarified water is acidified to about 1 % with nitric acid and can be aspirated in the atomic absorption spectrometer without further treatment if the concentration ranges present can be handled. Calibration standards made up as described earlier should also be acidified 1 % with nitric acid. If concentrations are too high an alternative line can be used or solutions are brought within the best range by dilution with deionized water also acidified with 1 % of nitric acid. If concentrations are too low, the sample may be evaporated to a lower volume, or the wanted metals extracted into an organic solvent as described on p. 148. Calcium and magnesium, relating to hardness values, can be determined in water without interference using the nitrous oxide acetylene flame. Potassium is added as ionization buffer.

An alternative extraction procedure using sodium diethyl dithiocarbamate was given by Platte[591] for iron, copper, nickel, zinc and cadmium. Tenny mentioned[728] that samples with a high suspended solids content must be digested by heating with nitric acid for 30 min. Ion exchange and extraction have been used[140] to determine silver down to 0.1 μg l^{-1}, while gold is determined in natural waters[841] after concentration by chelation. In this method the sample is stood overnight with bromine-water and hydrochloric acid and then passed through a column containing a polyschiff base. The gold is eluted with hydrochloric acid and hydrogen peroxide. After oxidation with potassium permanganate, the chelated gold is extracted with MIBK and this solution aspirated in the instrument. Molybdenum is determined in lake-water after extraction of the oxine complex with MIBK.[142]

Low levels of aluminium can be determined by extracting its oxinate into MIBK. Magnesium does not interfere because its oxinate is formed too slowly to be extracted at the same time.[233]

Sludges containing organic matter are wet or dry-ashed like other organic matrices and taken up in mineral acid. If the inorganic residue contains siliceous material, the silica can either be removed with hydrofluoric and perchloric acids or the residue completely solubilized by a method generally applicable to silicates. Alternatively, sludges and sediments can be extracted for heavy element analysis with hydrochloric acid/nitric acid mixtures. This is effective for several elements including cobalt, copper, chromium, iron, manganese, nickel, lead and zinc. For more complete analysis digestions may need to be carried out in a PTFE pressure vessel, when the most efficient extracting medium is hydrofluoric/perchloric/nitric acid.[40] It is essential, of course, to ensure that no organic matter is present if perchloric acid is employed. Somewhat less efficient extractants are boiling aqua regia, or hydrofluoric/hydrochloric/nitric acid,[211] but even with these, a pressure dissolution vessel of a specially strong type must be used.

Mercury is an increasing hazard from insecticides and fungicides and from the effluent of chemical plant and paper mills. The sensitive cold vapour method for mercury can be used to determine this element to an absolute sensitivity of about 5 ng. A cold digestion should be performed (p. 276) to ensure that mercury is liberated from methylmercury and similar compounds.

Again, many such extractions and concentration procedures are avoided if electrothermal atomization is used, most waters except salines being directly injected into the atomizer. Acidification with nitric acid helps both to remove any organic matter and to prevent the effects of chloride.

Solvent extraction has been used with electrothermal atomization to reach exceptionally low concentrations of some elements. Many workers use the APDC–MIBK system. Alternative approaches include silver by dithizone-carbon tetrachloride, lead by dithizone-chloroform[623] and beryllium with acetylacetone/chloroform from sodium EDTA at pH 7.[427]

Seawater

Although background correction is advisable, particularly at low wavelengths, the presence of $2\frac{1}{2}$–3 % of sodium chloride and other salts does not appear to affect the response of most trace elements in the flame. It does, however, preclude the concentration of seawater samples by simple evaporation by an effective amount. Billings[100] found direct atomic absorption to be sensitive enough to investigate major ion ratios in seawater, and some alkali and alkaline earth metals have been determined against standards containing sodium chloride. Chelating resins may be used[633] (see p. 151) to separate a number of metals from seawater. Chelex 100 is one of the most successful. The sample is brought to pH 3–5 and passed through a column of resin in the ammonium form. Elements such as cadmium, cobalt, copper, iron, manganese, nickel and zinc are retained and then eluted with 2 M nitric acid. Separation from alkali metals, considerable separation from calcium and magnesium, and concentration factors above $25 \times$ may be achieved.

APDC–MIBK extractions have been used extensively in oceanographic analysis. In particular, in the determination of cadmium, which has been reported quite widely from Japan, the APDC–MIBK method can be applied directly using large aqueous phase/solvent ratios[835] though some of the organic phase was not recovered due to its slight solubility in water. To avoid this, cadmium has also been coprecipitated with strontium carbonate as a preliminary to extraction with sodium diethyl dithiocarbamate. Large amounts of sodium chloride did not interfere.[573]

Nickel is determined by extraction of the dimethyl glyoxime complex into chloroform. For flame atomic absorption this extract must be taken to dryness and redissolved in 0.1 M nitric acid[622] but there seems to be no reason why the chloroform extract should not be injected directly into an electrothermal atomizer.

The presence of sodium chloride is undoubtedly a major problem in the analysis of sea waters with electrothermal atomization, both because of the background absorption interference and because of the chloride effect. Non-reproducible results may also be obtained if the sodium chloride recondenses at the ends of the graphite tube or at the electrical contacts between the tube and its supports.

These problems are avoided by the solvent extraction or ion exchange separations already mentioned.

In the case of cadmium, it is possible that, with rapid heating and close control of furnace temperature[464] the cadmium can be atomized and measured before the sodium chloride is vaporized.

The salt matrix background can, however, be much reduced, so that it can easily be handled by efficient background correction systems by the method of matrix modification first suggested by Ediger.[210] A concentrated solution of ammonium nitrate is added to the sample converting the sodium chloride to a matrix which is volatile below 500 °C. Campbell and Ottaway[127] have shown that zinc and cadmium can be measured directly in a 1:1 dilution of seawater provided a low temperature atomization, e.g. 1490 °C, is used as far as possible to separate the atomic absorption signals, which appear first, from background absorption. As there are depressive interferences from magnesium chloride and other sample constituents, the standard additions method of calibration must be used. Drying was at 100 °C for 45 s but there was no intermediate ashing step.

METALLURGICAL ANALYSIS

As atomic absorption is a technique for the determination of metals, it is not surprising that a very large number of its most valuable applications occur in the various metallurgical industries. Elwell and Gidley[1] pioneered atomic absorption methods for magnesium, zinc and lead as early as 1960. With the availability of hotter flames and a better knowledge of flame processes, however, a number of the interference effects reported and discussed between 1960 and 1965 are no longer of importance. In routine applications the technique frequently reduces to dissolution, dilution (usually with a spectroscopic buffer) and aspiration.

In metallurgical analysis, the need is always for the best accuracy and best speed. Control analysis is usually carried out under rigidly standardized conditions which have been designed so that the result is available in the shortest possible time. Check analyses are performed with the utmost care under less hurried conditions. Atomic absorption may be considered in both contexts. It will not, of course, compete in speed with the large direct-reading emission or X-ray spectrometers upon which most foundry control is now based. But smaller operators who cannot afford such equipment undoubtedly find a modified form of control with atomic absorption extremely valuable, particularly if a rapid method of dissolution can be evolved.

The potential accuracy and independent nature of atomic absorption analysis gives it an important place in check analysis and in the initial calibration of the working standards upon which the direct-reading spectrometers depend. That this is so is evidenced by the adoption of atomic absorption procedures by a number of national standardizing bodies.

A number of examples of the use of electrothermal atomization in metal analysis have now been reported. Sometimes preconcentration steps have been included in the methods which would have made the analysis perfectly feasible with the flame. For most elements the flame should first be considered for concentrations above 0.01–0.05% in alloys or 'pure' metals. At lower concentrations, where separation and/or concentration steps have to be included in flame methods, the use of electrothermal atomization may well render such steps unnecessary. The analysis then becomes simpler and less likely to incur contaminational errors.

Direct analysis by solid sampling is sometimes possible, but it must be remembered that because of the different atomization processes involved calibration of solid metal samples by solution standards will usually lead to erroneous results. Solution methods are therefore considered to provide better possibilities at the present time, unless appropriate solid standards are available.

Calibration graphs should be prepared with the matrix element present in the standards and halogen acids should be avoided as the principal component in the final solvent. Ashing temperatures should be sufficiently high to reduce the salts formed in dissolution to an oxide residue.

Light alloys

After the work of Elwell and Gidley[1] and the determination of several more individual elements, the determination of copper, calcium, manganese and zinc in magnesium alloys and the same elements plus magnesium in aluminium alloys was described by Mansell et al.[484] Wilson[816,818] used caustic soda dissolution and examined the capabilities of flames of three different orders of temperature in the determination of silver, copper, zinc, magnesium, zirconium and chromium, reporting that several interelement interferences were decreased in the hotter flames.

The analysis of aluminium alloys for nine elements after dissolution in hydrochloric acid was described by Bell,[91] and this has set the pattern for atomic absorption analysis of light alloys, dissolution in hydrochloric acid being much preferred to caustic soda for aluminium because of the lower final concentration of dissolved salts.

Other work on aluminium alloys has concerned the determination of particularly low traces of certain elements. Beryllium was determined without interference in the nitrous oxide acetylene flame by Peterson,[585] but Hirano et al.[321] had to sublime the aluminium off as aluminium chloride in order to determine calcium traces when using an oxyhydrogen flame.

The following simple preparation procedure will be found suitable for determining most elements in simple aluminium alloys:

Weigh out 0.5 g of sample, place in a 250 ml beaker and cover with a clock glass. Add 30 ml of 50% hydrochloric acid. When the initial reaction has subsided, cool the beaker then add 3 ml of hydrogen peroxide (20 vol). Evaporate to 15 ml, cool and filter quantitatively into a 100 ml flask and make up to the mark. Standards should be made up to contain 0–25 mg l^{-1} of copper, iron, magnesium, manganese, nickel, zinc and any other element in a 0.5% solution of high purity aluminium in 15% hydrochloric acid. Samples and standards for magnesium determination should be run with the nitrous oxide acetylene flame, or should contain 0.1% of strontium or lanthanum as chloride. The standard containing aluminium only gives the reference base line from which the other standards are measured. For concentrations higher than the ranges given, the sample solution is diluted with 0.5% aluminium solution and the final result multiplied by the dilution factor. Use the analysis lines and flames given in Chapter 9.

Calcium can only be determined in aluminium alloys using the nitrous oxide acetylene flame. The temperature of this flame eliminates most of the depressive interference of aluminium on calcium, but ionization occurs. Addition of an alkali salt, or dissolution of the sample in caustic soda followed by oxidation with nitric acid provides a solution with which flame ionization too is suppressed, and the detection limit of calcium is then about 0.0005% in the alloy.

The British Standard method for determining zinc in aluminium alloys in the range 50–200 μg g^{-1} (BS1728 Part 1 (1973)) also uses hydrochloric acid/hydrogen peroxide dissolution.

Similar methods to these may be used for the analysis of magnesium alloys. Magnesium interferes remarkably little with the determination of other elements, with the possible exception of silicon, and it may well be found to be unnecessary to add it to the standard solutions for magnesium alloys.

If silicon is to be determined in aluminium at low levels, a slightly different dissolution procedure is employed in order to ensure that the silica is dissolved and retained:[606]

> To 0.5 g of sample add 10 ml of water and, in small portions, 8 ml of hydrochloric acid. After allowing the reaction to subside add 15 ml of hydrogen peroxide (50 vol) in small portions and heat. Cool, add 5 ml of hydrofluoric acid and dilute to 100 ml. Polythene or PTFE ware must be used for this method. Standard solutions should contain 0.5% of high purity aluminium as well as a range of 0–250 mg l⁻¹ of silicon.

When silicon and other alloying element concentrations up to 10% or higher are known or suspected to be present, these can be determined by atomic absorption after a modified hydrofluoric acid dissolution.[515] Insoluble fluorides, particularly those of magnesium, are converted to soluble fluoroborates by the addition of concentrated disodium tetraborate (borax) solution, which acts as a convenient source not only of borate ions to complex the fluoride but of sodium ions to act as ionization buffer. Fluoroborate is an effective releasing agent (see p. 134) and this, together with the nitrous oxide acetylene flame, enables the major and minor elements normally present in aluminium alloys to be determined without interference. The procedure is as follows:

> Weigh 0.500 g of the sample into a 200 ml plastic beaker and add 25 ml of nitric acid solution (40%). Warm the beaker in a water bath until there is no further reaction. Cool to below 50 °C and cautiously add 5 ml of hydrofluoric acid (42%). Stand for 10 min then add 20 ml of borax solution (20%). Stand for a further 5 min then dilute to 100 ml in a polypropylene volumetric flask. This solution is aspirated directly to determine all elements in the range 0–1%. For concentrations between 1 and 10% dilute 10× with deionized water. For concentrations greater than 10% dilute the original solution 20×, but maintain the aluminium concentration at 0.5% by adding the appropriate amount of pure aluminium stock solution. Standards are prepared to contain 0.5% of aluminium and a range equivalent to 0.10–1.0% of silicon, chromium, copper, iron, magnesium, manganese, nickel and zinc in the alloy. A suitable range, e.g. 0.04–0.4%, of tin, lead, antimony and titanium can also be included in the same set of working standards. The standards are conveniently prepared by taking 0.5 g weighings of high purity aluminium, e.g. BCS 198f, through the above sample preparation procedure, adding appropriate aliquots of stock standard solution before making up to 100 ml.

Other workers[288] used a pressure vessel for the dissolution of aluminium–silicon alloys up to 28% silicon in hydrofluoric and nitric acids and hydrogen peroxide. Price and Whiteside also successfully applied the pressure dissolution method,[610] described on page 231 for siliceous materials, to the measurement of ten elements, including silicon, in aluminium silicon alloys.

Zinc alloys

Zinc base alloys present very few difficulties in routine analysis. The determination of a number of elements in both zinc and lead was detailed by Jimenez Seco[358] while zinc die-cast alloys were analysed by Smith et al.[694] using an internal standard method with a double channel instrument in order to achieve a high level of precision. In the latter method samples were prepared by dissolving in the minimum amount of hydrochloric acid, oxidizing with nitric acid, and then making up to a 1 % solution. Tin, lead, iron, aluminium, magnesium, copper and cadmium were determined.

A recommended procedure for zinc alloy is:

> Dissolve 2 g of the zinc alloy in 20 ml of 50% nitric acid, boil to expel nitrous fumes, transfer quantitatively to a 100 ml calibrated flask and make up with water. This solution is used for magnesium, iron and aluminium, and diluted by a factor of 2 for lead, cadmium and copper. Calibration standards are made by adding quantities of stock standards, of the concentration given in Table 8, to the amount of zinc solution (made by dissolving 100 g of pure zinc in 500 ml of nitric acid and making up to 1 litre) also indicated there, and making up to 100 ml with water.

TABLE 8. Calibration standards for zinc alloys

	Diluted stock solution (ppm)	Range added (ml)	Zinc solution 100 mg ml^{-1} (ml)	Equivalent %
Lead	500	0–10	10	0–0.5
Iron	50	0–12	20	0–0.03
Magnesium	20	0–15	10	0–0.03
Aluminium	200	0–12	20	0–0.12
Cadmium	100	0–10	10	0–0.10
Copper	500	0–20	10	0–1.00

In the analysis of high purity zinc, a number of trace elements, copper, silver, cobalt, nickel, indium and lead, can be extracted simultaneously with a mixture of sodium diethyldithiocarbamate and α-nitroso β-naphthol into MIBK.[299] With an original sample weight of 2 g and an MIBK volume of 10 ml, the above elements were determined at less than 1 ppm.

Copper-base alloys

Most minor and major elements in copper-base alloys can be determined by atomic absorption quite routinely.

A textbook on copper analysis[213] gives atomic absorption procedures for cobalt, lead, silver and zinc, and suggests that other elements could also be determined because the methods would be very similar. Cadmium, for example, could be done with the same flame conditions as the lead, but the same sample preparation would apply to most elements.

A few special methods have been suggested for individual elements, including tellurium and selenium,[699] low concentrations of lead[173] and arsenic,[817] and an ion exchange method was described[503] using Dowex-1 (the sample being dissolved in hydrochloric acid) to separate tin, cadmium and zinc from copper.

There are several methods for the separation of low-concentration impurities from pure copper. In one of these, after 20 g of copper are dissolved in nitric acid the impurities are coprecipitated on lanthanum hydroxide (adding lanthanum nitrate followed by ammonia), filtered and washed with ammonia. The residue is redissolved in nitric acid and diluted to volume for atomic absorption analysis with the flame. The standards are matched for lanthanum and acid contents. This is said to be better than coprecipitation on ferric hydroxide.[627]

Others have found it possible to determine impurities in the presence of copper[75] specifying separation of the matrix only if gallium, germanium and tin are to be determined at less than 1 % and if cobalt and iron are to be determined at less than 0.1 %.

Low levels of selenium can be extracted from copper or steels as the acetophenone complex from hydrochloric/perchloric acid mixture into chloroform.[842] The chloroform extract can be evaporated to dryness and taken up in MIBK for flame analysis, or presumably injected directly into an electrothermal atomizer for ultratraces.

The ferric hydroxide coprecipitation method is used by Mullen for determining selenium, tellurium, bismuth and antimony at levels below 0.5 ppm in 'pure' copper by electrothermal atomization. The precipitate is collected on a filter disc from which samples are punched for insertion in the atomizer. Arsenic was not determined by this method.[538]

A general scheme for flame analysis of copper alloys is made more difficult by the tendency of tin to precipitate as metastannic acid from nitric acid solutions and the insolubility of lead chloride at higher concentrations of lead in hydrochloric acid.

Two solvent/acid mixtures were investigated by Johns and Price,[361] viz. hydrochloric/nitric and orthophosphoric/nitric acids, each component acid being present at a concentration of 25 %. Nearly every type of copper alloy proved to be soluble in the former and, after the solutions are diluted to a working concentration, the tin remains in solution for at least twenty-four hours. At considerable dilution, a slight turbidity may be observed after eight hours in those solutions where 5 % of tin is present in the sample, but if the hydrochloric acid concentration is increased when the sample is further diluted no precipitation should occur.

Phosphoric/nitric mixture dissolved all copper alloys except those with more than about 5 % of tin. On dilution, the solutions were found to be stable for at least a week.

The hydrochloric/nitric mixture is recommended for routine analysis, as the need to retain the solutions for more than a few hours seldom arises, and this solvent can be used for the simultaneous analysis of materials of a much wider composition range.

The method which follows is based on the above,[361] the alloying elements all being measured in the single sample solution. Calibration solutions match the samples in both copper and acid content to ensure that minor interferences are compensated. Matching for other constituents is not necessary unless even the low level alloying constituents have to be determined with very high accuracy.

For most alloys, final nitric and hydrochloric acid concentrations of 5% will be found suitable. If a precipitate or haziness appears before the analysis of the sample solution has been finished, however, it may be necessary to use a somewhat stronger acid mixture.

Weigh 0.500 g of the sample into a 250 ml beaker and add 20 ml of 25% hydrochloric/ 25% nitric acid mixture. Warm until dissolution is complete, cool and transfer to a 100 ml volumetric flask. Make up to the mark with deionized water. This sample solution may be used directly for measuring all elements in the range 0–1%. For other concentrations dilute as follows with an 0.5% solution of pure copper in 5% hydrochloric/5% nitric acid: 1–5%: 5×; 5–10%: 10×; 10–20%: 20×; 20–50%: 50×; 50–100%: 100×. For the determination of copper itself, dilute the sample with 5% hydrochloric/5% nitric acid mixture only. Standard calibration solutions are prepared in 100 ml flasks from 1000 mg l^{-1} stock solutions of aluminium, antimony, arsenic, copper, iron, lead, manganese, nickel, silicon, tin and zinc according to Table 9. (Stock solutions are themselves prepared as described in the Element Tables in Chapter 9.) To these (except the copper standard) are added 20 ml of a 2.5% solution of copper in the 25/25 acid mixture (in order to bring the final copper concentration to 0.5% as in sample solutions) and the standards are then made up to 100 ml with water.

TABLE 9. Calibration standards for copper alloys

% element in sample	blank	0.25	0.50	1.00
Vol. of 1000 mg l^{-1} stock, ml	0	1.25	2.5	5.00

Aluminium, silicon and tin must be determined in the nitrous oxide acetylene flame. The other elements are normally measured in an air acetylene flame, but it may be more convenient to use nitrous oxide for these also, though there may be some loss of sensitivity. Copper appears to act as an ionization buffer at high flame temperatures, and this is another important reason for maintaining its concentration constant between samples and standards.

The instrumental conditions should be chosen to give the best noise-free response, particularly for major components. These may not be the same as for highest sensitivity, particularly the flow rates for the nitrous oxide acetylene flame.

Good results can also be obtained with the nitric/phosphoric acid solvent, in circumstances where this may be preferred. Calibration must be made with this solvent for, in addition to slightly lowered sensitivities, a greater 'blank' absorbance is observed in the air acetylene, though not in the nitrous oxide acetylene flame. Both effects are probably caused by the persistence of small agglomerates of undissociated copper pyrophosphate in the medium temperature flame.

Although the above recommended procedures give satisfactory results for a variety of different copper base compositions, a high acid content of the final solutions may cause corrosion with prolonged use in some types of spray chamber and burner systems.

Lead-base alloys and solders

Lead–tin alloys have long caused difficulties in wet analysis because of the tendency of tin to precipitate from nitric acid.

Lead metal with low tin content will usually dissolve completely in nitric acid. In order to determine traces of tin, Perry[583] dissolved 10 g of sample in 15 ml of nitric acid (sp. gr. 1.42) and then added 15 ml of 2% ammonium fluoride solution to complex the antimony which might be present. The solution was made up to 100 ml. With a multislot burner and air hydrogen flame, the limit of detection was 1–2 ppm of tin in the original sample.

Nitric acid and tartaric acid form the solvent for refined lead in another method for determining impurities.[572] The solution is aspirated directly into an air acetylene flame. Traces of silver and copper were separated from nitrate solutions of high purity lead and zinc at pH 1–1.5 by nitrogen bubble flotation after precipitation with dithizone in methyl cellosolve.[320] The atomic absorption method was compared with AC polarography by Beyer and Bond.[98]

The usual solvent for lead–tin alloys, and particularly solders, is hydrobromic acid/bromine mixture:

> Dissolve the sample in 3:1 hydrobromic acid:bromine. When dissolved, warm to remove the excess bromine. Dilute to known volume with 10% hydrobromic acid, with which all subsequent dilutions should be made.

Considerable success was claimed with the dissolution of lead–tin solders in a special mixture of fluoroboric and nitric acids, which produced a clear solution of the solders.[345] The solvent acid solution used was 70% nitric acid, fluoroboric acid and water in the proportions 3:2:5 by volume, and this was prepared afresh just before each analysis.

In a rapid dissolution technique for lead alloys[92] the solvent was 48% fluoroboric acid, 30% hydrogen peroxide and 0.2 M EDTA. Small samples (~ 1 mg) were dissolved on a steam bath in 5 min, though this was less effective for the richer solders (50% tin). The resulting solution was again aspirated into an air acetylene flame.

Fluoroboric acid and hydrogen peroxide dissolution also formed the basis of Powell's method.[614] Lead alloys were analysed for up to 0.2% of antimony and 0.4% of tin using the air and nitrous oxide acetylene flames respectively.

The fluoroboric/nitric acid dissolution method has been further developed in the author's laboratory and it is believed that certain limitations which have

TABLE 10. Standards for lead–tin alloys

| | Blank | Standard 1 | | | | | | Standard 2 | | | | | |
		Sn		Pb		Sb		Fe		Ni		Cu	
Middle and high % of range	—	0.4	0.8	0.4	0.8	0.2	0.4	0.02	0.04	0.02	0.04	0.05	0.1
ml of 1000 mg l⁻¹ stock	—	4	8	4	8	2	4	—	—	—	—	—	—
ml of 100 mg l⁻¹ stock	—	—	—	—	—	—	—	2	4	2	4	5	10
≡ mg element l⁻¹	—	40	80	40	80	20	40	2	4	2	4	5	10

| | Standard 3 | | | | | | Standard 4 | | | | | | | |
	As		Bi		Cd		Al		Ag		Zn		Au	
Middle and high % of range	0.2	0.4	0.05	0.1	0.1	0.2	0.1	0.2	0.02	0.04	0.02	0.04	0.05	0.1
ml of 100 mg l⁻¹ stock	20	40	5	10	10	20	10	20	2	4	2	4	5	10
≡ mg element l⁻¹	20	40	5	10	10	20	10	20	2	4	2	4	5	10

been mentioned in the literature are due to the fluoroboric acid not being freshly prepared. The recommended procedure, by means of which a number of elements may be determined, is as follows:

Reagents and stock standards
Fluoroboric acid: to 200 ml of 40% hydrofluoric acid at 10 °C, add 75 g of boric acid in small quantities. Allow to dissolve and store in a polypropylene bottle. Prepare freshly each day. Sample reagent solution: mix 45 ml of fluoroboric acid, 95 ml of nitric acid (sp. gr. 1.42) and 45 ml of 1% w/v tartaric acid solution in a 200 ml volumetric flask. Make up to 200 ml with water. Tartaric acid solution 1% w/v.
   Stock standard 1000 mg l^{-1} of the following elements: tin, lead, antimony, iron, nickel, copper, arsenic, bismuth, cadmium, aluminium, silver, zinc, gold (for directions see Element Tables, Chapter 9).
   Prepare stock standards, 100 mg l^{-1}, of the above elements as required by dilution with water.

Preparation of sample solution
Weigh exactly 1.000 g of finely divided sample and transfer to a 250 ml beaker. Cover with a watch glass. Add 40 ml of water, 6 ml of fluoroboric acid and 12 ml of nitric acid. Warm to dissolve, stirring if necessary. Cool to room temperature, add 5 ml of tartaric acid 1% and dilute with water to 100 ml in a volumetric flask. This 1% sample solution is suitable for the elements within the ranges indicated in the top line of Table 10. For elements within ten times this range prepare a 0.1% sample solution as follows: pipette 10 ml of this solution into another 100 ml volumetric flask, add 20 ml of the above 'sample reagent solution' and dilute to 100 ml with water. For elements at concentrations greater than 10× the basic ranges prepare a 0.01% sample solution by pipetting 10 ml of the 0.1% sample solution into a 100 ml volumetric flask, adding 20 ml of 'sample reagent solution' and diluting to 100 ml with water.

Calibration standards
Standards may be prepared to contain more than one element. Suitable groups of elements are suggested in Table 10 though other combinations may be used.
   Prepare the standards and blank in 100 ml flasks, add 6 ml of fluoroboric acid, 12 ml of nitric acid and 5 ml of tartaric acid 1% to each. Then add the amounts of the appropriate stock solution as indicated in the table, and make up to the mark with water.

Use nitrous oxide acetylene flame for aluminium and tin, air acetylene for the other elements. Reduce sensitivity for lead, if necessary by rotating the burner.
   It will be apparent from the fact that lead is present only with the major components that there are no interelement interferences. The method is suitable for lead–tin solders and for type metals containing antimony. A similar method can be used for the analysis of cadmium and cadmium alloys.

Iron and steel

All the metals for which iron and steel samples are analysed can be determined by atomic absorption. The non-metals, carbon and sulfur, cannot but phosphorus at low levels can be measured by electrothermal atomization. A considerable literature on ferrous analysis has built up since 1961 from which it is clear that there are many suitable procedures for a number of the elements, including several by which a number of the elements may be determined together. There remain problems both in the dissolution and in the various interelemental interferences.

A comprehensive review[669] of the applications of atomic absorption spectrometry to the analysis of iron and steel summarized earlier work and also made a valuable contribution to the determination of some trace elements.

It is clear that many common elements, particularly chromium, manganese, cobalt, nickel, copper and molybdenum, can be determined using an air acetylene flame after the sample has been dissolved in any one of several possible solvent acids. Thus Belcher and his co-workers[90,400-402] usually preferred sulfuric–phosphoric acid mixture while others use hydrochloric acid–nitric acid mixtures. The advantage of sulfuric–phosphoric mixture is that tungsten and other acid-hydrolysable elements are thereby retained in solution.

The outstanding problem encountered in the analysis of these elements is the considerable depressive effect of iron upon the sensitivity of chromium and molybdenum in the air acetylene flame. The effect has been overcome to a great extent by the addition of various spectroscopic buffers, thus enabling trace amounts of these elements to be determined with adequate sensitivity and accuracy. Quantitative accuracy at higher concentrations has been more difficult to achieve because of minor interferences which are not completely eliminated and which vary according to alloy composition. This situation, however, appears to be resolved by the use of the nitrous oxide acetylene flame[732] and a recommended general scheme is given at the end of this section after the discussion on individual elements. Given a method which is free from chemical interference effects, the precision required in most metallurgical analysis is achieved by great attention to details of operating technique, as outlined in the section on determination of major components.

Atomic absorption may be used as a sensitive means for determining low traces of usual and unusual elements. Sometimes this is possible by a direct procedure, but the elements to be determined may have to be concentrated after removal of the matrix elements.

For the purposes of an atomic absorption procedure, the elements to be determined fall into three groups. Manganese, nickel, copper, cobalt and lead (and probably magnesium) are all determined in an air acetylene flame; chromium, molybdenum (in absence of tungsten), titanium, vanadium, tin and aluminium require the nitrous oxide acetylene flame. All of these are best determined in a final solution which contains perchloric acid only and thus readily lend themselves to being determined together after a single sample dissolution.[733] The third group contains elements which require special dissolution procedures, e.g. niobium and silicon in the presence of hydrofluoric acid, and traces of lead, calcium and other elements where a matrix removal step may be necessary.

Because any general dissolution procedure for steel analysis is dictated by the requirements of chromium and molybdenum, these two elements are discussed first, The methods are detailed on p. 209.

Chromium and Molybdenum. The depressive effect of iron on the absorbance of chromium and molybdenum in the air acetylene flame, obviated to a large extent by the addition of ammonium chloride,[78,534] has been discussed in the section on interferences. Ammonium chloride was included in a general scheme based on dissolution in aqua regia,[604] which allows lower concentrations of both elements to be determined with the full sensitivity associated with an air acetylene flame. Small variations in response are liable to occur which made the method unreliable when extended to alloying concentrations of these elements. These variations are almost certainly due to a difference in oxidation state of the two elements between samples and standards. It is readily shown that a calibration curve obtained by using potassium dichromate [Cr(VI)] has a different slope from one using chromium metal dissolved in hydrochloric or even hydrochloric–nitric acid mixture [predominantly Cr(III)]. Oxidation of Cr(III) to Cr(VI) in the presence of hydrochloric acid or high concentrations of chlorides must be carried out with care otherwise chromium is volatilized off as chromyl chloride.

Nall has since shown[546] that 8-hydroxyquinoline is a very effective releasing agent for chromium in the presence of iron with an air acetylene flame. It also renders the response of chromium much less dependent upon observation height and on flame gas flow rates. At the same time, Nall confirmed that ammonium chloride is equally effective for molybdenum and was able to base a method for analysis of steels on the use of these materials as spectroscopic buffers.

Ammonium perchlorate has also been suggested as a releasing agent for chromium in presence of iron.[558]

Kirkbright *et al.*[409] found that the nitrous oxide acetylene flame overcame many interferences with the determination of molybdenum, and iron in this flame has an 'enhancing' effect on molybdenum.[619] Feldman *et al.*[227] reported the successful determination of chromium with a nitrous oxide acetylene flame, as also did Welz,[785] provided that calibration was effected with standard steels and not with synthetic standards having iron and nickel additions. The latter's findings again suggest a difference between samples and synthetic standards, probably in oxidation state of the chromium.

Provided that standards are taken through an oxidation step with the samples, the same response is obtained.[732] The oxidation is carried out with perchloric acid which, since it does not interfere with the chromium and molybdenum absorption response, forms the basis of the analysis solution. In perchloric acid solution and with the nitrous oxide acetylene flame, the influence of iron on both chromium and molybdenum is one of slight 'enhancement'. This is probably a reduction in ionization as it increases with the amount of iron present until plateau values are obtained at not less than 0.7 % of iron in the solution. In the method recommended for chromium and molybdenum, iron is always added to maintain a fixed level of about 1 %, whatever the sample dilution. In this way, chromium or molybdenum at any level can be determined. The method can also be extended to alloys other than steels.

The only limitation is when tungsten is also present in concentrations greater than 0.5%. This precipitates as tungstic acid when perchloric acid is the only solvent acid present, and some molybdenum coprecipitates giving low analytical results. Tungsten is retained in solution in the presence of phosphoric acid. This can be made the basis of an alternative procedure for molybdenum and tungsten, but is unsuitable for chromium as the latter again suffers interferences from the complex solution matrix.

Manganese. Many of the interference effects reported hitherto with various solvent acid mixtures were largely removed when perchloric acid was made the basis of the final solution,[335] though in an air acetylene flame interferences were experienced from high concentrations of cobalt, tungsten, chromium, nickel and molybdenum. These could be reduced by the addition of ethanol, which also helped to increase the sensitivity. Such interferences were not confirmed by Thomerson and Price[733] when the solutions prepared for chromium and molybdenum determinations were also used for determining manganese.

Nickel. Few workers have found chemical interferences in the determination of nickel. Most discussions have centred around the choice of acid solvent and the analytical line for measurement.

Sulfuric–phosphoric acid will retain tungsten in solution, but aqua regia is quicker. No chemical interferences are found in the presence of perchloric acid, and nickel is successfully determined in the solution prepared as recommended here for chromium and molybdenum.

Unless a monochromator of very good resolution is utilized, the nickel absorption line 232.0 nm may give a markedly curved calibration graph because of the breakthrough of non-absorbable radiation from a nickel line at 232.14 nm. The alternative line 341.5 nm is often measured instead.

Copper. No significant interferences have been reported whatever the acid solvent used or whatever major elements other than iron itself are present. Excellent results are obtained provided iron is incorporated in the standard. Copper can thus be determined by the method given. Traces of copper are extracted as diethyldithiocarbamate into methyl isobutyl ketone.

Cobalt. As with nickel, no significant interference effects have been found with cobalt in steels with any of the acid solvents. Iron increases the response of cobalt in perchloric acid solvent but the effect is overcome when the calibration standards contain the appropriate amount of iron as in the method given.

For determining traces of cobalt, Scholes[699] recommended the extraction of iron into isobutyl acetate.

Titanium. Titanium was not determined in steel until after the establishment of the nitrous oxide acetylene flame for routine use. Bowman and Willis[113] dissolved samples in aqua regia and fumed with sulfuric acid. The interference of iron depended upon both sulfuric acid concentration and flame conditions. Headridge and Hubbard[310] used hydrofluoric/nitric acid mixture and finally made up the solution to contain 50% ethanol. The titanium response was increased and, with iron added to the calibration solutions, no interelemental interferences were found. Mostyn and Cunningham[535] found that the addition of potassium chloride to all solutions based on nitric/hydrochloric acid solvent overcame all interferences, even that of iron, and so standards did not need to have iron added.

The same situation exists when the analysis solution is based on perchloric acid, although if mixed standards are made up according to the method given here iron will automatically be present. Titanium can therefore be determined under the same conditions as the other elements treated in the scheme without loss of sensitivity.

Tin. In the nitrous oxide acetylene flame there is a small increase in sensitivity in the presence of iron, but no other detectable interferences. A 1% sample solution in perchloric acid enables tin to be determined to about 0.01%. To determine lower levels of tin in steels and ferro alloys two possible separation methods have been suggested.[112] Precipitation of tin hydroxide using beryllium as a collector enables 0.0015% of tin to be measured, while down to 0.0008% can be determined after extraction of tin as its thiocyanate complex under defined conditions.

Vanadium. Vanadium has been determined in steel after dissolution in phosphoric/sulfuric acid mixture[128] and with perchloric acid in addition.[417]

In a final solution based upon perchloric acid, iron enhances the vanadium absorption and iron must therefore be present in the standards. No other interelement interferences are experienced in the nitrous oxide acetylene flame.

Aluminium. Aluminium was determined in steel by Amos and Thomas[56] shortly after the nitrous oxide acetylene flame was first introduced. 'Soluble aluminium' was included in a scheme based on hydrochloric/nitric acid dissolution given by Price and Cooke.[604]

In the perchloric acid-based analysis solution and nitrous oxide acetylene flame, iron increased the sensitivity of aluminium and was therefore added to the standards. No other interferences are found.

Soluble and insoluble aluminium are readily determined by both flame atomic absorption and electrothermal atomization. The sample is dissolved in nitric acid, e.g. 40%, or nitric/hydrochloric acid mixture and filtered. The solution is either aspirated in the flame or injected in the electrothermal atomizer. The residue is ashed and fused either with sodium carbonate borax (2:1) mixture[676] or with sodium carbonate and boric acid. The fusion is taken up in nitric acid and again either aspirated or injected. For flame determination of very low insoluble aluminium levels, the residue from the first dissolution can be treated with hydrofluoric and sulfuric acids to remove silicon prior to fusion with sodium carbonate and boric acid.[153] The melt is then extracted with hydrochloric acid and the iron extracted with isobutyl acetate before making up to volume and aspirating. Addition of ammonium sulfate to solutions used for electrothermal determination of aluminium is said to improve the signal.[578]

Lead. Lead can be measured directly at concentrations down to 0.002% in the perchloric acid method using an air acetylene flame. No interferences are found.

For lower levels, electrothermal atomization has been studied by Ottaway[568] and by Frech.[243] Levels down to 0.0001 % are determined by injecting 10 μl of a 0.25 % solution of the sample in hydrochloric/nitric acid solution.[244]

Scheme for the analysis of steels

The following scheme has been found to be successful for all steels except those containing tungsten. A single weighing and preparation procedure is needed for eleven common elements.

> Dissolve 1 g of sample in 10 ml of hydrochloric acid (sp. gr. 1.18) and 5 ml of nitric acid (sp. gr. 1.42). When the reaction has subsided add 10 ml of perchloric acid (sp. gr. 1.54). Evaporate slowly until the solution is fully oxidized and fumes of perchloric acid appear. (If the solution contains chromium it will probably turn red at this point.) Fume for 5 min, then cool and dissolve the soluble salts in 50 ml of water. Filter and dilute to 100 ml.
>
> Prepare calibration standards by adding appropriate volumes of stock solutions, to cover the basic ranges (Table 11) to 1 g samples of pure iron (B.C.S. 260/3 is suitable). Then proceed as for the samples. Chromium and molybdenum stock solutions in particular must be added before the fuming. Mixed standards, i.e. one set of standards containing ranges of any or all of the elements to be determined, can be prepared provided the iron content is maintained at 1%.

Further standard ranges can be prepared to cover alloy concentrations up to 5%, reducing sensitivity by one of the means described in Chapter 4 if necessary. To determine concentrations over 5% the sample is diluted up to 5 times, and sufficient stock iron solution is added so that the final iron concentration is always 1%. There appears to be no reason why other elements, including

TABLE 11. Calibration standards for steels

	Basic % range	Wavelength nm	Flame type
Manganese	0–1.0	279.5	air acetylene stoic.
Nickel	0–1.0	341.5	air acetylene lean
Chromium	0–1.0	357.9	N_2O acetylene stoic.
Molybdenum	0–0.2	313.3	N_2O acetylene* stoic.
Copper	0–1.0	324.8	air acetylene lean
Vanadium	0–0.2	318.4	N_2O acetylene rich
Cobalt	0–1.0	240.7	air acetylene lean
Titanium	0–1.0	364.3	N_2O acetylene rich
Tin	0–0.1	286.3	N_2O acetylene stoic.
Aluminium	0–0.1	309.3	N_2O acetylene rich
Lead	0–1.0	283.3	air acetylene lean
Tungsten	0–5	255.1	N_2O acetylene rich

* The acetylene flow rate for molybdenum must be carefully adjusted to give maximum height of red feather with no luminescence.

cadmium, antimony, bismuth, tellurium, calcium and magnesium, should not be included in this scheme.

Molybdenum in Presence of Tungsten. Tungsten is retained in solution in presense of phosphoric acid, even when perchloric acid is used in the final solution. The preparation of the samples is given below. Both molybdenum and tungsten have also been determined, using phosphoric/sulfuric/perchloric acid mixture, by Knight and Pyzyna.[417]

> Solvent acid mixture: to 300 ml of water add 100 ml of perchloric acid (sp. gr. 1.54), 100 ml of phosphoric acid (sp. gr. 1.75) and 100 ml of sulfuric acid (sp. gr. 1.84).
> Dissolve 1 g of sample in 50 ml of the solvent acid mixture and heat gently. When dissolution is complete, oxidize with dropwise additions of nitric acid and evaporate until the first fumes of perchloric acid appear. Cool, dilute, filter and make up to 100 ml. Prepare calibration standards by adding appropriate quantities of molybdenum and tungsten stock solutions to 1 g of pure iron and take through the dissolution procedure outlined above.

A composite scheme was given by Nall *et al.*[546] for manganese, nickel, copper, chromium and molybdenum using an air acetylene flame. The procedure is briefly as follows:

> Dissolve 1 g of sample in 10 ml of hydrochloric acid, oxidize with a little nitric acid and evaporate until nitric fumes disappear. Dissolve in 10 ml of hydrochloric acid, filter. After ashing treat the residue with hydrofluoric acid, remove fluoride by evaporation with hydrochloric acid and combine the solubilized residue with the main filtrate. Make up to 100 ml (solution A).
> *Copper, Manganese and Nickel.* Dilute solution A so that copper and manganese are in the concentration range 0–2.5 ppm and nickel 0–20 ppm with the final iron concentration always 0.2%. These elements may be contained in a single standard solution.
> *Molybdenum.* Dilute solution A so that molybdenum is in the range 0–20 ppm, incorporating 1% m/v ammonium chloride and maintain the final iron concentration at 0.2%.
> *Chromium.* Dilute solution A so that chromium is in the range 0–10 ppm, incorporating 0.25% 8-hydroxyquinoline and maintain the final iron concentration at 0.2%. Prepare standards in the same element ranges containing the same spectroscopic buffers.

Direct determination of other elements in steel

Phosphorus. The combination of electrothermal atomization and electrodeless discharge lamps overcomes the problems of sensitivity and noise associated with the conventional flame procedure. Although the available absorbing line of phosphorus at 213.6 nm has only one hundredth of the sensitivity of the principal (though usually unavailable) lines at 178 nm, concentrations of phosphorus in steel can readily be measured down to 0.01 %.[797] The sample is first made up in a 5 % solution.

> Dissolve 1 g of sample in 4.0 ml of hydrochloric acid and 3.0 ml of nitric acid. When the reaction has subsided add 1.5 ml of perchloric acid and warm until the perchloric acid refluxes down the sides of the beaker. Cool, add 5 ml of water, filter quantitatively into a 20 ml flask and make up to volume. Calibrate by method of standard additions,

injecting 20 μl of sample, drying, then 20 μl of water and carrying out the full atomization procedure. This procedure is repeated with 20 μl of 20, 40 and 60 ppm phosphorus solutions (as ammonium dihydrogen orthophosphate) ($\equiv$0.02, 0.04 and 0.06% P) respectively in place of the water. A suitable temperature programme would be dry 100 °C, 20 s; ash 300 °C, 30 s; atomize 2750 °C, 4 s; tube clean 2850 °C, 6 s.

Magnesium. Magnesium is determined to about 0.001 % in steel by the simple procedure of Belcher and Bray.[89] A 0.1 % solution of the sample in 1 % hydrochloric acid (or 0.5 % sample in 5 % hydrochloric acid) must contain about 0.5 % of strontium or lanthanum to overcome the effect of aluminium in the air acetylene flame. Other elements do not interfere.

Silicon. This is brought out of solution with the perchloric acid procedures described. Low concentrations of silicon in steel may be determined with the nitrous oxide acetylene flame using the following dissolution technique:

Dissolve 1 g of sample in 20 ml of 50% hydrochloric acid and carefully add 15 ml of hydrogen peroxide (50 vol). Boil, cool and dilute to 100 ml. Standards should contain iron as well as a range of silicon concentrations added as sodium silicate solution.

Higher silicon levels, e.g. as in cast irons, may be determined (Price and Roos).[606]

Dissolve 1 g of sample in 10 ml of hydrochloric/nitric (4:1) acid mixture in a PTFE cylinder or flask. Cool, add 2 ml of hydrofluoric acid and dilute to 100 ml in a PTFE cylinder or flask. If standards are made up with sodium silicate solution, a little sodium chloride solution should be added to both samples and standards to equalize the ionization buffering effect.

Niobium and Tantalum. The method given is that described by Thomerson[731] for niobium, and it enables tantalum to be determined in the same solution. The use of hydrofluoric acid necessitates the use of PTFE beakers and polypropylene flasks.

Add 5 ml of hydrochloric acid and 3 ml of nitric acid to 1.000 g of sample and warm gently. When the bulk of the sample has dissolved add 10 ml of hydrofluoric acid (40%) and evaporate the solution to a volume of about 10–12 ml. Cool, filter if necessary through a Whatman No. 541 paper and dilute to 50 ml in a polypropylene volumetric flask. Prepare niobium calibration and blank solutions by pipetting appropriate aliquots of niobium stock solution (1000 mg l^{-1} Nb) into PTFE beakers containing 1 g of pure iron, using the dissolution technique as described above. For some highly alloyed materials, with alloying elements in excess of 10%, the calibration solutions should be matched with these other major components when the greater accuracy is required.

Zirconium. The similarity of zirconium to niobium and tantalum suggests that a similar nitric acid–hydrofluoric acid dissolution technique as described above for these elements can be employed. Bond[109] has reported considerable enhancement effects from additions of ammonium fluoride which also suppresses other interference effects, thus the addition of equal amounts of this reagent to both sample and standard solution might be advantageous, especially for low levels of zirconium.

Arsenic, Selenium, Tellurium etc. Arsenic can be determined by extraction with 2-thenoyl-trifluoroacetone[510] into carbon tetrachloride after dissolution of the sample into nitric acid. Measurement is best made in the argon hydrogen diffusion flame. Tellurium was extracted into amyl acetate as the diethyl dithiocarbamate.[486] Down to 0.0005% is said to be determinable in an air acetylene flame.

These and other hydride-forming elements are determined using the hydride-generation method (p. 55). The steel sample is dissolved in perchloric/nitric acid mixture, fumed and cooled.[237] The residue is taken up in water for the determination of arsenic, bismuth, lead and tin, and in hydrochloric/nitric acid for antimony, selenium and tellurium. Solutions are then transferred to the hydride generating apparatus and treated with sodium borohydride in the usual way.

Arsenic, antimony and tin were measured by electrothermal atomization of a nitric acid solution of steel, but several elements were said to interfere.[616]

Other Trace Elements by Flame Atomic Absorption. Improvements in sensitivity may be attained by using stronger solutions than those already recommended. While some burner systems are able to accept comparatively high concentrations of total dissolved solids, others may be limited in this respect. It should also be remembered that a 1% solution of iron is a 3% solution of ferric chloride and a 6.3% solution of ferric perchlorate.

The discrete sample nebulization system (see p. 49) allows solutions of higher concentrations to be atomized than could normally be sustained continuously in a given system.[53] Using this method aluminium (0.0008%), arsenic (0.03%) and tin (0.007%) were measured in 200 μl aliquots of a 10% sample solution in hydrochloric/nitric acid.[739]

A group of trace elements, lead, bismuth, antimony and silver, are determined by extraction from hydrochloric acid, potassium iodide, ascorbic acid medium using the APDC–MIBK method,[726] while tri-*N*-octylamine is used to complex various elements for extraction into MIBK after the sample was dissolved in hydrochloric/nitric acid mixture, fumed with sulfuric acid and ammonium iodide added. Copper, zinc and cadmium were measured at the 0.1 μg g^{-1} level and lead, bismuth and antimony at 1 μg g^{-1}.[698]

Separation of the Matrix Elements. The following electrolysis procedure can be used to separate iron, nickel, chromium, cobalt, copper and tin:

> Dissolve 2 g of sample in 20 ml of nitric acid–sulfuric acid–water (2+1+4) mixture. Take to dryness and fume for 5 min. Alternatively, after dissolution in aqua regia, fume with perchloric or sulfuric acid. Make up to 20 ml with water, then electrolyse using a mercury pool cathode and platinum anode, connected to a DC supply giving 15 A at 5–6 V.

Trace amounts of aluminium, calcium, niobium, magnesium, vanadium and zirconium can then be determined, simply by concentrating the remaining solution, or performing further extractions or separations should these be found necessary.

Calcium was determined by Scholes[669] to 0.002% in maraging steel. After removal of the iron by electrolysis the solution was made up to 100 ml. Strontium chloride was added to prevent interference from phosphate, vanadium and titanium in the air acetylene flame. Calcium can also be determined in the presence of iron using the nitrous oxide acetylene flame, with the same order of sensitivity.

Removal of the iron itself by solvent extraction depends on the solubility of ferric chloride, in strongly acid solutions, in a number of organic solvents. According to Morrison and Freiser[532] 99.9% of iron is removed from 7.75–8.0 M hydrochloric acid solution by shaking for several minutes with an equal volume of isopropyl ether. Isobutyl acetate is also effective and is used much in iron and steel laboratories, a pH of about 1 usually being recommended for the aqueous phase. Amyl acetate or methyl isobutyl ketone may also be used. Most elements common in steels remain in the aqueous phase though vanadium (V), antimony (V), gallium (III) and thallium (III) are also extracted.

The aqueous phase may be analysed direct for elements which are depressed in the presence of iron, or concentrated to improve detection limits of trace elements. Further extraction can be made using the ammonium pyrrolidine dithiocarbamate–methyl isobutyl ketone system.

The potentialities of electrothermal atomization as applied to steel analysis have hardly been realized yet. The main problem of calibration has been overcome in solution techniques. There is no doubt, however, that even lower concentrations of many metals could be reached with solid sampling. Andrews and Headridge, for example, have had difficulty in keeping sensitivity low enough to enable work to be done on already calibrated solid standards.[63]

Further applications include the characterization of small metal particles by microsampling techniques and electrothermal atomization and also the examination of steel surfaces by electrographic sampling.[679] Here a paper impregnated with ammonium chloride forms a contact with the surface under examination, which is then stripped electrolytically.

Ferro-alloys

In general, the atomic absorption method can be used for both trace and major elements in ferro-alloys. Any convenient method of dissolution can be employed, provided the standards are made up to match the acidity and major components.

The following procedure may be used, for example, to determine aluminium and calcium in ferrosilicon, and could undoubtedly be extended to other elements:

Dissolve 1 g of sample in 10 ml of nitric acid and 10 ml of water in a PTFE beaker. Cover with a polyethylene lid and heat to boiling. Add 40% hydrofluoric acid dropwise through a hole in the lid until the initial reaction has subsided. Make further additions of hydrofluoric acid until all the silicon has been removed, washing down the sides of the beaker with water as necessary. Then evaporate to a volume of 5 ml and add 15 ml of 60% perchloric acid. Heat to fumes and continue fuming for 20 minutes. Cool the mixture, and add 50 ml of water, warming to dissolve. Cool again and make up to 100 ml in a calibrated flask. Aspirate this solution for the aluminium determination. For calcium, dilute 10 ml of this solution to 50 ml, adding in 4 ml of a solution containing 5% of iron as ferric chloride.

Standards should contain 100–500 mg l^{-1} of aluminium (equivalent to 1–5%), the presence of iron not being necessary, and 0–12 mg l^{-1} of calcium (equivalent to 0–0.6%) in 0.4% iron. The nitrous oxide acetylene flame is used for both elements.

Major components in ferro-alloys were determined by Smith *et al.*[692] Where possible the materials were dissolved by a solution procedure similar to that given above, though a sodium peroxide fusion was required for the insoluble residue of ferrochromium. The elements determined were vanadium, titanium, silicon, niobium, boron, manganese and chromium, and the nitrous oxide acetylene flame was used throughout.

Nickel and high temperature alloys

Andrew and Nicholls[62] were the first to analyse nickel for magnesium. With a low temperature flame, no interference was reported from aluminium and silicon. Nickel alloy was dissolved in hydrochloric acid and atomized in an air acetylene flame.[204] Standards were prepared from pure carbonyl nickel. Various flame conditions were investigated.

In a textbook on the analysis of nickel,[455] procedures are given for the determination of traces of zinc and lead after dissolving the sample in 8 M nitric acid and making up to a 2% solution.

Iron was determined in high temperature alloys by Cunningham.[165] The samples were dissolved in aqua regia but ammonium chloride was added to overcome the effects of nitric acid in an air acetylene flame. Cobalt and nickel did not interfere significantly. Welcher and Kriege analysed both nickel–base[784] and cobalt–base[782] high temperature alloys. Samples are dissolved in aqua regia (or, if they contain tungsten, niobium or tantalum, hydrofluoric/nitric acid mixture) and the nitrous oxide acetylene flame is used for all elements except iron. The tendency of cobalt to form a number of different complex ions may cause variations in response, and this may be overcome by evaporating both samples and standards twice to dryness with hydrochloric acid before finally making up to volume.

Details of the direct determination of a number of trace elements by flame atomic absorption are given by Hornick[330] and a preliminary separation from nickel alloys by a double ion-exchange method has also been given.[404]

The determination of trace levels of bismuth has been reported from a number of sources. Hydride generation after dissolution of the sample in nitric/hydrofluoric acid mixture and addition of EDTA[199] and electrothermal atomization after dissolution in the same acid mixture[783] are standard approaches. The latter has also been extended to lead, selenium, tellurium and thallium. A separation and concentration by electrolysis with copper prior to normal flame determination has also been described.[793]

Electrothermal atomization of the solid sample was shown by Welcher et al.[488] to be a practicable approach. 1 mg chip samples of nickel base and complex alloys were placed directly in a graphite tube atomizer to determine lead, bismuth, tellurium, selenium and thallium. Although the analysis time is shorter and no pre-atomization heating steps were necessary, accuracy and precision are less favourable than for solution methods.

The noble metals

All the noble metals except osmium show good sensitivity, though numerous interference effects have been reported when these elements have been determined in presence of each other. However, many of these interferences are overcome in very hot flames, though sensitivities are usually not improved.

Jewellery alloys normally dissolve in aqua regia when silver can be filtered off as silver chloride and other metals determined in the filtrate.

Gold can be determined in ores and alloys if these dissolve in aqua regia. Interferences reported in early work using low temperature flames are much reduced in air acetylene which gives similar detection limits.

For determining low concentrations of gold several extraction methods are available. Gold can be extracted virtually quantitatively from 3 M hydrochloric acid into methyl isobutyl ketone,[708] the distribution coefficient being greater than 1000. Samples digested in aqua regia and taken to dryness may be redissolved in hydrochloric acid, diluted and shaken with methyl isobutyl ketone[743,744] after the addition of a little hydrobromic acid.

Palladium is about as sensitive as gold and with the air acetylene flame relatively few interferences have been reported. Ginzburg et al.[280] used a chemical procedure for concentration, precipitating with lead and copper sulfide, but in the presence of these and other elements no other preparation was necessary. Erinc and Magee[217] extracted palladium from solutions of platinum alloys as the thiocyanate–pyridine complex from hydrochloric acid solution into MIBK, achieving a detection limit of less than 1 ppm in the extract.

To determine palladium in silver, Takeuchi[725] removed the silver by precipitation as chloride and then extracted the palladium as diethyl dithiocarbamate into methyl isobutyl ketone.

Rhodium is also quite sensitive but appears to be much affected by interferences, both from other noble metals and from other elements and acid radicals. There appears to be a complex pattern of depressions by copper, lithium and nickel and enhancements by zinc, magnesium, strontium and iron in a low temperature flame, and Deily[185] showed that in organic solvents the solvent effects are minimized in lean flames, but that the burner position is critical. Johns and Price[362] showed that there are residual interferences in an air acetylene flame similar to those reported by Strasheim[706] but that these are largely overcome in the presence of 1 % of sodium sulfate. Best results with sodium sulfate are obtained with a lean flame, but in air acetylene the calibrations tend to be non-linear.

This behaviour has been reported elsewhere.[378] With a nitrous oxide acetylene flame and no buffer, no interferences were detected, the sensitivities were maximal as with sodium sulfate and the air acetylene flame, and the calibration curves were linear over a greater concentration range.

Platinum was also determined by Strasheim *et al.*[706] who found that, in constrast to their experience with rhodium, the interferences in the low temperature flame were much reduced in presence of copper. Pitts *et al.*[589,590] investigated the behaviour of platinum in both air acetylene and nitrous oxide acetylene flames, following a method by van Loon[760] in which the serious interferences of palladium, gold and silver on platinum in the air acetylene flame were eliminated by the addition of 1% lanthanum chloride. Pitts *et al.* found that in the presence of lanthanum with the air acetylene flame, most interferences from other noble metals, from base metals and from several common anions were virtually eliminated. The releasing action of lanthanum in this context was discussed. If the nitrous oxide acetylene flame is used, as has already been seen for rhodium, no interferences are detected, though it is stated that the sensitivity of platinum is lower than in air acetylene by a factor of about 4.

Iridium, Ruthenium and Osmium were determined by Ashy and Headridge in rhodium sponge by dissolving in hydrochloric acid with sodium chlorate at 250 °C under pressure in a sealed Pyrex tube. These metals were then extracted with methyl triphenyl phosphonium chloride in chloroform. The extract was evaporated and the residue dissolved in acetonitrile, lithium perchlorate being added as spectroscopic buffer. This solution was then aspirated in an air acetylene flame for iridium, nitrous oxide acetylene for osmium, and either for ruthenium.[65] In order to avoid loss of osmium when the sealed tube was opened, the contents had to be frozen with solid carbon dioxide.

Silver is dissolved in nitric acid, but for the determination of other precious metals van Loon[760] used the following dissolution procedure:

> Add 5 ml of nitric acid to the sample, heating to leach the silver. Evaporate to 0.5 ml and add several ml of hydrochloric acid, continuing small additions until no further gases are evolved. Transfer to a volumetric flask. Add sufficient lanthanum chloride solution to make the final lanthanum concentration 1% and make up to the mark with 6 M hydrochloric acid.

Silver itself is assayed in fine silver bullion[317] after dissolution in nitric acid by precipitating the bulk of the silver with standard sodium chloride solution, then determining the remaining silver in the supernatant by atomic absorption.

Studies of the determination of noble metals by electrothermal atomization were made by Knoop[38] using a large graphite furnace tube and by Everett[219] using a miniature pyrolytically coated graphite tube. Both found that the sensitivities of all elements investigated improved up to the maximum temperature that could be employed. It appears to be preferable, however, to use a lower temperature than this, as the lower sensitivity is compensated by better reproducibility, and longer furnace tube life. Although rhodium, platinum and iridium were shown to have little mutual interference when present in binary mixtures, a more complex interference pattern is again experienced when a number of the elements are found together. It is therefore recommended that for mixtures of unknown composition the method of standard additions should be used. For palladium, platinum, rhodium, ruthenium and iridium, absolute detection limits of between 2.5 pg and 0.5 ng were found, and for osmium, 10 ng.

The 'new' metals

Zirconium and 'Zircalloys'. Two basic methods were given by Elwell and Wood[214] for dissolving zirconium alloys for atomic absorption analysis. For determining calcium and lithium, zirconium must be separated by ion exchange before the sample is aspirated.

> To 2 g of sample in 30 ml of water in a platinum dish, add 5.5 ml of hydrofluoric acid and allow to dissolve, then oxidize with a few drops of nitric acid. Add a solution of strontium chloride ($\equiv 6$ mg Sr) and pass through an Amberlite IR:120H (20–50 mesh) column. Wash the column with four 30 ml portions of 1% hydrofluoric acid. Then elute with three 10 ml portions of 60% hydrochloric acid allowing the second of these to remain in contact with the resin for 20 min. The combined eluates are used for atomic absorption.

This procedure can be used for determining less than 10 ppm of calcium in the alloy.

For copper and sodium, basically the same procedure is advised, though the ion exchange separation is omitted.

> Dissolve 2 g of sample in 4 ml of water in a platinum dish to which hydrofluoric acid is added dropwise till the dissolution is complete. Then add nitric acid dropwise until clear and make up to 10 ml in a polythene measuring cylinder and transfer back to the platinum dish.

These elements are all determinable in an air acetylene flame.

The other dissolution method involves fluoroboric acid and does not therefore require platinum ware.

> Dissolve 0.5 g of sample in 5 ml of 10% sulphuric acid and 2 ml of fluoroboric acid. Clear with a few drops of nitric acid if necessary, cool and make up to 50 ml.

Although this method is given specifically for the determination of zinc, there would appear to be no reason why other elements should not be measured in the same solution.

Low traces of cadmium (0.1–1 ppm) were extracted by Mizuno *et al.*[527] after dissolution in sulfuric, hydrofluoric, boric and nitric acids and adjusting to pH 8–10 with ammonia–ammonium citrate buffer, into chloroform as dithizonate. The cadmium was re-extracted back into hydrochloric acid for atomic absorption determination.

Zirconium aluminium alloys may be dissolved in sulfuric acid and ammonium sulfate. Aluminium is then determined in the aqueous phase after zirconium has been extracted with bis (2-ethyl hexyl) phosphate into chloroform.[677]

Titanium Alloys. Two dissolution methods based on fluoroboric acid were given by Elwell and Wood.[214] The first is recommended for determining magnesium, though presumably other trace elements could thereby be determined.

> Dissolve 0.5 g of sample in 15 ml of hydrochloric acid (30%) to which fluoroboric acid is added dropwise until the dissolution is complete. Make up to 25 ml.

For higher concentrations and alloying elements, specifically zinc:

> Dissolve 0.25 g of sample in 10 ml of 50% hydrochloric acid, adding fluoroboric acid dropwise until dissolution is complete. Make up to 250 ml.

Titanium is also dissolved by hydrofluoric/nitric acid mixtures, and the method for determining sodium in zirconium may also be used for titanium.

Niobium and Tantalum Alloys. The determination of copper and zinc in niobium is similar to the copper method given for zirconium, according to Elwell and Wood.[214]

> Dissolve 2 g of sample in 5 ml of hydrofluoric acid in a platinum dish. Add nitric acid dropwise until dissolution is complete, cool, make up to 10 ml in a polythene measuring cylinder and then return to the original platinum dish. Aspirate this solution for atomic absorption measurements.

Kirkbright *et al.*[407] determined copper in niobium and tantalum after extracting the 8-hydroxyquinoline complex from fluoride solution at pH 4.5 into ethyl acetate. Titanium and molybdenum in one-hundredfold excess interfered with the extraction but were masked by the addition of hydrogen peroxide. Either air acetylene or air propane flames could be used.

Cobalt and zinc may be extracted into chloroform as thiocyanate diantipyridylmethane ion association complexes from solutions of niobium, tantalum, molybdenum and tungsten. The solutions should be adjusted to pH 3.25 in a citric acid medium and approximately 1.2 M in sodium thiocyanate.

Niobium alloys have been dissolved in sulfuric acid, ammonium sulfate and hydrogen peroxide[660] prior to determination of molybdenum and tungsten in the percentage range, using nitrous oxide or air acetylene flame.

Niobium has also been attacked by electrolytic dissolution in ethanol ethylene glycol mixture, saturated with ammonium chloride prior to the determination of copper and iron in the 0.01–1.0% range.[485]

Both niobium and tantalum have been analysed for trace elements by direct electrothermal atomization of the powdered solid sample.[352,551] The sample, 1–50 mg of tantalum powder in one example, was weighed into a tantalum boat which was then introduced into a graphite tube furnace.

Uranium Alloys. Magnesium is determined in uranium without separation after dissolution (0.5 g sample) in 25 ml of hydrochloric acid.[338] Hydrogen peroxide, if required to assist dissolution, is decomposed when the solution is evaporated to dryness and redissolved in hydrochloric acid. To avoid chemical interference effects, the method of standard additions should be employed.

The bulk of the uranium matrix can be extracted into carbon tetrachloride with tributyl phosphate, allowing some concentration to be made[373] without affecting burner performance. Baudin et al.[83] described an analytical procedure for iron and aluminium (50–500 ppm and 500–1000 ppm respectively) based upon a sulfuric acid dissolution without separation. 1 g of the finely divided sample is attacked in the cold with 25 ml of sulfuric/nitric acid mixture (15 + 5, + 15 volumes of water). The solution is then evaporated to white fumes and the residue redissolved in water and made up to 100 ml, when it is about M/20 with respect to sulfuric acid. Iron is determined with an air acetylene flame and aluminium with nitrous oxide acetylene.

Complex alloys produced by nuclear fission have been analysed for molybdenum, ruthenium, palladium and rhodium by Scarborough.[665] The alloy is dissolved in hydrochloric/nitric acid mixture to which a little hydrofluoric acid was also added. An air acetylene flame was used for all the elements with a multislot burner. The absorbances given by mixed standards containing uranium were essentially the same as standards for individual elements with uranium added. Uranium thus eliminates the mutual interferences between these elements and the method may be used for a wide range of alloy compositions.

Rare earths in the 0.01–0.1 % range were determined in perchloric acid solutions of uranium alloys[271] but lead in the ppm range had to be separated from the sample in 2 M hydrobromic acid on a Dowex 1-X8 column and then eluted with 6 M hydrochloric acid for measurement in a flame.[425]

A graphite furnace device suitable for determining trace elements in uranium and plutonium has been described by Cox.[158]

Electroplating solutions

With samples already in the form of solutions the control of electroplating baths is an ideal application for atomic absorption spectroscopy. In order to ensure an electrodeposit of the correct quality, concentrations of the major metals, metal-containing additives (used to provide characteristic plating properties) and trace impurities should all be checked. Impurities are introduced through bath to bath transfer of workpieces, in the commercial grades of the salts used to make the bath, in some superficial dissolution of the workpieces themselves, and in the continued use of hard waters.

If the electrodeposit is itself an alloy its metal/metal ratios may have to be checked. Plate thickness can also be determined by atomic absorption spectroscopy.

When both major and trace elements are measured in one sample, the full concentration range may be as high as 50 000:1 and consequently a number of dilutions may be required. For the major elements a dilution of 5000 is sometimes necessary. Alternatively the less sensitive absorption lines may be employed or the burner rotated.

Some plating solutions contain several major components, and therefore standards for trace elements should normally be made up in a matching matrix, unless it is proved that it does not interfere. Alternatively trace elements can be determined by the method of standard additions.

Plating solution analyses are not generally required to a very great degree of accuracy: impurities to $\pm 10\%$, additives to $\pm 5\%$ and main elements to $\pm 2\%$ is not unusual. On occasion it is even sufficient to report that the traces are below a certain level and that the major elements are above a prescribed minimum.

Copper Plating Solutions. Cyanide copper-plating baths contain 10–50 g l^{-1} of copper as sulfate and 5–35 g l^{-1} of sodium cyanide, while in a typical acid bath there may be 200 g l^{-1} of copper sulfate in 3% sulfuric acid.

Iron, lead and zinc are the main impurities and unless they are present at very low concentrations (<40 ppm) good results are obtained with a $20 \times$ dilution of the sample and comparison with simple aqueous standards. For very low levels where smaller dilutions must be used the matrix compositions of standards and samples should be matched.

Copper at major levels may itself be determined if the recommended steps to reduce sensitivity are taken.[429]

Nickel Plating Solutions. A typical nickel bath may contain, per litre: 300 g of nickel sulfate, 50 g of nickel chloride or sodium chloride, and 40 g of boric acid. Nickel itself is thus 50–60 g l^{-1}, and copper, zinc, iron, lead, chromium, calcium and magnesium may all be in the ppm range.

Shafto[675] used the standard addition method for some of these traces, while Whittington and Willis[799] determined copper and zinc in Watts-type nickel solutions, using purified Watts solutions as a basis for standards. In the control of nickel plating baths, Parker[575] determined iron, lead, copper, calcium and magnesium in a $10 \times$ dilution of the sample solution, comparing with simple aqueous standards. To determine zinc and chromium, standards had to be made up in the nickel–base solution because of enhancement of the former and depression of the latter in an air acetylene flame.

Concentrations of zinc between 0.03 and 10 ppm have been extracted with tri-octylamine hydrochloride, then re-extracted back into 1 M nitric acid for measurement in the air acetylene flame.[349]

To determine the nickel itself, use of the alternative line, 352.5 nm, and rotation of the burner allows a much smaller dilution factor to be used.

Chromium Plating Solutions. These are prepared by dissolving 200–500 g of chromic oxide per litre of water. The resulting solution is sufficiently acid to dissolve small amounts of copper, iron, zinc and nickel from the workpieces. The impurities are harmful only at the g l^{-1} level and the solutions can therefore be analysed after a 100 or $200 \times$ dilution. All elements except iron (slightly depressed by high concentrations of chromium) can be determined by comparison with simple aqueous standards. Iron standards must contain the correct level of chromium. Chromium itself is not normally determined by atomic absorption in this case, as it is also required to know the concentration of Cr(III) (Parker).[575]

Zinc Plating Solutions. Alkaline zinc cyanide baths contain 30–55 g of zinc, 80–150 g of sodium cyanide and 20–35 g of sodium hydroxide per litre. Acid zinc baths contain 20–35 g of zinc, 20–25 g of sodium chloride and 12–15 g of boric acid per litre. Zinc itself may be determined in either solution by comparison with simple aqueous standards, using the alternative line 307.6 nm. Copper, lead, iron and tin can be allowed to build up to 0.1 g l^{-1} without detriment, and when determined by atomic absorption do not suffer interference from zinc in an air acetylene flame.

Lead/Tin Tinning Baths. Silver, copper and palladium have been determined by the method of standard additions, the lead–tin alloy having been dissolved in a mixture of hydrofluoric and nitric acids.[119]

Cadmium Plating Solutions. Alkaline baths contain 20–25 g of cadmium, 80–120 g of sodium cyanide and 15–35 g of sodium hydroxide per litre. Copper in the range 1–10 ppm and nickel (1–1000 ppm) require control. Whittington and Willis[799] found that, while copper showed no interference, cyanide suppressed the response of nickel. The complex must thus be degraded with sulfuric/perchloric acid and standards must be made up from a 'pure' plating solution. Cadmium can be determined without interferences in an air acetylene flame after suitable dilution of samples.

Silver Plating Solutions. These usually contain the double potassium–silver cyanide, with silver 5–30 g, potassium cyanide up to 50 g per litre and small additions of potassium carbonate. Some chemists prefer to determine silver by a Volhard titration if great accuracy is required, however, silver does not exhibit interference effects in the air acetylene flame and simple standards may be used. Only copper among impurity elements is occasionally determined.

Gold Plating Solutions. Different types of gold bath usually have gold in the range $1–10 \text{ g l}^{-1}$ as gold cyanide, with a small excess of potassium cyanide. Additions of cobalt and nickel cyanides may be made at the $1–2 \text{ g l}^{-1}$ level. Copper, iron and zinc are allowed to accumulate up to 1 g l^{-1}, but chromium and lead are not tolerated above 10 mg l^{-1}.

Cobalt, copper, nickel, iron and zinc may be determined in an air acetylene flame with no interference effects. Chromium and lead show interferences and the standards must be based upon a 'pure' plating solution, though Kapetan[381] found also that lead was interference-free.

Although none of the trace impurities interferes with the determination of gold in the air acetylene flame, cyanide itself appears to give rise to differences in the response of gold. The effects are not overcome in the hotter nitrous oxide acetylene flame. Correct results are obtained, however, if the cyanide complexes are destroying by fuming with sulfuric/perchloric acid mixture and standards contain the same acids. Alternatively, standards may be prepared to contain the other major constituents in the correct concentration.

Rhodium Plating Solutions. A typical rhodium bath contains 2.5–5.0 g of rhodium as sulfate per litre of 5% sulfuric acid. Accumulations of up to 1 g l^{-1} of cobalt, iron and nickel may be tolerated in used baths. Over this amount they have a detrimental effect on the deposit. These elements can be determined without interference in the air acetylene flame after dilution of the samples.

Rhodium suffers fairly complex interference effects[362,378] which are largely overcome in an air acetylene flame in the presence of 1% of sodium sulfate. In the nitrous oxide acetylene flame the interferences are not evident, but because of ionization effects, the concentration of sodium (which is often present as a trace impurity in the bath) must be equalized in the standards. In fact it is much simpler to add an excess, e.g. 1%, of sodium sulfate to both samples and standards.

Analysis of Electrodeposited Films. Both composition and thickness of films can be determined by atomic absorption. Deposits may be removed from the base metal by selective stripping, mechanical removal or partial etch, and then dissolved in aqua regia. Kometani[421] applied the method to the determination of film thickness by assuming that the plate density is equal to the bulk density. Knowing the area from which the deposit has been removed, the thickness is calculated from the weight determined by atomic absorption spectrometry. The bulk density of gold for example is 19.3 g cm^{-3} and the thickness is calculated from the formula:

$$\text{thickness} = \frac{\text{weight of deposit } (\mu g)}{\text{area (cm}^2) \times \text{density (g cm}^{-3}) \times 100} \, \mu m$$

The majority of electrodeposited films may be examined by the above principles, simply by removing the film with mineral acid attack.

Deposited alloy compositions may also be determined. The copper/zinc ratio of brass which has been coated on to iron wire is arrived at with adequate accuracy after stripping the sample with ammonium persulfate and dilute ammonia solution. It is important to acidify the sample solution with hydrochloric acid before aspiration and to compare with standards containing the same major ingredients.

The analysis of both evaporated and electrodeposited films was also discussed by Woolley,[831] who examined combinations of metals such as nickel–chromium, nickel–cobalt, nickel–iron as well as single element films of chromium, gold and aluminium.

Ferrites

The problems of ferrite analysis are very similar to those of alloys once the material has been brought into solution. This can usually be achieved using a pressure technique such as will be described in more detail under silicate analysis. Alternatively sealed tubes can be used.

> Weigh 50 mg samples of the finely divided ferrite material into strong Pyrex test tubes. Add 5 ml of 6 M hydrochloric acid and seal off the mouth of the tube. Heat to 210 °C

until dissolution is complete. Cool, open and make up to 100 ml with water. Dilute further as necessary for individual elements and compare with simple aqueous standards.

Little interference between components should be experienced after concentrations have been brought within the working range by dilution.

INORGANIC ANALYSIS

This section includes analyses in the field of pure chemicals, ores and other mining applications, and siliceous materials, whether natural or man-originated. It should be recognized that this can at best be a guide to some of the ways of approaching the many types of analysis which are now possible.

Pure chemicals

The problems encountered with pure inorganic chemicals and particularly metal salts are, when these are dissolved, similar to those met with in metal analysis. There are very many cases where the methods are so simple, e.g. dissolution in an appropriate solvent, that it would be impossible to detail them here. The importance of running 'blank' determinations (reagents, solvents etc. but no sample) is again emphasized, in flame as well as in electrothermal methods.

It is often necessary to devise special separation procedures for individual matrices, if the general separation procedures already mentioned are ineffective.

Some materials, e.g. high purity inorganic acids, can be evaporated in a silica crucible and the residue taken up in a much smaller amount of hydrochloric or nitric acid. This is a useful preliminary to electrothermal atomization. Electrodeposition of impurity elements from metal salt solutions on to a graphite electrode is also a useful means of separation where the matrix is one of the less readily deposited metals. The plated material may either be dissolved off the electrode, or the electrode graphite crushed and introduced quantitatively into an electrothermal atomizer.[730]

Impurities in graphite powder are also determinable by introducing the sample directly into the atomizer.

Langmyhr[438] measured a number of trace elements in sulfuric and hydrofluoric acids and ammonia solution. For the acids the use of small glassy carbon boats was advocated, facilitating the removal of the acid fumes from the sample, which was dried before being introduced into the graphite furnace.

High purity sodium metal was analysed by Scarborough et al.[666] The sample was dissolved with argon saturated in water vapour and then neutralized with hydrochloric acid. The wanted impurities are separated by the simple expedient of co-precipitating with lanthanum as hydroxides by bubbling ammonia, diluted with argon, through the solution. The precipitate is then redissolved in hydrochloric acid.

Mining and geochemical applications

It is often difficult to differentiate between the problems encountered in these fields of application. Only the uses to which the results are put (and sometimes the degree of accuracy required) are different. Geochemical analyses, including mining applications, have been the subject of a book[3] and methods of attack for many types of geochemical sample are given there.

In the determination of trace elements, particularly of trace precious metals, it is usually convenient to remove the silicon at the dissolution stage, hence many materials are attacked with mixtures of hydrofluoric acid and other acids, usually hydrochloric and nitric.

For our purpose, however, materials in this category may conveniently be divided into siliceous and non-siliceous types, for the latter usually incur greater problems in preparation. In this section essentially non-siliceous materials are discussed, as silicate materials in general are treated as a separate subject in the section which follows.

More than ever, the principle applies that if a sample can be brought into solution it can be analysed by atomic absorption, and in the majority of cases this applies to electrothermal atomization as well as to the flame. In mineralogical analysis, the dissolution method must be chosen with the analyte elements in mind, particularly if these are relatively volatile or form volatile compounds.

Since Strasheim et al.[705] determined copper in ores in 1960, atomic absorption has been used to determine both major and minor wanted elements, as well as trace impurities in many different types of ore. Ores to be assayed for copper were leached by these workers in a hot mixture of hydrofluoric and nitric acids. This was said to be more successful than either sulfuric/nitric or hydrochloric/nitric mixtures, and others have used hydrochloric/nitric acid[222] and perchloric/nitric acid[689] when copper, zinc and lead were to be measured. Both flotation products for copper, and ores and concentrates dissolve for analysis in hydrochloric/nitric acid mixtures.

Lead ores may also be dissolved in hydrochloric/nitric acid mixtures[84] but the determination of tin in cassiterite requires a more difficult separation.[311] The ore must be ground with ammonium iodide and heated to 500 °C. The tin iodide which volatilizes may be dissolved in hydrochloric acid. Addition of ammonium iodide to the final solution increases the sensitivity of the tin.

Molybdenum has been determined in siliceous ores[504] and in slags.[593] The molybdenum is leached with hydrochloric acid, and the residue digested with nitric, hydrochloric and perchloric acids, fumed and taken up in hydrochloric acid. Aluminium chloride is added to overcome the effects of interfering elements on molybdenum. The releasing action of aluminium chloride rises to a plateau at about 1000 ppm of aluminium.

The presence of mineral deposits may be discovered through the analysis of plants for metallic elements. This form of biogeochemical prospecting was described by Hornbrook[329] who determined molybdenum in soils and plants, the latter after an ashing procedure similar to those discussed in the biological section.

Gold may be assayed in ores by methods similar to those mentioned on page 215. In most cases an attack on the ore material with aqua regia is satisfactory and the gold may then be concentrated by one of several methods. Gold may be extracted directly into MIBK[450] provided conditions are chosen which give a clean extract in the presence of emulsion-forming insoluble residues. p-Dimethyl-aminobenzal rhodanine extracts gold from an aqua regia dissolution into amyl acetate[625] after the solution has been taken to dryness and redissolved in hydro-chloric acid. The absolute sensitivity of the method is 0.2 μg of gold and no interferences were detected from any metal likely to be present. Gold was precipitated, with tellurium, in a reduction method[749] in which methanol was added up to 10% to improve sensitivity.

For measuring gold in lead and zinc sulfides[472] and other geological samples[682] similar direct extractions have been used. Sulfides are dissolved in hydrochloric/nitric acid mixture with added bromine, while geological samples are attacked conventionally with hydrofluoric, hydrochloric and nitric acids with additions of hydrobromic acid. Gold is then extracted into MIBK and determined in an electrothermal atomizer. Gold may also be extracted with 0.2 M dibutyl sulfide into toluene.[651]

Column separation of gold has also been used,[718] the ore being digested with hydrochloric/nitric acid mixture, evaporated and then redissolved in hydro-chloric acid. After dilution, the gold was adsorbed on an Amberlite XAD-7 column, and after washing with 1 M hydrochloric acid, eluted with acetone/1 M hydrochloric acid 4:1 and measured in an air acetylene flame.

The determination of platinum in basic rocks was described by Simonson.[685] After digestion with nitric/hydrofluoric acid and evaporation, the residue is redissolved in hydrochloric acid and extracted with dithizone into methyl isobutyl ketone. At this stage, only otherwise interfering elements are extracted. Platinum itself is not extracted until reduced to Pt(II). The aqueous phase is therefore reduced with stannous chloride and again extracted into methyl isobutyl ketone. This platinum-containing extract is then aspirated into a lean air acetylene flame. An alternative extraction is possible.[723] After digestion first with aqua regia and then with hydrochloric acid, the samples were made up to volume and filtered. (Many base metals not present as silicates may be determined in this solution.) To an aliquot was added potassium iodide solution and the platinum was extracted into methyl isobutyl ketone. After further addition of the potassium iodide solution and two periods of shaking, the organic solution was aspirated into a lean air acetylene flame.

Rhodium was determined in chromite concentrates after precipitation with tellurium or separation into the gold bead resulting from fire assay[668] although the latter results in incomplete recovery of the rhodium.

Soils, rocks and ores were analysed for bismuth[344] after digestion with nitric acid, and filtration. An air acetylene flame was used and standards were simply made up in nitric acid, few metals other than calcium giving detectable interference.

Methods for the atomic absorption determination of potassium, rubidium and strontium for geochronological purposes were given by Gamot et al.[268] Caesium and lanthanum were used together as a buffer. Accurate results were obtained for potassium by operating in a purified atmosphere. Strontium[323] and rare earths[398] have been determined in phosphate rocks. Strontium was extracted from the sample with hydrochloric acid, and phosphate removed by passage of the resulting solution through a cation exchange column, the cations being subsequently eluted with hydrochloric acid. Barium (10–300 μg g^{-1}) is measured after acid attack or fusion and separation from the extract by ion exchange using Dowex 50-X8 (200–400 mesh protonated form).[241]

Many rare earth elements, with the exception of cerium, are determined in a hydrochloric acid solution aspirated into a nitrous oxide acetylene flame, sodium chloride being added as ionization buffer. Main interferences are from silicate, fluoride and aluminium. Samples are prepared by fusing with sodium peroxide, followed by dissolution of the alkali-insoluble residue in hydrochloric acid. The rare earths are then precipitated with oxalic acid, ignited and redissolved in hydrochloric acid.

Some of the more sensitive (and less common) rare earth elements were determined in various rare earth minerals by van Loon et al.[761] Zirconium rare earth silicates were decomposed with hydrofluoric acid, while the calcium mineral was subjected to decomposition by alkaline carbonate fusion. The wanted elements were precipitated as hydroxides, dissolved in nitric acid and made up to a known concentration with lanthanum buffer. Solution detection limits ranged from 0.1 ppm for ytterbium to 10 ppm for lutetium.

There are many examples of the application of atomic absorption analysis where sample dissolution presents very few problems, as in zinc, manganese and iron ores, limestones etc. Some more difficult materials, however, include sulfide ores and concentrates which are first fused with sodium peroxide[370,636] particularly for the subsequent determination of the hydride-forming elements. Otherwise hydrochloric/nitric acid mixture is an effective solvent.[652]

Fly ash has been fused in lithium borate[574] prior to its analysis by electrothermal methods for trace elements.

The possibilities of direct solid sampling in the electrothermal analysis of geological materials have been reviewed by Langmyhr.[436] Sample preparation is one of the more obvious difficulties. Small samples lead to errors through heterogeneity, and while it is relatively easy to grind this type of material to pass a 270 mesh sieve, to go much further, e.g. to a 400 mesh sieve, takes a very long time. Some materials, particularly mica, are difficult to prepare, and a volatile element like mercury may be lost during grinding.

Most theoretical treatments of electrothermal atomization processes assume oxides as being the starting material (and oxysalts which produce oxides at the ashing step). Many ores are sulfide based. Di- and polysulfides give off sulfur when heated to between 500 °C and 1000 °C, and at 1500 °C monosulfides decompose or react with the tube graphite. Iron-containing materials such as pyrites leave a residue, probably Fe_4C_3.

Siliceous materials

A number of types of material contain silica or silicates as one of the major constituents. Often the other major elements include calcium, magnesium, aluminium and perhaps iron or manganese, or sodium and potassium. Among such materials are silicate rocks, lavas and minerals, slags and refractories, ceramics, glasses, cements, ashes etc.

From the point of view of atomic absorption they have several problems and features in common. They require special methods to bring them into aqueous solution, but when in solution, all constituents (including traces of phosphate if present) can readily be measured. There are, however, a number of interference effects to be overcome before accurate determinations of either major or minor constituents can be made.

An examination of the reported methods for the atomic absorption analysis of siliceous materials reveals that they fall into two distinct groups: those in which the sample is brought into solution by way of acid attack and those in which the sample is fused with an alkali fusion salt or mixture. A further sub-division of both types of method is into those where the silicon is retained in solution and determined by atomic absorption and those where it is removed, either completely by volatilization, or precipitated as silica, and perhaps determined by gravimetry.

The main disadvantage of fusion methods is that they result in a prepared solution of high dissolved-solid content, which can usually be tolerated in the determination of major constituents where a greater dilution factor is applied, but which generally requires separation before minor or trace elements can be determined. On the other hand, with fusion methods it may be easier to ensure that all silicon is present in solution.

Fusion Methods. By far the most common fusion medium is lithium metaborate, which may be used to attack silicate rocks, slags, cement and coal ash. Some workers substitute a mixture of lithium carbonate and boric acid, or prepare lithium metaborate from these beforehand.

Typical procedures have been given by van Loon and Parissis[762] and by Boar and Ingram.[104] The rock or ash sample, 0.2 or 0.5 g, ground to at least BSS 250 mesh, is mixed with 2 g of lithium metaborate in a platinum or graphite crucible, and heated to 900 °C in a muffle furnace for 15 min or until a clear melt is obtained. If there is more than 15% of iron as magnetite, 50 mg of ammonium vanadate is added to flux the iron. The melt is then cooled and placed in a beaker containing 8 ml of nitric acid and 150 ml of water. The mixture is stirred continuously until dissolution is complete. If the dissolution is not allowed to take place at room temperature the silica may form polymeric hydrated silicic acid which does not dissolve.

Aliquots are taken and diluted as necessary. Silicon, aluminium and titanium must be determined with the nitrous oxide acetylene flame, but no spectroscopic buffer is added for these elements, or for calcium and magnesium if these are measured in the same flame. Calcium and magnesium determined in an air acetylene flame require the addition of 1 % of lanthanum to the final solution to overcome interference by silicon, aluminium, borate and titanium. It is necessary to match the lithium, borate and acid concentrations of prepared samples and standard solutions in order to equalize degrees of ionization of the readily ionized elements in either flame.

In an alternative approach for routine analysis of slags[621] the sample is fused with sodium carbonate and leached with hydrochloric/nitric acid mixture. The precipitated silica is filtered off. Aluminium, calcium, magnesium, manganese and iron were determined in iron, copper and lead slags, and silicon was determined in a separate sample.

Acid Attack. Most solvent acid mixtures contain hydrofluoric acid, but, depending on the way in which it is used and the other component of the acid mixture, it is possible to remove the silicon completely by vaporization as silicon tetrafluoride, or to keep the silicon in solution for determination by atomic absorption. The choice can therefore be made at the outset as to whether the silicon is to be determined with the other elements or perhaps gravimetrically in a separate sample. Even where the silica content is more than 25 % it is now possible and worthwhile to determine it by atomic absorption.

Langmyhr and his co-workers[437,447] found that hydrofluoric acid is a suitable reagent for the dissolution of siliceous material and showed that no silicon is lost by volatilization provided that an excess of the acid is used and that the solutions are not too strongly heated. This principle was used in the later work by Langmyhr and Paus[440] who distinguished between two possible groups of materials. The first includes those which can be decomposed easily below 112 °C, the boiling point of the azeotropic mixture of water and hydrofluoric acid, using relatively simple equipment. The second group comprises materials which are more difficult to decompose, requiring to be attacked at higher temperatures inside a pressure dissolution vessel.

Three possible methods of attack can be employed, depending on the material being analysed. For decompositions which take place at room temperature or temperatures up to 30–40 °C, plastic beakers or platinum dishes, with lids to prevent loss by spray action, can be used provided that the volume of solution is not allowed to reduce appreciably. Where decomposition must take place at higher temperatures, e.g. on a boiling-water bath, plastic bottles or Erlenmeyer flasks with screw stoppers are employed. If still higher temperatures are necessary a special vessel is required. Examples are given of all three methods.

The principle of the retention of silicon in hydrofluoric-based solutions in open vessels at lower temperatures was confirmed by Price and Roos[606] who used hydrofluoric acid to prevent the precipitation of silica in the atomic absorption analysis of some metals and alloys as well as cement. The dissolution itself must be carried out in either a platinum or a PTFE vessel. Subsequent dilutions can be made either in plastic volumetric ware which is now available, or the excess hydrofluoric acid may be complexed with boric acid allowing glass volumetric ware to be used.

A suitable procedure[642] for preparing many cements for atomic absorption analysis is as follows:

> Weigh 0.500 g of powdered sample into a 100 ml polythene or PTFE beaker. Wash down the sides of the beaker and add 10 ml of hydrochloric acid (sp. gr. 1.18). Break up gritty particles with the end of a stirring rod. When the sample has dissolved (apart from remaining or precipitated silica) rinse down and remove the rod, then add 1.0 ml of hydrofluoric acid (40%). Swirl until all silica is dissolved, add 50 ml of boric acid solution (4%) and mix. Transfer to a 200 ml calibrated flask, add 20 ml of lanthanum chloride solution (5%) and dilute to the mark with water. This solution is used for the determination of aluminium, iron, manganese, silicon, sodium, strontium and zinc.
>
> Transfer 10 ml of this solution to a 100 ml calibrated flask, add 5 ml of hydrochloric acid and 9.0 ml of lanthanum chloride solution (5%) and dilute to the mark with water. This solution is suitable for determining calcium, magnesium and potassium.

Calibration ranges for standards and brief operating conditions are summarized in Table 12.

Decomposition on a boiling-water bath may be necessary for certain kinds of cement, industrial silica, glasses, quartzite, sand, sandstone, coal ash, feldspars and steelmaking slags and sinters. The decompositions are most conveniently carried out in polythene or polypropylene bottles with screw stoppers.

> Weigh out 0.2 g of finely divided sample into the bottle, moisten with 1 ml of water, add 5 ml of hydrofluoric acid (40%), stopper the bottle and place on a boiling-water bath.[440,441,444] If a clear solution is obtained add 50 ml of 2% boric acid solution,

TABLE 12. Calibration standards for cements

| Element | Wave-length | Flow rates (l min⁻¹) | | Composition of standards | | |
		Acetylene	Oxidant	Conc. range: ppm	Boric acid %	Lanthanum %
Al	309.3	4.0	N₂O 5	0–125	1.0	0.5
Ca	422.7	1.4[a]	air 5	0–200	0.1	0.5
Fe	248.3	1.4	air 5	0–125	1.0	0.5
Mg	285.2	1.4[a]	air 5	0– 12.5	0.1	0.5
Mn	279.5	1.4	air 5	0– 2.5	1.0	0.5
K	766.5	1.0	air 5	0– 2.5	0.1	0.5
Si	251.6	4.3	N₂O 5	0–750	1.0	0.5
Na	589.0	1.0	air 5	0– 12.5	1.0	0.5
Sr	460.7	4.0	N₂O 5	0– 12.5	1.0	0.5
Zn	213.9	1.0	air 5	0– 0.5	1.0	0.5

[a] Sensitivity is reduced by rotating burner or using an emission (short path length) head.

mix and transfer quantitatively to a calibrated flask. Make up to volume after adding an appropriate quantity of lanthanum chloride solution or other buffer required.

A useful variation of this technique was given by Reid et al.[628] for the analysis of steelmaking slags. For blast furnace slags, hydrofluoric acid alone is satisfactory, but sinters, OH and BOS slags need hydrochloric acid as well. The higher pressure caused by this acid may be accommodated by placing the bottles in a domestic pressure cooker which, at its 15 psi (1.05 bar) setting, gives a temperature of about 121 °C. The weighed sample is transferred to the bottle and the solvent acid added. For sinters this is 5 ml of 9 + 1 hydrofluoric/hydrochloric acid and for OH and BOS slags, 5 ml of a 7 + 3 mixture of the same acids. After a suitable time in the pressure cooker, which may vary from $2\frac{1}{2}$ minutes for BF slag to 20 minutes for sinters, the pressure is released, the bottles are rapidly cooled, and an excess (e.g. 100 ml of 1%) boric acid is added. As no solution is lost by evaporation, all additions may be made by pipette, and time is not wasted in making up to volume finally in a calibrated flask. In Reid's method the nitrous oxide acetylene flame was used for all elements (calcium, magnesium, silicon, manganese and aluminium) except iron, and sodium chloride was the only buffer used, to reduce ionization of calcium and magnesium.

If the decomposition results in the formation of a precipitate, the decomposition vessel must be closed again after the addition of the boric acid and heated again as before until a clear solution is obtained.

Materials which may require attack at higher temperatures may be treated in a special pressure dissolution vessel. Such a device would be constructed from stainless steel and contain a PTFE inner vessel or liner, as described by Langmyhr and Paus,[442] Bernas[97] and others. Bernas' device is available commercially, and others are marketed by Parr Instrument Co.[577] in the USA and by S. & J. Juniper (UK).

Bauxite was analysed for silicon, aluminium, iron, calcium, titanium and trace constituents[442] by the general method outlined above, but after decomposition under pressure for 30 minutes. Ferrosilicon required the addition of 1 ml of concentrated nitric acid[443] and some iron ores and slags required the addition of 1 ml of aqua regia.[445] Sulfide minerals and ores,[446] 0.2 g samples, were decomposed with 2.5 ml of hydrofluoric acid and 2 ml each of nitric and hydrochloric acid. The temperature for all these decompositions was $110° \pm 5$ °C. A pressure vessel was operated at 135 °C by Sheridan[678] who digested samples of mineral slag and wool for 40 minutes with 48% hydrofluoric acid plus a little aqua regia, prior to determining silicon, aluminium, magnesium and calcium. Before transferring the digest, the excess of hydrofluoric acid was complexed with boric acid.

If a residue of undecomposed material remains, it may be necessary to repeat the operation, the sample having been ground to a finer powder, with a prolonged heating time at maximum temperature.

It is always essential that the excess hydrofluoric acid should be complexed by addition of boric acid as soon as possible after the decomposition vessel has been cooled and opened. This allows subsequent handling to be done in glassware.

A General Method for Siliceous Materials. Interference problems between elements determined in silicates have frequently been mentioned. The use of reference standards for calibration, or of synthetic solutions with compositions very similar to the samples being analysed, may avoid this problem when the actual composition of the sample is approximately known.

A quite general method of analysis for silicate and similar materials has been developed which can be applied to a very wide range of materials without prior knowledge of the sample composition. Interferences can normally arise both in sample preparation and in the flame, and in order to achieve high analytical accuracy even minor interferences must be avoided. The addition of boric acid to a cool solution of fluorides, for example, may not convert all fluorides to fluoroborates. When lanthanum is added lanthanum fluoride may therefore be precipitated, coprecipitating other elements and leading to low results. Complete conversion of fluorides to fluoroborates takes place if a second heating step is inserted in the procedure after addition of the boric acid.[609,610] Complete conversion to fluoroborate almost completely removes interelement interferences, especially if the nitrous oxide acetylene flame is used, because fluoroborates are readily dissociated at flame temperatures. There is, for example, no depression of either the magnesium or aluminium signal in the presence of up to ~ 3000 ppm of silicon. It is therefore unnecessary to add lanthanum as releasing agent. Either caesium chloride or potassium chloride should be used as ionization buffer, however, depending upon whether or not potassium itself is to be determined. The procedure is as follows:

> To 0.2 g of sample, ground to pass a 200 mesh sieve and transferred to a pressure dissolution vessel, add 5 ml of water, 2 ml of aqua regia and 2.5 ml of hydrofluoric acid (40%). Close the vessel, heat for 30 min at 160 °C, cool, open and quickly add 10 ml of boric acid (4%). Close the vessel and re-heat for 20 min at 160 °C. Cool, transfer quantitatively to a 100 ml calibrated flask. Add 5 ml of either potassium or caesium chloride solution (2%) and dilute to volume. Standard solutions are prepared from the 1000 ppm stock solutions described under individual elements in Chapter 9 and Table 13 gives suggested ranges and final sample dilutions.

This method can readily be miniaturized to handle microsamples of weights in the region of 10 mg:

> Transfer the accurately weighed sample to the pressure vessel, add 2 ml of water, 0.5 ml of nitric acid and 0.5 ml of hydrofluoric acid (40%). Close the vessel. Heat for 15 min at 160 °C, cool, open and quickly add 3 ml of boric acid (4%). Close the vessel, reheat for 10 min at 160 °C, cool, open, and transfer the solution quantitatively to a 10 ml calibrated flask. Add 1 ml of either caesium or potassium chloride solution (2%) and dilute to volume.

Excellent agreement with figures given for a number of chemically analysed reference standards of Portland cement, blast furnace slags, firebrick, etc. has been obtained with both the macro- and micro-versions of this method.

Variability with the micro-version is only slightly less good (e.g. $\sim 1\%$ instead

TABLE 13. Calibration standards for siliceous materials

Element	Conc. range in solid %	Standard solutions ppm	Sample solution dilution, %
Silicon	0–50	0, 25, 50	0.01
	0–5	0, 25, 50	0.10
Calcium	0–50	0. 25, 50	0.01
	0–5	0, 25, 50	0.10
Magnesium	0–5	0, 2.5, 5.0	0.01
Aluminium	0–50	0, 25, 50	0.01
	0–5	0, 25, 50	0.10
Iron	0–50	0, 25, 50	0.01
	0–5	0, 25, 50	0.10
Manganese	0–5	0, 2.5, 5.0	0.01
Titanium	0–5	0, 25, 50	0.10
Sodium	0–5	0, 2.5, 5.0	0.01
Potassium	0–5	0, 2.5, 5.0	0.01

of 0.2–0.5%) and this is almost certainly attributable to the greater difficulty in making a very small sample representative of a larger bulk.

The micro-version is, however, a valuable procedure for quantitative analysis of microsamples, e.g. glass fragments, forensic samples, archeological and mineralogical materials. It can be extended to elements other than the nine listed above (nickel, chromium and zinc for example) and also to other high silicon matrices such as aluminium silicon alloy.

Pressure dissolution methods have also been used as a preliminary to trace element determination by both flame and by electrothermal atomization. Welz dissolved granites in hydrochloric/hydrofluoric acid mixture under pressure at 150 °C prior to measurement of trace elements by electrothermal atomization.[786]

Paus dissolved ores in hydrochloric/nitric/hydrofluoric acid mixture under pressure[580] and, after addition of boric acid and adjustment to pH 3, extracted iron with acetylacetone into chloroform and other trace metals with APDC into chloroform. The pressure dissolution method is a good way of preventing loss of mercury prior to reduction in determination by the cold vapour method.[116] The sample is attacked with sulfuric/hydrofluoric acid mixture and potassium permanganate. When the vessel is opened, the permanganate is decomposed with hydroxylamine chloride and the mercury released in the cold vapour accessory with stannous chloride. Traces of beryllium (0.2 μg g^{-1}) were extracted from the pressure digest at pH 7 with acetylacetone into chloroform.[426]

Gallium can be extracted from acid digests, made up to 6 M in hydrochloric acid, directly into organic solvents.[561,776]

The determination of ultra-trace impurities in high purity inorganic materials can be limited by the 'blank' value of solvent acids used in decomposition, which can often be considerably higher than the value to be determined. Woolley has described an enclosed all PTFE apparatus for decomposition of 100–200 mg amounts of sample by acid vapours generated from acid contained within a different compartment within the apparatus.[832] High purity glasses were powdered and decomposed at 110 °C overnight using a mixture of 6 ml of nitric acid and 4 ml of hydrofluoric acid (50%). After vapour-phase attack the sample remained as a dry powder in the sample compartment, consisting of nitrates of all the elements present, the silicon tetrafluoride formed having been absorbed into the acid mixture when the vessel was cooled. Iron and copper added deliberately to the solvent acid up to 250 ppm were not transferred to the sample, confirming that contamination of the sample by impurities in solvent acids can be avoided by vapour phase attack.

Removal of Silicon. The foregoing methods were evolved with the definite object of retaining silicon quantitatively in solution, but if this element is not to be determined for any reason within a general scheme the methods of attack become somewhat simpler. The silicon is then either volatilized off completely by fuming with hydrofluoric acid and sulfuric or perchloric acid, or is dehydrated with either of these acids and separated by filtration.

This former principle can be applied to the analysis of glass.[37] A sample weight of 0.1 g is moistened with 2–10 ml of water and 1 ml of perchloric acid and 5–20 ml of hydrofluoric acid added. The mixture is digested on a steam bath and then fumed to near dryness. After a second fuming with 1 ml of perchloric acid the residue is dissolved in perchloric acid and made up to volume. A spectroscopic buffer is added to the aliquot used for determining magnesium, calcium, strontium and barium. Both Belt[93] and Thompson[735] analysed silicate rocks by this type of procedure, and performed a statistical analysis of results obtained on two standard rocks. Belt's procedure was to attack 0.5 g of sample with 2 ml of water, 20 ml of hydrofluoric acid and 10 ml of nitric acid. After standing for two hours, 2 ml of perchloric acid were added and the mixture fumed until no more fumes were evolved. The residue was then dissolved in hydrochloric acid and made up to volume. Lanthanum was added to overcome the effect of aluminium on magnesium and calcium in an air acetylene flame, but no buffer was required for sodium, potassium, manganese or iron. Thompson's procedure was similar except that iron was determined in the solution to which lanthanum had been added. Acid and buffer concentrations in samples and standards were carefully matched. Coefficients of variation quoted by both authors for a common sample NBS W–1 are remarkably similar, though Belt's reproducibility figures were obtained on one day, while Thompson's were obtained from repeated recalibrations over a period of some months (see Table 14).

Many slight variations of this method have since been published, but it is perhaps most useful as a preliminary to trace element determination in cases where a full quantitative analysis is not required.

After attack with hydrofluoric/perchloric acid mixture for example, barium was separated by ion exchange on Dowex 50-X8;[241] cadmium (from 40 ng g^{-1})

was extracted with Amberlite LA-2 in xylene after the evaporation residue had been dissolved in hydriodic acid;[681] copper was separated on a Dowex A-1 resin column;[753] and tin was separated with hypophosphite/iodide mixture into benzene.[729]

TABLE 14

	Content % (as metal)	Coeff. of variation % Belt[93]	Thompson[735]
Calcium	8.1	1.6	1.9
Magnesium	4.12	1.9	1.1
Iron	7.6	0.7	2.1
Sodium	1.57	1.9	2.5
Potassium	0.50	5.0	5.2
Manganese	0.136	1.0	1.2

A number of elements were measured by Fuller[255] in lead silicate glasses using two methods of this type. In one, the sample is treated with hydrofluoric acid, evaporated, then evaporated with nitric acid. The residue is redissolved in nitric acid and, after addition of a buffer, iron, copper, nickel and cobalt are extracted with bathophenanthroline into MIBK. Hydrofluoric/sulfuric acid mixture is an alternative solvent: after evaporation to dryness, the residue is taken up in hydrochloric acid, from which solution iron, copper, nickel, cobalt, manganese and chromium are extracted with sodium diethyl dithiocarbamate into MIBK. These impurities, all in the range 0.02–1 μg g^{-1} were then measured by electrothermal atomization.

ORGANIC MATERIALS

Determinations of trace and major concentrations of metals in organic matrices are often some of the easiest analyses to carry out by atomic absorption. In this section types of sample that are not soluble in aqueous media but which may form solutions with other organic liquids, samples from which it may be necessary to remove the organic matrix by oxidation, as well as those which are simply soluble in aqueous media, are discussed. Biochemical samples, too, could be classed in this latter general category, but for convenience are discussed separately.

As organic solvents usually increase the sensitivity of many metals, due to the more highly reducing nature of the resulting flame, and perhaps to the greater rate of vaporization of the solvent, direct nebulization of an organic solution usually results in improved sensitivities. Care must be taken to provide non-aqueous standards which are known to give the same response in the flame as the sample itself.

Apart from such direct methods, the wanted metals can sometimes be extracted from the organic matrix with mineral acids, the matrix itself can be removed by wet or dry oxidation, or small samples can be effectively attacked by the oxygen flask method.[497]

Among the many types of non-aqueous samples analysed by direct procedures are petroleum and petroleum products, materials encountered in the paint, oil and colour industry, plastics, fibres, etc.

Petroleum products

Crude Oils and Feedstocks. The determination of trace metals in crude oils is necessary before such oils are refined and submitted to catalytic 'cracking', because some metals, in particular nickel, copper, iron, sodium and vanadium, may poison the catalyst and reduce the efficiency of the plant. These metals are often present, at least in part, as volatile organometallic compounds which accompany the hydrocarbons during fractionation.

The sources of oils, too, may often be identified by the relative concentrations of some trace elements; rapid determination is therefore useful not only to oil companies but also to those engaged in forensic and environmental pollution investigations.

Barras[82] diluted gas oil feedstocks $1+2$ with *n*-heptane, comparing with standards made by dissolving NBS organometallic compounds in the same solvent. Molecular absorption was corrected by measurement at a non-absorbing wavelength.

Trent and Slavin[750] used the method of standard additions, adding increasing amounts of nickel cyclohexane butyrate, dissolved in 'pure' lubricating oil and diluted in *p*-xylene, to thin oil samples also diluted with xylene. An air acetylene flame with auxiliary air was used. Kerber[386] used a $1+4$ dilution in *p*-xylene, constructing a calibration curve from synthetic standards made from the same materials, and correcting for background absorption. A detection limit of 0.05 ppm of nickel was obtained.

Moore *et al.*[530] found that dioxane was a better solvent for crude and fuel oils from the point of view of solvent power and burner characteristics. They determined copper and nickel using the standard addition method.

Vanadium was determined by Capachi-Delgado and Manning[128] with a nitrous oxide acetylene flame. With a $1+2$ dilution in *p*-xylene, a detection limit of $0.05 \mu g \ ml^{-1}$ was obtained. Standards were made by dissolving bis-(1-phenyl-1,3-butane diono)oxovanadium (IV) in oil, and diluting with the same solvent. This compound has become widely used as a standard for vanadium, although it has been criticized because it may give a different response from vanadyl tetraphenyl porphyrin, which typifies the form of vanadium present in natural petroleum.[672] Different organo-nickel compounds dissolved in xylene are also said to give different responses even in the nitrous oxide acetylene flame.[435]

In the analysis of crude oil, the sampling procedure itself requires some care.

Before the analytical samples are measured out, the bulk sample should be thoroughly shaken, and heavy crudes warmed to 60 °C in a water bath to improve mixing.

A tentative procedure for measuring sodium and nickel at levels above 1 μg g^{-1} and vanadium above 10 μg g^{-1} in crude oils and residual fuel oils has now been recommended by the Institute of Petroleum.[350] The basis of this is as follows:

> 0.25 g of sample (warmed to 50°–60 °C if viscous) is weighed accurately into a 25 ml calibrated flask. A little solvent mixture (10% isopropyl alcohol, 90% white spirit) is added to dissolve the sample with agitation, and the volume is then made up to 25 ml.
>
> Standards are prepared from sodium tetraphenyl boron, nickel 4-cyclohexylbutyrate and bis-(1-phenyl 1,3-butanediono)oxovanadium using the same solvent mixture. For sodium and nickel, an air acetylene flame is used, and for vanadium, nitrous oxide acetylene. The line 352.5 nm is recommended for nickel.

Lubricating Oils and Greases. A tentative method has also been published[351] for barium, calcium, magnesium and zinc in unused lubricating oils which are free from additives containing aluminium. The samples are again simply diluted in an organic solvent, but an ionization buffer has to be added.

> 0.5 g of sample (or sufficient sample to contain not more than 100 μg ml^{-1} of the prepared solution) is weighed accurately into a 25 ml calibrated flask, and 10 ml of ionization buffer (2500 mg of sodium or potassium as sulfonate per litre of white spirit) is added. Further white spirit is added to dissolve the sample, and the volume then made up to 25 ml. Standards are prepared from suitable organometallic compounds of the above four elements, dissolved in white spirit and containing the same concentration of ionization buffer ($\equiv 1$ mg ml^{-1} of sodium or potassium). Subsequent dilutions of the prepared sample solution should always contain 1 mg ml^{-1} of sodium or potassium. Each element is determined with an air acetylene flame.

Mixed solvent systems have been proposed to enable aqueous stock standard solutions to be incorporated in working standards made from the same solvents that are used to dilute samples in quality control. A solvent mixture consisting of cyclohexanone: butan-l-ol: ethanol: hydrochloric acid: water (10: 6: 4: 1: 1) was successfully used for measuring calcium and zinc,[326] these elements being introduced into the standards as chlorides in the aqueous components. A simpler solvent, 2-methylpropan-2-ol: toluene $(3 + 2)$ was subsequently used for barium either alone or in presence of calcium.[327] This had good burning characteristics in the nitrous oxide acetylene flame, improving both accuracy and sensitivity. It was thought that this solvent could also be used for calcium and zinc.

Calibration standards of lubricating oil containing 500 ppm of various metals are prepared from 4-cyclohexylbutyrates (or other organometallic compounds given in Chapter 9) by the following method recommended by Hopkin & Williams Ltd:

> Transfer to a weighed 200 ml flask an amount of cyclohexylbutyrate containing 50 mg of the metal, add the amounts of xylene and 2-ethylhexanoic acid specified in Table 15 and heat on a hotplate (avoid charring) with swirling until a clear solution or gel forms.

Add 6-methylheptane-2,4-dione or 2-ethylhexylamine (if specified in the table) and continue heating and swirling until a clear solution is obtained. To the hot solution add 80 to 90 ml of lubricating oil and gently shake the flask to mix the contents. Allow to cool to room temperature, dilute to 100 g with lubricating oil and mix.

Greases may be treated by one of two methods: dissolution in a suitable solvent, or, if the elements to be determined are at too low a concentration level, dry-ashing followed by dissolution of the residue in hydrochloric acid. Mostyn and Cunningham[537] used the latter method even for determining the molybdenum sulfide content of lubricating greases, finding it to be both more rapid

TABLE 15. Solubilizers for amounts of metal standards equivalent to 50 mg of the respective metals

Metal	Xylene ml	2-Ethylhexanoic acid ml	6-Methylheptane 2, 4-dione ml	2-Ethylhexylamine ml
Barium	3	5		
Cadmium	2	2[a]		4
Calcium	3	5		
Chromium	3	3		
Copper	2	2[a]		4
Iron	3	4	3	
Lead	3	5		
Magnesium	3	5		
Manganese	3	5		
Nickel	3	5		
Silver	2	2[a]		4
Tin	5			
Vanadium	3	1[a]	3[b]	
Zinc	3	5		

[a] Add just before adding the lubricating oil.
[b] Plus 3 ml of bis(2-ethylhexyl)amine.

and more convenient. Ashing would have to be carried out at a carefully controlled temperature to avoid loss of elements by volatilization. In particular, if molybdenum is being determined, the temperature should not exceed 600 °C. The ash was extracted with ammonia and the samples and standards were made up to contain equal concentrations of ammonium chloride.

Wear Metals in Lubricating Oils. An increase in the metal content of crank-case and circulated lubricating oils signifies a source of potential trouble in an engine. The actual metal may indicate the location of the trouble. For example, the presence of lead, tin or antimony might mean wear in a bearing; iron, chromium or nickel would be present because of piston wear; while contamination by sodium or boron could arise through a leak in the cooling system. Periodical analysis of the lubricants enables such troubles to be located and corrected before they cause a major breakdown.

Wear metals are different from naturally occurring trace metals, in that they are probably present in the form of fine metallic particles, or even of a colloidal suspension. They may therefore be expected to give a different response from the same metals present in the form of soluble organic salts as in oils with additives or in working standards. Experience has shown, however, that the particle size is frequently so small that, with hot flames, the metal is completely atomized and hence good results can be obtained in relation to synthetic standards made up from organometallics.

Means and Ratcliffe[506] successfully determined iron, silver, copper, magnesium, chromium, tin, lead and nickel in jet engine lubricants. After a $1+4$ dilution with p-xylene the oil was aspirated into an air acetylene flame. Standards were of metal naphthenates dissolved in 'unworn' oil, also diluted with p-xylene. Methods for preparing metal naphthenates were given. Burrows[122] in a similar method used methyl isobutyl ketone as the solvent for railway engine oils and Sprague and Slavin[701] described a rapid method for aircraft engine oils. By utilizing a composite standard solution for seven elements, iron, nickel, chromium, lead, copper, silver and magnesium and a digital concentration readout system, a precision of 5% was claimed for an analysis rate of fourteen seconds per element per sample. A mixed solvent containing 85% methyl hexyl ketone, 13.5% ethanol, 0.5% water and 1% hydrochloric acid has been recommended by Russian workers.[768]

One of the particular problems in analysing this type of oil is again in taking a truly representative sample. The metallic particles gradually separate on standing, and settle to the bottom of the container as a sludge. The main sample should therefore be taken from the engine quickly after shut-down, and should be rewarmed and thoroughly mixed again before the working samples are measured out. A dilution of $1+4$ with 10% isopropanol/90% white spirit mixture is a satisfactory procedure, and standards are made by dissolving organometallic compounds in the same solvent. It is probably unnecessary to add unused oil to the standards for the viscosity difference is negligible, and in any case the important feature of this analysis is that it enables concentration *trends* to be quickly recognized over specified periods of time. Provided the figures are correct relatively, an actual error of 5% or even more is usually unimportant.

A General Method for Fuel, Lubricating and Worn Oils Using Dry-Ashing. There is always a possibility that errors will be caused by differences in response between the forms of metal in the sample and standard when simple dilution methods are used; e.g. different organometallic complexes, or particulate and dissolved forms. This applies particularly to the non-routine situation where the constitution and metal content of the oil may not be known before the analysis.

An alternative approach free from the possibility of such errors arises from the comprehensive report by Gorsuch[289] on the recovery of trace elements after wet and dry ashings. It has been shown[767] that all the wanted metallic elements are retained after dry ashing at 500 °C provided a suitable ashing 'aid' such as *p*-toluene sulfonic acid is employed. The metals are extracted from the residual ash with nitric acid, and caesium chloride is used as ionization buffer. Details of the method (as modified slightly in the author's laboratory where some twenty elements have been measured) are as follows:

> Ash a suitable weight of sample (10–50 g) in a platinum dish at 500 °C in a muffle furnace with 0.5 g *p*-toluene sulfonic acid per 10 g of sample. When ashing is complete (20–30 min) warm the residue gently with 10 ml of nitric acid (25%), stirring and crush-in the ash with a glass or PTFE rod. Filter the slurry into a 50 ml calibrated flask, wash with a further 10 ml of nitric acid (25%), add 5 ml of caesium chloride solution (1%) and dilute to the mark.
>
> Calibration solutions are prepared from stock standards (see Chapter 9) to contain the same concentrations of nitric acid (5%) and caesium chloride (0.1%) as the sample solutions.

Further dilutions of sample solutions should also contain the same concentrations of nitric acid and caesium chloride.

Analysis of Fuels. Trace elements in both fuel oils and gasolines often require to be determined, particularly lead, copper and zinc, as they accelerate the oxidative deterioration of refined products or otherwise reduce their storage stability, and vanadium which, when present in oil destined for boiler firing, leads to corrosion and produces a fine toxic ash.

Scott and Killer[670] used methyl isobutyl ketone as the solvent for direct determination of vanadium in fuel oil, diluting the sample $1 + 9$. They suggested that the dilution technique provides sufficient sensitivity for trace elements at concentrations down to about 1 ppm. Below this they recommended that the trace metals be separated and concentrated by wet-ashing, chromatography or extraction. Acid extraction would appear to be a useful technique, as 90% or greater recoveries of vanadium, nickel and iron were obtained when 250 ml of artificial blends were shaken with 10 ml of 2 M hydrochloric acid.

The petroleum analysis causing the greatest problems in atomic absorption is the determination of lead, added in the form of tetraethyl lead (TEL) or tetramethyl lead (TML) as an anti-knock. Its concentration is usually between 200 ppm and 1500 ppm and it may be present as either TEL or TML or as a mixture of both. Standards produced by dissolving TEL in gasoline generally give a calibration graph of different slope from that produced with TML. It must therefore be known which compound is present, but if both are suspected a correct result is unlikely to be obtained. Mostyn and Cunningham[537] confirmed that this problem persists, though, in limiting their work to materials containing only TEL, they obtained some very satisfactory results in spite of a 'memory' effect in which TEL standards take more than 90 s to reach a maximum reading. The problem was overcome by using a carefully timed operating sequence of 15 s aspiration of the gasoline sample or standard, followed by $1\frac{1}{2}$ min interval wash of pure iso-octane. A coefficient of variation of 1.6% was obtained at a concentration level of 3.5 ml of TEL per imperial gallon of gasoline. 2 ml of sample are diluted to 100 ml with iso-octane. Standards are prepared from TEL and iso-octane.

The only entirely satisfactory answer to this problem, enabling each of the various lead alkyls present to be determined, involves a gas chromatographic separation, using atomic absorption as a lead-specific detector. Such methods have been described, the lead alkyls being decomposed to form atomic lead either in a conventional electrothermal atomizer[673] or in a special silica tube furnace in the resonance beam of the atomic absorption spectrophotometer.[143] Ebdon[207] has also coupled a gas chromatograph outlet directly to the burner of an atomic absorption spectrophotometer which was supporting an air hydrogen diffusion flame. Accuracy in the determination of lead tetramethyl and lead tetraethyl was shown to be superior to that obtained by GC using an electron capture detector, and sensitivity was adequate.

Total lead in gasoline may be determined after a simple and rapid extraction with iodine monochloride.[360] Lead is removed from both alkyls, the final solution is aqueous, and comparison standards consist simply of dilute solutions of lead nitrate. The method is as follows:

> Pipette 5 ml of gasoline into a 100 ml separating funnel containing 10 ml of 1.0 M iodine monochloride. Add 25 ml of iso-octane, stopper the flask and shake for 3 min. Transfer the lower, aqueous, layer to a 100 ml calibrated flask. Wash the remaining organic layer with 10 ml of water by shaking for a further 1 min and add the washings to the flask. After diluting to volume with water, aspirate into an air acetylene flame. Compare the absorbance with that given by aqueous standards of lead nitrate containing 5–50 ppm of lead. Good sensitivity is obtained by using the lead resonance line at 217.0 nm.

Iodine monochloride is readily prepared for this determination:

> Add 112 ml of 25% potassium iodide solution to 112 ml of hydrochloric acid sp. gr. 1.18, in a 500 ml beaker, and cool to room temperature. To the cooled solution slowly add 18.75 g of potassium iodate crystals and mix until a clear orange coloured solution is obtained. Dilute to 250 ml with water.

It is reported that lead, present as TEL, TML, or mixed alkyl leads, can be determined in a premixed air acetylene flame against a single alkyl lead standard if 3 mg of iodine are added to the 1 ml petrol sample.[383] This is then diluted to 50 ml with methyl isobutyl ketone. It was stated that there are no interferences and the method is suitable for determining from 0.16 to 0.8 g of lead per litre of gasoline. In the ASTM procedure[67] also iodine is added to overcome response differences between the various lead alkyls.

Electrothermal Atomization of Petroleum Products. Electrothermal atomization methods for the measurement of trace elements have been applied to most types of material dealt with in the petroleum industry. In many cases it is only necessary to dilute the sample with a suitable solvent, e.g. xylene or MIBK in order to ensure that a suitable amount can be accurately injected into the atomizer. Light oils and gasolines can be injected directly. The addition of magnesium sulfonate, prior to injection, has been recommended[218] to prevent loss of other metals during pyrolysis.

Phosphorus may also be determined in gasoline and petroleum products, but has been found to be lost during the extraction or digestion of samples.[200] This is overcome by injecting the sample on top of a 5 % solution of lanthanum nitrate in ethanol already placed in the furnace. Tricresyl phosphate additives can be converted to magnesium pyrophosphate before dissolution and injection on top of 1 % lanthanum nitrate solution.[707]

For some elements a prior wet ash may be necessary, e.g. for vanadium to improve sensitivity, and for elements such as antimony and selenium to avoid vaporization losses before atomization.

Before embarking on such work using micropipettes, the analyst is referred to the section on p. 181 on the micropipetting of non-aqueous solvents.

Paints, oils and colours

Metallic elements occur as principal constituents in inorganic pigments and in additives to drying oils, but may be present in smaller amounts in pesticides and antifouling compounds. Trace elements may be undesirable because of their toxic properties or because they affect the colour or stability of the product.

Regulations are laid down in some countries on the toxic metal content of paint for toys and domestic appliances. In the United Kingdom for example, the 'soluble' lead content of a dry coating of paint on a toy must not exceed 5000 ppm and arsenic, antimony, barium, chromium and cadmium must each not exceed 250 ppm. The soluble metal content is that which can be extracted in a 0.25 % solution of hydrogen chloride under specified conditions, which approximates to the dissolving power of human gastric juices.

The problems encountered in analysis of paints[603] are not unlike those in the petroleum industry, except that the vehicle may consist of natural or synthetic polymerizable fluids and the pigment is a suspended solid of high concentration. Many of the solvents used in the analysis of petroleum products, viz. methyl isobutyl ketone, n-heptane, iso-octane or white spirit, can equally well be used to dissolve paint vehicles. White spirit (which is also used as a thinner in the paint industry) is preferred for reasons already given in the section on petroleum.

Direct Methods. Oils or thinners to be analysed for trace elements may be diluted with white spirit by a factor of ten, and compared with standards of metal naphthenates also dissolved in white spirit.

Drying oils themselves often contain metal naphthenates, especially those of cobalt, lead, manganese, zinc or zirconium, and these metals are all determinable by a similar method, though a greater dilution factor would be required.

Fuller recommends direct dissolution in MIBK for determining cobalt, magnesium and lead in paints and driers by electrothermal atomization.[256]

Pretreatment Methods. To conform to the official specification,[702] 'soluble' lead, arsenic etc. are determined in the aqueous phase after a dried and weighed quantity of the sample is shaken for one hour at 16 °C with 1000 times its weight of hydrochloric acid solution containing 0.25 % by weight of hydrogen chloride.

Standards would be made up to contain the same concentration of hydrochloric acid.

Metals can be extracted from paints with strong acid under pressure. Nitric acid at a temperature of 145° or 150 °C was used for alkyd and latex paints[596] and other materials.[325] Others recommend dry ashing at temperatures around 500 °C. The ash may then be extracted with nitric acid and ammonium acetate for flame determination of lead and cadmium,[215] or fused with carbonate borax mixture and re-extracted for complete analysis of automobile paints,[201] or dissolved in sulfuric acid and potassium permanganate in a pressure vessel for chromium determination.[552]

The Delves microsampling cup has been used for determining lead and cadmium. Scrapes of pencil paint may be transferred directly to the cup[522] or the paint suspended in water or MIBK and dried in the cup[314] for determining lead. Two procedures have been given[449] for lead and cadmium: suspension of the paint in water or toluene dimethylformamide mixture followed by transfer of a 10 μl aliquot to the Delves cup; and preliminary ashing of a 10 mg sample at 500 °C and transfer to the flame. Calibration is by the method of standard additions.

Pigments may be separated to enable either the pigment itself or the vehicle to be analysed, by dissolution of the vehicle in an organic solvent followed by centrifugation. Inorganic pigments are then dissolved in a suitable acid mixture, or if necessary first fused with an alkali salt, and then leached out with acid. The vehicle is analysed by a direct method as described above unless the metal concentrations are too low to be determined directly.

Chromous oxide has been measured in pigments from works of art[94] by fusion with sodium peroxide, sodium carbonate and potassium carbonate mixture and extraction with hydrochloric acid.

Toxic lead, chromium and barium were determined in magazine colours by drying and grinding the sample and extracting with 0.1 M hydrochloric acid at 37.5 °C for 4 h.[518] Potassium chloride was then added before the elements were measured in a nitrous oxide acetylene flame.

A useful means for bringing organic materials into aqueous solution in readiness for atomic absorption analysis is the oxygen flask method[497] which has been applied to the analysis of commercial plastics. More usually, however, the organic constituents of paints can be analysed for very low concentrations of trace metals by a preliminary wet oxidation step. Powder paints, too, with organometallic constituents, and antifouling paints, which contain compounds of copper and mercury are best treated by wet oxidation. Phenyl mercury compounds are volatile, however, and if these are present the procedure must be carried out carefully to avoid loss of mercury. Suitable wet oxidation methods for organic materials with or without mercury were given by the Society for Analytical Chemistry.[365]

With the analyte in aqueous solution it is sometimes desirable to separate the trace elements, either from an excess of alkali fusion salt or from a major pigment element. One of the separation techniques described in an earlier section may then be used. Probably the ammonium pyrrolidine dithiocarbamate–methyl isobutyl ketone system is the most useful.

Major Constituents. Separated inorganic pigments can in principle be checked for major or impurity elements. Many pigments dissolve readily in one or other of the mineral acids. Titanium dioxide may cause some difficulties, however, and is dissolved in sulfuric acid after fusion with potassium bisulfate and glucose.

The main pigment elements include aluminium, antimony, arsenic, barium, calcium, cadmium, copper, chromium, mercury, iron, lead, magnesium, manganese, sodium, potassium, silicon, titanium, zinc and zirconium. The sensitivity of many of these may need to be reduced by burner rotation or measurement of an alternative resonance line. Operating conditions may be inferred from the details given in Chapter 9.

Textiles, fibres and plastics

Both natural and synthetic fibres have been analysed for trace elements, but synthetic fibres may also contain residues of catalysts, stabilizing agents, etc.

Aluminium was determined in wool by Hartley and Inglis[307] after dissolving the sample in constant boiling mixture hydrochloric acid. Although hydrochloric acid tends to depress the response of aluminium in a nitrous oxide acetylene flame, the presence of wool protein aminoacids enhances it. Standards were therefore prepared to contain the appropriate acid concentration and hydrolysed wool protein. Hair and hide could also be analysed by the same method.

Some elements, e.g. copper,[684] are readily determined in textile materials after simple acid extraction. Synthetic fibres may need to be ashed and the residue dissolved in hydrochloric acid. The resulting solution can be analysed directly, or an extraction can be carried out.

Trace metals in synthetic fibres and polymers remaining from previous processing may affect, either adversely or favourably, their stability and other characteristics. Olivier[562] evolved methods for the examination of four types of material: soluble polymers; insoluble polymers; wool, cotton and cellulose; polymers containing volatile metallic compounds. Polymers (0.5 g sample) are dissolved, where possible, in an organic solvent. Appropriate solvents are given in Table 16. After warming if necessary the solution is made up to 25 ml with the solvent and aspirated direct.

It is important to ensure that, if organic solvents are to be aspirated directly, they will not dissolve or otherwise damage parts of the sample-handling section of the atomic absorption spectrometer that are themselves constructed from plastic or polymerized materials.

Polymers which do not dissolve in an organic solvent must be wet-ashed. The sample is heated with 2–3 ml of concentrated sulfuric acid, and hydrogen peroxide added dropwise until the organic matter has been destroyed. The solution is then made up to the required volume with water.

Wool, cotton and cellulose fibres are hydrolysed, again taking 0.5 g samples. Wool is treated with 15 ml of 5% sodium hydroxide solution, 'dissolution' being complete in about 30 min. If the sodium is likely to interfere, hydrochloric acid may be used instead, but this takes longer. Cotton and cellulose are shaken for 30 min with 6–7 ml of 72% sulfuric acid. Finally, in all cases the solution is made up to 25 ml with water.

Volatile metal compounds are determined after the polymer has been dissolved in any possible solvent. If the solvent is unsuitable for both flame and electro-thermal atomic absorption analysis (organic halides, cresols—see p. 129—concentrated acids, etc.) a double precipitation into aqueous medium is carried out. The organic solution is dropped slowly into water with vigorous stirring,

TABLE 16. Organic solvents for polymers etc.

Polymer	Solvent
Polystyrene	Methyl isobutyl ketone
Cellulose acetate	Methyl isobutyl ketone
Cellulose acetate butyrate	Methyl isobutyl ketone
Polyacrylonitrile	Dimethyl formamide
Polycarbonates	Dimethyl acetamide
Polyvinyl chloride	Dimethyl acetamide
PVC/PV acetate	Cyclohexanone
Polyamides (nylon)	Formic acid
Polyethers	Methanol
Wool	5% Sodium hydroxide
Cotton and Cellulose	72% Sulfuric acid

and after the second such precipitation the metals are transferred quantitatively to the aqueous medium. For lead the aqueous medium should be dilute ammonium acetate, for silver dilute ammonia, and for antimony dilute hydrochloric acid. A further concentration of most metals can then be made with the ammonium pyrrolidine dithiocarbamate–methyl isobutyl ketone system, or with cupferron for aluminium extraction.

Plastic materials may be treated either by the oxygen flask method of Matsuo[497] or ashed.

The following wet-ashing procedure has been found to be suitable for determining calcium and magnesium in polyvinyl chloride:

To 1 g of sample add 20 ml of perchloric acid. Heat gently and then evaporate to fumes. Add 20 ml of 5% lanthanum chloride solution, filter, and make up to 100 ml. Compare with standards containing 0–10 ppm of calcium and magnesium and 1% lanthanum chloride, aspirating into an air acetylene flame.

If any residue remains on the filter paper, ash the paper and fuse with 5 g of potassium hydroxide. Extract the fusion with 5 ml of hydrochloric acid, add 20 ml of 5% lanthanum chloride solution and make up to 100 ml.

If calcium or magnesium are present, compare with standards containing 0–10 ppm of calcium and magnesium, 1% of lanthanum chloride and 5% of potassium chloride.

Lanthanum chloride is added when an air acetylene flame is used, as there is a possibility that titania or alumina fillers may pass into solution to a sufficient extent to cause some depression of the calcium absorption response.

A number of applications of electrothermal atomization to the measurement of trace elements have been reported, and in most of these the solid sample has been used directly. Kerber *et al.*[387] used this technique for determining copper, iron, manganese and silicon in paper, and aluminium, copper and iron in plastics. Henn[315] also used solid samples directly, transferring them to the furnace with a tantalum spoon, or alternatively extracted the polymer with 0.16 M hydrochloric acid, injecting 50 μl aliquots to determine chromium, copper and iron in the range from 0.01 μg g^{-1} upwards.

The concentration of lead in plastic containers for pharmaceutical products has been checked by direct electrothermal atomization.[281] A washed (but not dried) sample (1–3 mg) cut from the container was analysed for lead and 'bracketed' with a suitable standard. The ashing time required varied with the type of plastic material and the ashing step usually had to be carried out in two or even three stages. Examples are:

polyethylene, 30 s at 650 °C, then 30 s at 800 °C;
polypropylene, 120 s at 600 °C, then 60 s at 800 °C;
polyvinylchloride, 60 s at 380 °C, 60 s at 620 °C, then 600 s at 800 °C;
polystyrene, 60 s at 800 °C.

Atomization was generally between 1600 °C and 1800 °C.

Organometallic compounds

The analysis of organometallic compounds may include the determination of the metallic components by atomic absorption. For independent analyses the compounds are wet-ashed and the resulting aqueous solutions are compared with synthetic aqueous standards.

A typical procedure suitable for the determination of arsenic in triphenyl arsine and phenyl arsonic acid, and for the determination of barium, calcium, cadmium, cobalt, manganese, lead and zinc in organometallics is:

Dissolve 0.5 g samples in 20 ml of 1+1 nitric perchloric acid mixture. Digest gently for four hours and dilute to 100 ml with water. Make a further dilution, e.g. 1+19 for the arsenic compounds mentioned and compare with standards containing 10–200 ppm of arsenic or barium or 1–20 ppm of other elements in perchloric acid of the same concentration as the prepared sample solution.

This procedure is not suitable for the determination of tin as metastannic acid is precipitated. To determine tin in, for example, dipropyltin dichloride or hexabutylditin:

Warm carefully 0.1 g samples in 15 ml of 1+1+1 nitric/hydrochloric/phosphoric acid mixture under a reflux condenser. When dissolution is complete, dilute to 100 ml with water. Compare with standards containing 10–200 ppm of tin in an acid mixture of the same final concentration.

Analyses of materials containing known stable organometallic compounds may be done by a direct procedure using standards carefully made up from the compound itself. The method can be used, for example, in the control of the composition of silicone fluid mixtures and blends. Samples may be diluted in ethanol or a suitable ethanol water mixture, while standards are made up by dissolving the silicone fluid in the same solvent.[202]

In the context of organometallic compounds, flame atomic absorption is a useful detector in high speed liquid chromatography.[367] The example given is separation of chromium complexes of acetylacetone, 2′-hydroxyacetophenone and hexafluoro-acetylacetone on a Porasil-A column, the solvent being a 0.5% solution of pyridine in toluene.

Pharmaceuticals and toiletries

Both major and unwanted trace elements have to be controlled in pharmaceutical products, the individual components or raw materials. The major pharmacopoeias include some atomic absorption methods.

Direct analysis is possible if the material is soluble in aqueous or non-aqueous media, and standards are made up in the same solvent and with the same major constituents present. Vitamin tablets have been ground and dissolved in 10% hydrochloric acid for measuring copper and manganese[139] and aluminium was determined in pharmaceutical products simply by extracting with 4 M hydrochloric acid, filtering and diluting.[389]

Concentration of trace elements, if necessary, is achieved by the methods already described, solvent extraction, dry or wet oxidation, the latter always being relied upon if an entirely independent method is needed. Lead, arsenic, mercury, copper, iron and zinc may have strictly enforced upper concentration limits, and these elements may be introduced in the raw materials or through contamination from the equipment used in the preparative processes. Lead, arsenic and mercury do not have good sensitivities and the special methods given for these elements usually have to be used. Leaton[452] described procedures for the determination of metals in drugs, dealing with liquids, suspensions, tablets, powders and ointments. Prepared samples were diluted with either 2% nitric acid or 1% lanthanum chloride solution.

The determination of zinc in crystalline zinc insulin and its preparations was described by Spielholtz and Toralballa.[697] 5 mg of sample is suspended in 10 ml of water and dissolved by the addition of one drop of hydrochloric acid. This solution is then made up to 50 ml. Alternatively 1 ml of protamine insulin solution is diluted to 50 ml, after acidifying with 1 drop of hydrochloric acid if necessary. Standards in the range 0.2–3.0 ppm of zinc were prepared simply by diluting a stock zinc solution with water. The addition of albumin had no effect on the zinc absorbance. Zinc insulin is used in therapeutic medicine, but zinc and other heavy metals which would be determinable by a similar method are used in X-ray studies of the structure of insulin.

Vitamin B_{12} (cyanocobalamin) in pharmaceutical dosage forms is determined indirectly by its cobalt content.[96] About 20 tablets are dissolved in hot water, the solution filtered and cobalt determined in the filtrate. Comparison should be made with a standard solution of cyanocobalamin.

The determination of aluminium, bismuth, calcium and magnesium in antacid preparations presents an almost classical example of stable compound interference. Aluminium itself must be determined in a nitrous oxide acetylene flame, and bismuth in a lean air acetylene flame. The interference of aluminium on magnesium and calcium in the air acetylene flame is completely overcome by addition of lanthanum. The procedure is as follows:

> Dissolve 0.5000 g of powdered sample in 10 ml of water, 5 ml of hydrochloric acid and 5 ml of nitric acid. Digest on a hotplate until dissolution is complete. Transfer to a 100 ml flask, washing with 20 ml of M hydrochloric acid and dilute to the mark. Dilute this solution further as indicated in Table 17 incorporating lanthanum chloride solution in the final analysis solution to give a concentration of 1% lanthanum where appropriate in both samples and standards.

TABLE 17. Standards and sample dilutions for antacid analysis

	Further dilution of sample solution	Standard range	Lanthanum conc. in final soln.	Equiv. % in sample
Aluminium	5×	0–100 mg l^{-1}	—	0–10% Al
Bismuth	—	0– 80 mg l^{-1}	—	0– 1.6% Bi
Calcium	50×	0– 25 mg l^{-1}	1%	0–25% Ca
Magnesium	500×	0– 2 mg l^{-1}	1%	0–20% Mg

The further dilutions of the first solution are necessary to accommodate the different sensitivities of these four elements. Silicon is another element which is present in some preparations in major concentrations, either as 'tri-silicate' or even as silicone oil. In the latter case, unfortunately, homogenization with ethanol-based diluents (as mentioned for silicone oil blends, p. 246) does not appear to be successful, and therefore preliminary cold wet oxidation (avoiding loss of silicone) must be applied before taking up in a hydrofluoric acid-based solvent. Such a solvent is also used, in methods similar to those described for siliceous materials, for trisilicate materials.

A feature of many modern drugs is that they form extractable complexes with particular metals. This is the basis of a number of indirect methods of drug analysis by atomic absorption. These are therefore mentioned here rather than in the section on Indirect Methods on p. 278.

Copper complexes are formed and measured for the determination of isonicotyl hydrazine,[390] flufenamic acid,[517] chlorpheniramine maleate,[393] barbituric acid derivatives[526] and *p*-aminobenzoic acid.[394] Benzyl penicillin is extractable as its cadmium complex,[392] brucine and strychnine as molybdenum complexes,[683] and noscapine is measured by the chromium response of its complex with Reinecke salt.[516] Disodium edetate was measured as its nickel complex in streptomycin[343] while silicon could be determined directly in the same material.[342]

Very little work seems to have been reported from the pharmaceutcal field on electrothermal atomization of drugs, etc., though this should be a very appropriate use of the technique.

THE ANALYSIS OF BIOLOGICAL SAMPLES

In this section the 'biological' label is intended to cover all materials—organisms, plants, animal products—that have been formed as a result of, or are necessary to support, various forms of life. Medical and pathological applications of atomic absorption are the subject of the next section. The book by Christian and Feldman[8] is an excellent guide to the literature relevant to the total biological field up to about mid-1969.

The major problem in the analysis of biological samples, whatever their origin, is the variety of forms in which the organic matrix may exist. This may be overcome in a widely applicable method in which all biological samples are either dry- or wet-oxidized and the residue taken up in some mineral acid. After addition of a spectroscopic buffer, e.g. lanthanum chloride if an air acetylene flame is to be used, potassium or lithium chloride if nitrous oxide acetylene, the solution would be made up to volume and compared with standards containing the same anions and the same buffer. This is the basis of very many methods and where an unusual analysis has to be carried out, without preliminary research, the principle can be relied upon to give accurate results.

More direct methods, e.g. dissolution of the sample where possible, acid extraction of the wanted metals, are less time-consuming but have to be proved to give correct results for each different matrix, and may not give adequate sensitivity for trace elements without further operations.

We consider biological samples according to the type of matrix for, in accordance with principles already established, it is invariably possible to determine more than one element, if desired, in a given prepared sample.

Liquid samples—waters, juices, beverages

Table and mineral waters would be analysed by methods already given for natural waters, except that it may be necessary to remove dissolved gases such as carbon dioxide as these may interfere with the regular flow of sample through the nebulizer capillary.

Fruit Juices. These are analysed for trace elements absorbed from the environment or for elements such as iron and tin which derive from the cans in which juices are stored and distributed. The problem of iron and tin was studied by Price and Roos[607] who showed that, for orange and pineapple juices, correct results are obtained by acidifying the juices and separating suspended matter by centrifugation. The method was extended[643] to other metals in juices: calcium, magnesium and potassium—which are present in relatively major concentrations—and manganese, zinc and sodium. A 5× dilution of the sample is necessary to overcome the viscosity effects caused by the presence of sugar, while hydrochloric acid both extracts metals from suspended solid material and also standardizes the acidity of the prepared solution. The procedure is as follows:

To 20 ml of the juice in a 100 ml volumetric flask add 10 ml of hydrochloric acid, then make up to the mark with water. Shake well, transfer sufficient quantity to a dry centrifuge tube and centrifuge. This solution is aspirated for iron and tin, but further dilutions as shown in Table 18 may be needed for the other elements. Make dilutions so that the final solution always contains 5% of hydrochloric acid, and the solution for calcium contains 0.5% of lanthanum as lanthanum chloride. Prepare standards containing the same acid and lanthanum concentrations as the samples. For tin use the nitrous oxide acetylene flame, and for all the other elements air acetylene.

TABLE 18. Preparation of fruit juices

	Approx. level ppm	Dilution	Standard range ppm
Tin	0–100	none or 5×	0–50
Iron	0– 20	none or 5×	0–10
Manganese	5	5×	0– 5
Zinc	5	5×	0– 5
Calcium	100	10×	0–25
Sodium	20	10×	0– 2.5
Magnesium	200	200×	0– 5
Potassium	1000	200×	0–20

The method was tested both by subjecting a number of samples also to wet oxidation, comparing with 'pure' aqueous standards, and by determining recoveries of known amounts of each element added to the samples. Interferences were detected in the air acetylene flame between iron and the organic constituents of the juices, citric acid and sugar. In test experiments this was completely overcome by the addition of 60 ppm of phosphoric acid, less than is naturally present in the juices. There is thus no need to make special additions to the juices or standards to overcome this effect. None of the other elements is affected in this way and there is no interference with tin in the nitrous oxide acetylene flame. Lanthanum is added to prevent depression of the calcium absorbance by phosphate.

Arsenic has been determined in fruit juices and soft drinks by direct hydride generation using zinc and sulfuric acid as the reductant.[620]

Wines and Beers. The significance of traces of some metals in wines and beers is well recognized. The stability of beers, and also the quality of the foam, is associated with the presence of some metals, while in the UK the concentration of lead, copper and zinc is the subject of legislation. Zinc-deficient worts may have to be supplemented with zinc ions in order to stimulate fermentation.

Probably the first application of atomic absorption to the analysis of wines was made by Zeeman and Butler[839] who dehydrated the samples with sulfuric acid and then ashed at 500 °C before taking up again in acid. Frey *et al.*[246] determined a number of trace metals in beer, wort and yeast either direct or after extraction and subsequently[247] investigated the effect of several of these metals on the fermentation processes. While it is possible to aspirate alcoholic beverages directly in order to determine copper,[709] it may be found necessary to add ethanol to the standards, bringing them to either 86° or 100° proof in order to determine iron in beverages of similar proof strength.[513] Ethanol should be added to wine samples, normally of 40° proof, in order to conform with the composition of the standards.

The determination of copper, iron, zinc, lead and magnesium in both wine and beer for routine quality control was described by Weiner and Taylor.[780] These elements are mostly determined by direct aspiration, though lead and iron may have to be extracted from beer and some table wines.

The following methods will be found to be satisfactory for beer and table wines:

> *Direct method.* Decarbonate beer by pouring rapidly from one beaker to another a number of times and allow to settle so that the foam collapses back into the liquid. Aspirate beer and wine samples without further treatment unless dilution is seen to be necessary. Prepare standards by adding up to 1 ppm of copper, 2 ppm of zinc, 2 ppm of magnesium, 2.5 ppm of iron and 0.5 ppm of lead to decarbonated beer or wine of known low trace metal content. Aqueous standards are prepared for wines; these should contain 10% of ethanol.
>
> *Extraction method.* Transfer 50 ml of wine or decarbonated beer to a 250 ml beaker, add 5 ml of glacial acetic acid and boil for 2 min. Cool and transfer quantitatively to a 100 ml centrifuge tube. Add 2 ml of 1% aqueous ammonium pyrrolidine dithiocarbamate solution and, by pipette, 10 ml of methyl isobutyl ketone. Shake for about 5 min and centrifuge at 3000 rev. min⁻¹ for 10 min. Aspirate the upper organic layer direct from the tube. Prepare standards by pipetting 0–5 ml of standard iron (50 mg l⁻¹) and standard lead (10 mg l⁻¹) solutions into 100 ml calibrated flasks, and make up to the mark with water. Pipette 50 ml of each solution into 100 ml centrifuge tubes, add 5 ml of glacial acetic acid, 2 ml of 1% ammonium pyrrolidine dithiocarbamate solution and, by pipette, 10 ml of methyl isobutyl ketone. Continue as for the samples. These standards are equivalent to 0–2.5 mg l⁻¹ iron and 0–0.5 mg l⁻¹ of lead in the original samples. If only iron is to be determined the addition of acetic acid and the boiling can be omitted.

The response of both copper and iron may be enhanced in the presence of ethanol. This is offset by use of 'pure' beer or wines where possible in preparation of standards for the direct methods and, of course, is eliminated in the extraction method. This finding is consistent with those of Meredith et al.[513] and beverages of higher ethanol content should be compared with standards carefully matched with respect to ethanol when the direct method can be employed.

Boiling with acetic acid breaks down possible lead complexes in beer, so that small amounts can be completely recovered. There is no doubt that other extractable elements can also be determined in the organic phase in the second procedure. Cadmium and lead are for example determined down to 1 ng g^{-1} and 0.1 μg g^{-1} respectively using a simple APDC–MIBK extraction.[60]

Electrothermal atomization is readily applied to beverages, it usually being possible to transfer an aliquot of the acidified sample direct to the atomizer. Mack[473] simply diluted wine and fruit juice 1:10 and calibrated by standard additions for measuring lead. Organic tin, present as di(n-octyl) tin, can be extracted with benzene after the sample has been diluted with hydrochloric and acetic acids.[511]

For measuring trace elements in milk by electrothermal atomization, the best procedure appears to be to ash after evaporation with nitric acid. The residue is taken up in nitric acid.[765]

Calcium and copper have been measured in milk by flame atomic absorption after deproteinization with trichloracetic acid.[306,566]

Food and feedingstuffs

Convention has it that human beings eat food while animals feed. Both food and feedingstuffs consist mostly of an organic matrix containing traces of various metals. In feedingstuffs these are often added in the form of mineral mixes while in foods they can be present as desirable minerals or undesirable toxic elements. Either way their concentration has to be known and controlled, but in atomic absorption the analytical problems are identical. The metals must be separated from the solid matrix and this may be achieved either by removing the matrix by wet or dry oxidation or by extracting the elements into mineral acid.

A collaborative study on the determination of zinc in foods was reported by Rogers.[637] Both wet- and dry-ashing procedures were examined and 100% recoveries were obtained in each case. Copper and lead in meat and meat products are determined in samples dry-ashed at 500 °C and the residue dissolved in nitric acid.[174] The ranges were 1–5 ppm and 1–10 ppm respectively. The lithium contents of some consumables—salt, lettuce, potatoes, etc.—have also been measured.[337]

There are interesting examples of non-ashing methods. Food grade phosphates were analysed for lead after dissolution in hydrochloric acid and extraction with diethyl-ammonium dithiocarbamate into xylene.[369] Zinc was determined in sugar products and molasses solutions after further dilution with citric acid solution by Mee and Hilton.[507] This procedure gave the same results as dissolution after ashing.

Sugar matrices have been removed in two ways prior to electrothermal atomization for measuring trace elements. Morris *et al.*[531] simply fermented overnight with yeast at 40 °C, pH 4.5, to give a much simpler matrix. In another case inorganic chromium was measured directly, but for organic chromium the matrix was oxidized with an O-plasma before transfer to the furnace.

Hoover *et al.*[328] determined lead in plant and animal products by digestion with nitric, sulfuric and perchloric acids followed by co-precipitation of the lead on strontium sulfate. After conversion to lead carbonate by agitation with ammonium carbonate, the precipitate is dissolved in nitric acid and the lead determined in an air acetylene flame.

For routine analyses of both foods and feedingstuffs the dry-ashing method has two particular advantages: it requires less operator time than wet-ashing and it leaves a residue containing the minimum amount of unwanted ions. Such a residue is readily dissolved in hydrochloric acid. Dry ashing procedures must be carefully carried out to guard against the possibility of loss of wanted elements by vaporization. An ashing temperature of not more than 500 °C will ensure that most elements, with the exception of mercury, arsenic and some other metalloids, remain in the residue. For the exceptions mentioned, wet oxidation is essential, and that for mercury either in the cold, or in an enclosed vessel.

Ashing Procedures. A dry-ashing procedure is used for the analysis of raw materials and for the control of formulated feedingstuffs for added copper, zinc and magnesium.[634] Recovery experiments indicate that none of these elements is lost during the procedure described, and the accuracy of the method was established by comparison with standard colorimetric methods. Coefficients of variation of 1–2% were obtained for copper and magnesium and 3% for zinc, all in the ppm range.

The procedure is readily extended to other elements, e.g. calcium and manganese. Recommended operating details are as follows:

> Grind the sample if necessary so that it passes a B.S.S. No. 16 sieve. Weigh 2.0 g of sample (or other convenient quantity) into a silica evaporating basin, place on a hot-plate and allow to 'smoke' until completely charred. Transfer the basin to a muffle furnace at a temperature of 470 °C and ash.
>
> When ashing is complete, cool and extract with the minimum amount of hydrochloric acid and evaporate to dryness. Extract again with 10 ml of 25% hydrochloric acid, boil and filter into a 100 ml calibrated flask. Wash the filter through with warm 1% hydrochloric acid solution, make the contents of the flask up to 100 ml with water and mix.
>
> This solution is used for the determination of copper, magnesium, manganese and zinc. To determine calcium, dilute the solution by a further factor F to give a calcium concentration in the range 1–10 mg l^{-1}, adding 2 ml of 5% lanthanum solution per 100 ml of final solution. An air acetylene flame is used for all these elements. Multi-element standards are prepared to contain 0–10 mg l^{-1} of copper (equivalent to 0–0.05% in an original 2 g sample), 0–2 mg l^{-1} of magnesium (0–0.01%) and 0–5 mg l^{-1} of both manganese and zinc (0–0.025%).
>
> Standards for calcium should contain 0–10 mg l^{-1} of calcium and 0.1% lanthanum as lanthanum chloride. The calcium content of the original sample would be $VF/100\,W\%$ where V is the reading in mg l^{-1} from the calibration graph, F is the further dilution factor, and W is the weight in g of the original sample.

This method has been applied successfully to all types of animal feedingstuffs, oilseed cakes, and to milk, bonemeal and other products of animal origin. There is no obvious reason why it should not be applied to any material with a largely organic matrix and in this sense it may justifiably be described as universal.

Dry ash procedures have therefore been used by many workers, though volatile elements, e.g. mercury and the arsenic group, present problems. Arsenic has, nevertheless, been determined after the sample was dry-ashed with magnesium nitrate at 500 °C.[824,843] After subsequent digestion arsenic was measured down to 5 μg g^{-1} by the hydride generation method. It has been confirmed that no losses of arsenic, cadmium or lead occur, even after ashing at 450 °C for sixteen hours with magnesium nitrate, though mercury still requires special treatment.[248] Alternatively, there are many references to the determination of arsenic after the sample has been wet oxidized with nitric, sulfuric and perchloric acid mixture. In particular, Fiorino et al.[232] after wet ashing the sample, used the hydride method for sequential measurement of arsenic, selenium, antimony and tellurium at trace levels of 10–20 ng per 1 g sample.

In general, the measurement of the arsenic group elements by electrothermal atomization is also preceded by wet ashing. Selenium is reduced and precipitated with ascorbic acid.[347,348]

Tin may be determined in foods by electrothermal atomization, either directly of the solution resulting from acid digestion, or of the chloroform extract of the stannic neocupferron complex.[322]

Low trace levels of certain elements such as cadmium and lead are also determined in the flame after wet ashing and extracting with APDC–MIBK.

A suitable general wet ashing procedure is given under Plant Materials on p. 255.

Wet ashing of food materials should not require the use of a pressure dissolution vessel, except possibly for mercury determinations (q.v.). If such a vessel is used, however, only one built to withstand the extra pressures generated by organic materials should be employed, and perchloric acid should *not* be used in it with organic materials.

Acid Extraction. For the vitamin–mineral mixes actually added in the preparation of feedingstuffs, a simple acid extraction technique may be used. The following may be used in order to determine cobalt, copper, iron, magnesium, manganese and zinc in this type of sample:

> Extract 1 g of sample with 20 ml of hydrochloric acid at 65 °C for 30–60 min, adding further acid if necessary. Then add a little water, boil and filter into a 50 ml calibrated flask and make up to volume. Compare with standards made up in 5% hydrochloric acid, using an air acetylene flame.

Vitamin B$_{12}$ has been determined indirectly[798] by extraction and measurement of cobalt.

Acid extraction may also be used to determine trace elements in water-insoluble matrices. A good example of this type of sample is butter or edible oil. The latter are often produced in a process which involves a nickel catalyst. Both the nickel and copper contents usually have to be known. The simple acid extraction procedure may well result in an emulsion which it is difficult or impossible to break but carbon tetrachloride dissolves the organic phase, preventing formation of the emulsion and forming a lower layer. The aqueous layer containing the extracted elements remains on the top, ready for direct aspiration. Details (Price *et al.*)[608] are as follows:

> Dissolve 10 ml or 10 g of sample in carbon tetrachloride in a 50 ml centrifuge tube and add 10 ml of 10% nitric acid. Stopper and shake continuously for 10 min. Centrifuge for 2 min at 2500–3000 rev. min^{-1}. Aspirate the upper layer, and compare with standards containing 0–50 ppm of nickel and copper in 10% nitric acid.

Determination of Mercury in Foodstuffs, etc. Fractions of a part per million of mercury in foodstuffs are determined using the cold vapour method described on p. 56. The sample must first be either wet oxidized in the cold, using the method described for tissue material (see p. 275) or at higher temperatures under pressure. Nitric/sulfuric acid mixtures have been effectively used to digest fish tissues at 140 °C in a PTFE pressure vessel.[68,582] Care should be taken with this kind of method as mentioned above.

Mercury has also been measured by electrothermal atomization, although the likelihood of its being lost before the atomization stage in the temperature programme is greater than that for the arsenic group elements. Low temperature oxidation of the sample, combined with sulfide treatment to stabilize the mercury at 'ashing' temperatures up to about 300 °C are the necessary sample pretreatment.

Plant materials

As early as 1958, David[181] determined zinc, iron, copper and magnesium in plant material. Although his equipment was less sophisticated, his preparation procedure could hardly be improved today. One or two grams of the dried plant material was digested with 4 ml of sulfuric/perchloric acid $1+7$, and nitric acid was added in small amounts until destruction of the organic matter was complete. The mixture was then fumed until all nitric and perchloric acid was removed, and the remaining 0.5 ml of sulfuric acid solution was made up to 10 ml with water. Many further procedures originated in David's laboratory including that for calcium[180] and strontium[177] which attains good sensitivity by removal of phosphate by ion exchange.[176]

Dry-ashing and sulfuric acid digests were employed by Bradfield and Spencer[114] to determine magnesium, zinc and copper when investigating leaf analysis as a guide to the nutrition of fruit crops. The supply of calcium, manganese, strontium and zinc to plant roots growing in soil has been investigated[304] in chamber experiments by determining these elements both on plant digestate and in soil extracts.

Since that time many examples of wet ashing procedures have appeared in the literature, all based on nitric acid with either sulfuric acid or perchloric acid, or both.

Plant materials should be dried at 40 °C, crushed to pass through an 0.5–0.7 mm sieve and then oven-dried at 105 °C. When plants containing higher concentrations of manganese are digested and diluted, a brown suspension of manganese dioxide may appear, which is redissolved by addition of a drop of sulfurous acid.

The following typical wet ashing procedure was given by Allen and Parkinson[52] and still proves satisfactory after a number of years:

> *Acid digestion.* To 200 mg of finely ground plant material in a 100 ml Kjeldahl flask, add 0.5 ml of sulfuric acid, 1.0 ml of perchloric acid (60%) and 5 ml of nitric acid. Allow the digestion to proceed slowly at first and then increase heat until sulfuric acid refluxes down the side of the flask. When destruction of the organic matter is complete, cool, transfer to a 50 ml calibrated flask and make up to volume with water. In the determination of calcium and magnesium, include 0.05% of lanthanum.

The wet oxidation procedure is also used as a preliminary for measuring arsenic by the hydride evolution method. A. J. Thompson[736] preferred nitric/ sulfuric acid mixture for this purpose.

Dry ashing of plant material has become less popular of recent years, perhaps because of the possibility of loss of the more volatile elements. It has been shown, for example, that cadmium can be lost in the ashing of tobacco leaves at 450°–500 °C.[819] Where such a risk is not incurred, however, the following procedure, again from Allen and Parkinson[52] may be used:

> *Dry-ashing.* Place 200 mg of finely ground sample, contained in a silica or platinum dish in a cool muffle furnace. Allow the temperature to rise to 450 °C, and ash at this temperature for 3 h. Cool, dissolve the residue in 5 ml of 5 M hydrochloric acid. Add a few drops of nitric acid and evaporate to dryness on a water bath. Redissolve in 5 ml of 5 M hydrochloric acid, warm, filter into a 50 ml calibrated flask and make up to volume with water.

Since it was shown by Premi and Cornfield[598] that trace amounts of a number of elements are quantitatively extracted from plant material and organic residues simply by boiling with hydrochloric acid, and that complete destruction of the organic matter is not important, a number of analysts have used this type of procedure. Lead, zinc, copper, nickel, chromium, cobalt and cadmium were measured in marsh sediments after extraction for 90 min at 90 °C in 1:1 hydrochloric/nitric acid mixture. The method is speeded by using extraction under pressure. Varju[763] preferred to digest with 6 N hydrochloric acid at 110 °C in a closed vessel, stating that this gives similar results to the dry ashing method but is much simpler. Murphy *et al.*[540] recommend that pasture samples be heated in borosilicate tubes fitted with PTFE screw caps for 4 h with 2 M hydrochloric acid. These are then filtered and made up to volume for the determination of calcium, magnesium, sodium, potassium, copper, manganese and zinc.

The basic method, which Premi and Cornfield used for copper, zinc, iron, manganese and chromium, is as follows:

> *Acid extraction.* Weigh 0.5 g of dried and finely ground sample into a 250 ml beaker and add 25 ml of 6 M hydrochloric acid. Boil gently for 15 min, allowing the volume to decrease to about 5 ml, but not to become dry. Add 5 ml of hot water, boil, filter quantitatively into a 50 ml calibrated flask and make up to volume with water. Prepare standards for each element in 0.05 M hydrochloric acid.

Grass, straw, sewage sludge and farmyard manure were all successfully analysed by this method. It should be noted that potassium, sodium, calcium, magnesium and manganese are all extracted quantitatively in the cold when dried, ground samples are shaken in M hydrochloric acid for 24 hours,[303] but that under these conditions only 60–70% of copper, zinc, iron and chromium are extracted.

Ashing and acid extraction methods may be used as a preliminary to concentration by solvent extraction of elements which are normally too low in concentration to be measured directly in a flame. Alternatively, such elements can be measured by electrothermal atomization. It is noticeable, however, that comparatively little work on the use of electrothermal atomization techniques in the analysis of either plant material or soils has been published, probably because amounts of sample are very rarely limited. The technique has been applied to the measurement of phosphorus in plant material, however,[388] the phosphorus first being oxidized to phosphate and then stabilized by addition of lanthanum. The latter also minimizes effects of interference from calcium, sodium, potassium and magnesium. An electrodeless discharge lamp makes possible the use of the phosphorus line at 213.6 nm (see also p. 210) and a pyrolytic graphite tube furnace was found to give lower sensitivity than a standard tube. Phosphorus has been determined directly in 20 μl aliquots of soybean and rapeseed oils diluted 1:1 in MIBK.[599] In this case, addition of lanthanum was not beneficial.

Boron may be extracted from aqueous solution with 2-ethyl-1,3-hexanediol into methyl isobutyl ketone or into chloroform. This principle forms the basis for determining water soluble boron[508] and acid-soluble or total boron.[779] The nitrous oxide acetylene flame is required but there are no significant interferences.

There are very many papers in the literature in which atomic absorption methods are recorded as having been used as the means of carrying out a particular investigation in the fields of ecology, agronomy, etc. Many can be found in the *Journal of the Association of Official Analytical Chemists* and others can be traced through the classified abstracts and *ARAAS*.[32]

Soils

Extractable Elements. The soil chemist may require to determine either extractable nutrients, the total trace element content, or even the 'ultimate soil nutrient potential'.

Many buffer solutions are used in agriculture to release exchangeable or mobile ions and these have been described by Black.[101] In general, the buffer and extracted elements may be nebulized in the atomic absorption spectrometer, with dilution as necessary. There are many examples of the direct application of flame atomic absorption to the buffer extracts, though electrothermal methods could be used equally well.

Exchangeable sodium, potassium, calcium and magnesium were determined by David[179] after extraction with M ammonium chloride solution. 0.5 M acetic acid was used[474] to extract calcium and magnesium, strontium being added to the extract to overcome interference by phosphate. Acetic acid extracts were used by Ure[757] to determine cobalt, which was present in the range 1–50 μg. This corresponds to a detection limit of 0.05 ppm if an original soil sample of 20 g is shaken with 400 ml of 0.5 M acetic acid, evaporated twice with nitric acid and made up to 10 ml in ethanolic hydrochloric acid. Tamm's reagent (0.1 M oxalic acid + 0.1 M ammonium oxalate) has also been used[434] to determine free silicon in soils, and[102] to determine extractable iron and aluminium. In this method, 10 g of soil are heated for half an hour with 200 ml of the reagent and the resulting solution used for analysis. The extract may have to be acidified to prevent precipitation of oxalates.

Modified Morgan's solution (0.625 M ammonia + 1.25 M acetic acid, pH 4.8) may be employed to determine magnesium values, nebulizing after dilution and adding strontium chloride solution if aluminium or phosphorus are shown to lower the magnesium absorbance. Lead and nickel are leached from soils with M ammonium acetate in soil pollution studies.[305,505] The resulting solution is equilibrated with 0.1 M calcium chloride solution, centrifuged and filtered for analysis. Nadirshaw and Cornfield[544] examined four extractants for assessing particular functions or potential availability of manganese in soil: M ammonium acetate (pH 7.0) for exchangeable and water-soluble manganese(II); the same, containing 0.2% of hydroquinone to extract in addition 'active' manganese and manganese(III); Morgan's reagent (0.5 M acetic acid, 0.75 M sodium acetate, pH 4.8); and 0.5 M acetic acid, intended as a replacement for Morgan's solution which tended to clog a single slot burner. Results were obtained by comparison with standards made up in each extractant and agreed well with those obtained by colorimetric procedures.

Typical extraction procedures for mobile and exchangeable ions, using ammonium acetate or acetic acid, have been developed in conjunction with S. E. Allen[51] at the Institute of Terrestrial Ecology for the determination of calcium, magnesium, manganese, potassium and sodium. Soil samples should be air dried at 40 °C before extracting. Lanthanum is added to overcome interference from aluminium, silicon and phosphorus in the determination of calcium and magnesium.

> Transfer about 5.0 g of air dried sample, accurately weighed, to a 250 ml polythene bottle and add 125 ml of either (a) M ammonium acetate or (b) 2.5% acetic acid. Fit a

polythene screw cap onto the bottle and clamp in an end-over-end shaker. Shake for 1 h. Filter the mixture through a Whatman 44 filter paper and retain the filtrate (solution A). Perform this extraction in duplicate and also make up a reagent blank with each batch of samples.

Manganese, sodium and potassium. Determine directly in solution A. To calibrate, prepare standards containing 0, 2.5 and 5 mg l^{-1} of each of these elements in presence of (a) M ammonium acetate or (b) 2.5% acetic acid.

Calcium and magnesium. Pipette 20 ml of solution A into a 100 ml flask, add 20 ml of lanthanum chloride solution (0.4% La^{3+}), 5 ml of sulfuric acid (20%) and either (a) 20 ml of 4 M ammonium acetate solution or (b) 8 ml of 25% acetic acid. Make up to volume with water. Process the reagent blank similarly. If greater dilutions of the working solution are required these can alternatively be prepared using the volumes indicated in Table 19. To calibrate, prepare standards containing 0, 5, and 10 mg l^{-1} of calcium and 0, 0.25 and 0.5 mg l^{-1} of magnesium in presence of 800 mg l^{-1} of lanthanum, 1% sulfuric acid and either (a) M ammonium acetate or (b) 2.5% acetic acid.

TABLE 19. Further dilutions for calcium and magnesium determination

Total dilution of solution A (F)	Solution A or blank (ml)	(a) 4 M amm. acetate (ml)	OR (b) 25% acetic acid (ml)	0.4% La^{3+} solution (ml)	20% H_2SO_4 (ml)	Final volume (ml)
$10\times$	10	22.5	9	20	5	100
$20\times$	5	24	9.5	20	5	100
$50\times$	2	25	10	20	5	100

Calculate the concentration of each extractable metal in mg 100 g^{-1} of soil, correcting for oven dry weight as follows. For manganese, sodium and potassium

$$\text{mg } 100 \text{ g}^{-1} = \frac{(C_s - C_b) \times 1.25 \times 10^3}{W_{40} \times (100 - L)}$$

and for calcium and magnesium

$$\text{mg } 100 \text{ g}^{-1} = \frac{(C_s - C_b) \times 1.25 \times F \times 10^3}{W_{40} \times (100 - L)}$$

where C_s is concentration in mg l^{-1} found in final sample solution, C_b is concentration in mg l^{-1} found in reagent blank solution, F is total dilution factor (5, or if further dilution table is used, 10, 20 or 50), W_{40} is weight of soil dried at 40 °C (5 g approx.) and L is % loss, at 105 °C, of soil previously dried at 40 °C. This value should be determined in a separate experiment.

The accuracy of the figure obtained depends to some extent on the element and the sample but is normally of the order of 5–10%.

Other trace elements can be determined by conventional atomic absorption procedures either after extraction of the kind described above, or after wet oxidation or fusion if a total value is required. Arsenic was measured in soils by hydride generation after pyrosulfate fusion and acid digestion[736] and cadmium and other elements are readily determined by APDC extraction into MIBK for flame,[661] or into chloroform for electrothermal atomization.[756] For this kind of method, an effective digestion acid is 4 M nitric, M hydrochloric acid, used[331] particularly for silver, copper and zinc.

For the determination of mercury by reduction and measurement by the cold vapour method the sample was first digested at 60 °C with potassium dichromate and nitric acid solution[238] or less conventionally, released from the soil by radio frequency heating and collected in acid permanganate solution.[433]

Total Content. The total mineral content of soils is not often required, particularly by the ecologist. When it is, the preparation of the initial solution requires fusion and/or hydrofluoric acid digestion procedures, depending on the elements to be determined. After the organic material has been destroyed, the methods employed would be identical with those described for the analysis of rocks and minerals, though for calcareous soils with low clay content, digestion with hydrochloric acid is generally adequate to bring nutrient elements into solution.

Biological processes

Atomic absorption is an ideal technique for the control of nutrient trace metals in processes where metal-dependent enzymes or bacteria are active. Frequent checks of a number of such metals may well be required with a good degree of accuracy so that their concentrations may be kept manually or automatically within certain closely defined limits.

A further example of this sort of application is the check made by Weiner and Taylor[780] on the magnesium content of mashing liquors and worts in beer making. The changes occurring in the magnesium content of the wort during the fermentation process were also investigated, and sufficient magnesium to ensure continued growth and metabolism of the yeast was found to be contributed by the malt itself.

PATHOLOGICAL AND MEDICAL

Pathological and clinical samples are essentially little different from other types of organic matrices, but in many cases, and particularly the more routine determinations, considerable effort has been made to reduce preparation procedures to a minimum. This is possible because the number of types of sample is limited to serum or whole blood, urine, cerebro-spinal fluid and tissue, and also because instrumental sensitivities in the flame mode are now good enough for many of the interference effects encountered in earlier work to be 'diluted out'. Method simplification is desirable because the technicians in an average clinical laboratory have little time for lengthy or intricate procedures. Indeed, the simplicity and reliability of atomic absorption make possible many analyses which with other methods would be both time-consuming and uneconomic.

Similar methods to those described here are applied in the field of veterinary pathology.

The elements routinely measured in flame atomic absorption are the 'electrolytes': calcium, magnesium, sodium and potassium; and also iron, copper and zinc. These are all readily determined with an air acetylene flame, on which most clinical laboratories now standardize.

TABLE 20. Normal levels of some elements in body fluids, etc.

	Blood		Urine	Tissue
	Serum	Other fraction[a]		
Calcium	2.3–2.6 mM l^{-1}	—	2.5–11 mM day^{-1}	
	9.3–10.3 mg 100 ml^{-1}		100–450 mg day^{-1}	
Magnesium	0.9 mM l^{-1}	—	4–12.5 mM day^{-1}	bone ash
	2.15 mg 100 ml^{-1}		100–300 mg day^{-1}	0.6%
Potassium	3.1–5.5 mM l^{-1}	P 16.07 mg 100 ml^{-1}	53–91 mM day^{-1}	
	12–22 mg 100 ml^{-1}		2120–3640 mg day^{-1}	
Sodium	140±7 mM l^{-1}	—	40–156 mM day^{-1}	
	322±17 mg 100 ml^{-1}		920–3590 mg day^{-1}	
Iron	male 125 μg 100 ml^{-1}	'iron binding cap'	} up to 1 mg day^{-1}	
	female 90 μg 100 ml^{-1}	300 μg 100 ml^{-1}		
Copper	male 105 μg 100 ml^{-1}	W 92 μg 100 ml^{-1}	} up to 50 μg day^{-1}	
	female 114 μg 100 ml^{-1}	W 97 μg 100 ml^{-1}		
Zinc	120 μg 100 ml^{-1}	W 880 μg 100 ml^{-1}	300–600 μg day^{-1}	
		P 300 μg 100 ml^{-1}		
Aluminium	4 μg 100 ml^{-1}			
Arsenic	3.5–7.2 μg 100 ml^{-1}	W 6–20 μg 100 ml^{-1}	< 100 μg day^{-1}	
Beryllium			< 2 μg l^{-1}	liver 0.012
				μg 100 g^{-1}
Boron				tissue 0.5–1
				μg g^{-1}
Cadmium		W ~0.5 μg 100 ml^{-1}	up to 2 μg day^{-1}	kidney ash
				2–6 mg g^{-1}
Chromium	4 μg 100 ml^{-1}	Cells 2.3 μg 100 ml^{-1}		
Cobalt	0.02 μg 100 ml^{-1}	P 1.2 μg 100 ml^{-1}	1–10 μg day^{-1}	
		W 1–12 (av. 4.3)		
		μg 100 ml^{-1}		
Lead		W 20–30 μg 100 ml^{-1}	10–75 μg day^{-1}	
Lithium	1 μg 100 ml^{-1}			
Manganese	1–2 μg 100 ml^{-1}	W 2.4 μg 100 ml^{-1}	0.8 μg day^{-1}	
Molybdenum	0.4 μg 100 ml^{-1}			
Nickel	2 μg 100 ml^{-1}		30 μg day^{-1}	
Selenium	1 μg 100 ml^{-1}			
Strontium	5 μg 100 ml^{-1}			bone
				0.01–0.25%
Vanadium	1 μg 100 ml^{-1}			

Figures taken from Christian and Feldman[8] and various sources.

[a] P = plasma; W = whole blood.

Lead, cadmium and mercury are also frequently determined in the screening of individuals exposed to the dangers of poisoning by these metals and in actual cases of poisoning, though for these, direct flame atomization is not sufficiently sensitive. Many other elements are now being measured in clinical samples using electrothermal atomization.

The latter technique enables normal and pathological levels of many elements to be established that have never been reached by other methods. Thus a whole new line of investigation has been made possible. A comprehensive essay on the clinical significance of some trace metal measurements has recently been published by Delves.[189] Normal levels of a number of elements are summarized in Table 20.

The presence of protein in whole blood or serum sometimes leads to difficulties in trace element analysis: interference in the flame, blockage of poorly designed burners and high backgrounds and involatile residues in electrothermal atomizers. Deproteinization is then carried out, usually with trichloracetic acid. When sensitivity is a problem, even after deproteinization, solvent extraction can be used, or alternatively the organic matter can be destroyed in a wet ashing process. The resulting solution may then be satisfactory for analysis, or better still, sensitivity can be achieved by ashing a larger sample and separating and concentrating the element to be measured with solvent extraction.

Dry ashing of biological specimens should be undertaken with some care. Although losses of iron, cadmium, chromium and zinc through volatilization at ashing temperatures below 600 °C may be very low, retention by the crucible can be serious.[420]

Some of the earliest applications of atomic absorption were to the analysis of clinical samples. Willis, in particular, published a series of papers [801-8] and a comprehensive monograph[809] in 1963. Dawson and Heaton[183] also gave flame emission and absorption methods for analysis of clinical samples and a textbook on biological applications appeared in 1970.[8] Reviews on the application of atomic absorption to clinical analysis since that time include those by Berman,[95] Sunderman[721] and Price.[602]

Calcium and magnesium

The relative concentrations and sensitivities of these two elements in serum make it convenient to determine them together in the same dilution, although interfering constituents such as phosphate and protein do not have the same effect in the air acetylene flame.

Willis[804] obtained results by simple dilution with water ($1+9$ or $1+24$), dilution in the same ratio with sodium EDTA solution (1%) and also after deproteinization with trichloracetic acid, finding them comparable. Gimblet et al.[279] also found trichloracetic acid deproteinization and EDTA dilution comparable but Pybus[611] deproteinized with perchloric acid, adding strontium to remove interferences from phosphate and other anions.

Zettner and Seligson[840] developed a single diluent, containing lanthanum chloride, butanol and octanol, for the determination of calcium, in order to overcome phosphate interference and to improve the response in the flame. Trudeau and Freier[751] simply diluted the serum $1 + 49$ in 0.1% lanthanum solution (as $LaCl_3$). More recently, exceptional accuracy was claimed by Pybus et al.[613] using a double channel atomic absorption spectrometer with strontium as internal standard reference element. The serum was diluted $1 : 50$ in a solution which contained 50 mM hydrochloric acid, 10 mM lanthanum chloride and 0.12 mM strontium chloride.

The determination of magnesium, in both serum and urine, has been shown to be largely uninfluenced by the presence of phosphate or protein. Dawson and Heaton[184] diluted samples of serum, plasma, urine, and ashed materials with 0.1 M hydrochloric acid.

The need for special diluents for calcium determinations suggests a stronger bond between calcium and protein in the serum. In investigating the atomic absorption method for calcium, Cooke and Price[155] believed that, in a reliable method, the correct figures must be obtained, whatever the dilution factor employed and in spite of small variations in flame conditions. Phosphate and sulfate both tend to suppress the calcium absorption, as do low concentrations (< 60 mg%) of protein. As the protein concentration increases, however, there is an enhancement in absorption, due to the more reducing nature of the resulting flame. Dilution with water or hydrochloric acid gives the correct result only when the dilution factor is such that the effect of protein counterbalances that of the interfering ions and thus cannot be relied upon as a generally applicable method. The use of lanthanum as a releasing agent does not appear to fulfil the conditions laid down above, but disodium EDTA solution does.

The releasing effect of disodium EDTA solution rises to a plateau when its concentration is about 0.75%. Recoveries of added calcium are also good, showing that serum-calcium and non-serum-calcium respond in the same way.

Calcium and Magnesium in Serum. Details of the method, which can be used for both calcium and magnesium determinations, are as follows:

> Dilute a small volume, e.g. 200 μl, of serum $(1 + 49)$ or $(1 + 99)$ with disodium EDTA solution, 0.75%. Aspirate into an almost luminous air acetylene flame (or better, nitrous oxide acetylene flame). Prepare standards containing 0–15 mg 100 ml^{-1} of calcium and 0–4 mg 100 ml^{-1} of magnesium and dilute similarly with 0.75% disodium EDTA solution.

The accuracy of this method should be in the order of 1% of the calcium and 0.6% of the magnesium contents.

Calcium in Urine. Calcium is determined in urine by the following procedure:

> Dilute the urine so that its calcium concentration lies between 2 and 10 mg l^{-1}, including 1 ml of 5% lanthanum solution (as $LaCl_3$) per 10 ml of final solution. Prepare standards containing 0–10 mg l^{-1} of calcium in 0.5% lanthanum solution.
>
> If the results are required in mmoles of calcium per litre, multiply by the factor 0.02495.

Magnesium in Urine

> Dilute the specimen with water so that its magnesium concentration lies between 0.2 and 1 mg l^{-1}. The dilution factor will be between 50 and 1000. Aspirate into an almost luminous air acetylene flame and compare with standards containing 0, 0.5 and 1.0 mg l^{-1} of magnesium as magnesium chloride. Multiply the results obtained by the dilution factor and by 0.04114 to convert to mmoles of magnesium per litre.

Calcium and Magnesium in Solids and Other Specimens. Removal of organic matter by wet or dry oxidation is the surest approach. Comparable results should be obtained by the three basic methods using dry-ashing, digestion with nitric/perchloric acid mixture and extraction of the wanted metals from dried tissue with 0.5 M nitric acid. It is usually necessary to add releasing agents to the final solution, however, to remove interference from phosphate, nitrate, perchlorate and protein.

Most other methods given for the examination of solid specimens, e.g. skeletal muscle, tissue, hair, faeces, etc., are simply variations of this or the wet- and dry-ashing procedures already given in the section on plant materials.

Dry-ashing is also the most suitable method for preparing samples of milk. The residue from 1 g of milk should be dissolved in hydrochloric acid, and diluted to 200 ml so that the final solution also contains 0.5 % of lanthanum.

Sodium and potassium

The fact that few authors have discussed the determination of these alkali metals in serum or urine in any detail since Willis[805] testifies to the reliability of the method. These elements are often determined by flame emission but the principal advantage of the absorption method is that interference on sodium by the calcium–OH band emission is readily overcome with instruments having modulated sources. The same preparation can be used for either technique.

Sodium and potassium mutually interfere through the shifts which they induce in each other's ionization equilibria. As potassium is usually present in lower concentrations in both serum and urine, however, these effects are manifest as a considerable enhancement in potassium response (as compared with a solution containing no sodium) but a very small enhancement of the sodium. Response variations which affect analytical accuracy are countered either by adding the correct amount of the second element to the standards used for determining the first, or by ensuring that an excess of the second element is present in both samples and standards for the first, particularly if the concentration of the second element is variable.

Serum. The normal serum electrolyte levels fall within fairly restricted ranges and the variation of sodium within its range produces a negligible variation of the potassium response. It is therefore necessary, in determining potassium, simply to ensure that the sodium concentration in the standards is comparable to that of the water-diluted samples. A dilution factor of 250 × is suitable. When

TABLE 21. Electrolytes in serum

	Ca+Mg	Na	K
Sample diluent	0.75% EDTANa$_2$	water	water
Sample dilution factor	100	1000	250
Standard range:			
mM l^{-1}:	Ca 0–0.04; Mg 0–0.015	0–0.15	0–0.04
mg l^{-1}:	Ca 0–1.5; Mg 0–0.4	0–3.5	0–1.6 in 0.56 mM l^{-1} (=12.9 mg l^{-1}) Na solution
Air acetylene flame	almost luminous	lean	lean
Pathlength	10 cm	rotated if necessary	10 cm

sodium is determined a greater dilution factor is necessary (1000 ×) and at such dilutions the potassium concentration is so low that its effect on the sodium response is negligibly small. Dilutions of the sample are made with water and the standards do not need to contain potassium.

Urine. Because of the possible wide variations in concentration, an excess of one element should be added to the samples and standards used for determining the other. In practice this is necessary when potassium is being determined, but as very low sodium concentrations rarely occur water is used as the diluent in this case.

Tables 21 and 22 summarize the preparation of serum and urine samples for determining the four electrolyte elements.

Other Materials. When sodium and potassium are to be determined together in the same diluted sample, it is necessary to add another alkali element as ionization buffer. Because of its low ionization potential, caesium is the best, but rubidium or lithium—the latter in higher concentrations—may be used, and lithium salts specially free from sodium and potassium are now available for this purpose.

TABLE 22. Electrolytes in urine

	Ca	Mg	Na	K
Diluent for sample and standards	0.5% La^{3+}	0.5% La^{3+} or water	water	0.1% Na
Sample dilution factor	20–2000	50–1000	500–2000	500–2000
Standard range:				
mM l^{-1}	0–0.25	0–0.04	0–0.1	0–0.025
mg l^{-1}	0–10	0–1.0	0–2.5	0–1
Air acetylene flame	almost lum.	almost lum.	lean	lean
Pathlength	10 cm	10 cm	rotated if necessary	rotated if necessary

Sodium and potassium may be determined in solid materials, after wet- or dry-ashing or acid extraction by one of the procedures already described for calcium and magnesium.

Sodium and Potassium by Emission. The conditions given above are also suitable for the measurement of these elements in emission, when the analytical sensitivity is usually found to be slightly greater.

Lithium

Normal levels of lithium are extremely low, about 0.01 ppm in serum,[8] and these are best reached by electrothermal atomization. Lithium therapy is used in the treatment of manic depressive illnesses and the progress of the treatment is monitored, ensuring that toxic limits are not exceeded by analysing blood and urine. In these cases, serum lithium levels of about 7 ppm are maintained.

Lehmann[454] diluted serum 1:20 with 0.1 M hydrochloric acid, obtaining mean recoveries of 101%, though others used 1:10 dilutions of serum and 1:50 dilutions of urine in water. Pybus[612] too used water as the diluent for serum except where high lithium levels were encountered, when sodium and potassium were also added. Direct dilution with water is still often favoured[156,830] though greater dilution factors, e.g. 20× or 50×, can now be used.

Deproteinization, when carried out with trichloracetic acid, ensures that lithium is displaced completely from the protein, and for this reason is probably the better method. Details are as follows:

> Into a dry centrifuge tube pipette 0.5 ml of serum, 2.0 ml of water and 2.5 ml of 10% trichloracetic acid solution. Mix well and centrifuge at 3000 rev. min^{-1} for 10 min. Aspirate the supernatant fluid into a lean air acetylene flame. Prepare standards with up to 2 μg ml^{-1} of lithium in a solution also containing 5% of trichloracetic acid and 322 μg ml^{-1} of sodium as sodium chloride. Lithium can then be measured either in absorption or in emission.

Iron

Serum. Abnormal serum iron concentrations are found in various types of anaemia, in pregnancy and in the presence of malignancy. Iron can be measured in the flame with very great precision with a 5× or 10× dilution of the sample, instrumental sensitivities normally being good enough to render unnecessary the special deproteinization procedures with minimal dilution formerly employed. Aqueous standards are usually employed, though it is acknowledged that this may incur a very small though constant analytical inaccuracy. Care must be taken to avoid haemolysis of the original samples. If this is believed to have occurred both the protein and protein-bound iron must be removed by precipitation with hot trichloracetic acid before the residual true serum iron is measured.

It has been observed[680] that the sodium chloride absorption band in the region of the 248.3 nm iron line is enhanced considerably by trichloracetic acid and hydrochloric acid. To avoid interferences when these acids are used, therefore, a background correction technique should be employed.

If serum iron is measured by electrothermal atomization, the sample should first be diluted $1+4$ with 0.1 M hydrochloric acid, then $1+1$ with trichloracetic acid and centrifuged. A 20 μl aliquot of the supernatant is injected directly into the atomizer. Ash at 1100 °C for 20 s and atomize at 2700 °C for 5 s. With this method, calibration is made directly with aqueous standards.

Other Samples. Little progress seems to have been reported in the standardization by atomic absorption of iron levels in whole blood since van Assendelft[759] in 1968 pointed out some difficulties with the method.

Iron is determined in urine, depending on the level, by deproteinization with trichloracetic acid, followed by direct aspiration into the flame or possibly following an APDC–MIBK extraction for very low levels.

Copper and zinc

Serum. These elements can be measured in a $10\times$ or even $20\times$ dilution of serum by flame atomic absorption, effectively eliminating interference by proteins and avoiding the need for deproteinizing before aspiration. Recoveries after deproteinization were found by some workers to be very minimally better than by the direct dilution method ($99.9\pm0.9\%$ instead of $99.3\pm1.4\%$)[385] but it was also affirmed that the more complex diluents that have been employed in the past, e.g. containing *n*-butanol, do not improve recoveries.

Because of the low wavelength of the zinc absorption line, sodium chloride may be added to the aqueous standards to match the scatter effects, but the use of background correction is strongly advised. The flame method for copper and zinc by simple dilution is as follows:

> Pipette 1 ml of serum into a 10 ml calibrated flask and make up to the mark with water. Alternatively prepare a $1+9$ dilution with a suitably accurate diluter. Prepare copper calibration solutions containing 10 μg and 20 μg of copper per 100 ml (corresponding to 100 μg and 200 μg respectively in 100 ml of sample) from stock copper nitrate solution. Use water as the blank. Prepare zinc calibration solutions containing 0, 10, 20 μg of zinc per 100 ml of 14 mM sodium chloride solution (corresponding to 0, 100 and 200 μg of zinc per 100 ml of sample respectively) from stock zinc nitrate solution. If background correction is used for zinc, the sodium chloride can be omitted from the standards.

Serum copper and zinc are readily measured by electrothermal atomization using a sample dilution of $1+9$ or $1+19$ with water. Calibration is again with aqueous standards, and 20 μl aliquots are injected into the atomizer. For both elements, ash at 800 °C for 20 s and atomize at 2700 °C for 5 s. Such dilutions enable samples to be measured out accurately with a micropipette.

The significance of abnormal copper and zinc levels in serum has been comprehensively discussed by Delves.[189]

Electrothermal atomization has also been used by Delves to measure the copper content of plasma protein fractions.[190] The albumin, α_1, α_2, β and γ globulins are separated from a 2 μl sample by cellulose acetate membrane electrophoresis. Pieces containing these fractions are cut from the electrophoretic strip and atomized separately. Most of the serum copper is shown to be associated with the α_2 fraction. Zinc and copper have both also been measured in serum protein fractions by flame atomic absorption by the expedient of coupling a liquid chromatograph with a DEAE cellulose column directly to the nebulizer of an atomic absorption spectrophotometer.[192]

Urine. Excretion of zinc is usually at a higher level than that of copper, and accounts for zinc being a normal contaminant of domestic sewage. Low copper concentrations in urine, present in cases of neoplasm, may be determined after extraction with ammonium pyrrolidine dithiocarbamate into methyl isobutyl ketone. High levels of copper and zinc can usually be determined by direct aspiration.

Other Materials. For flame atomic absorption, solid specimens require wet- or dry-ashing, followed by further concentration, if necessary, by solvent extraction. Backer[69] claimed that chloric acid digestion was simpler than perchloric/nitric acid mixtures, and is suitable for hair and brain tissue samples.

Dental samples are atomized in a graphite furnace directly from the solid, after crushing and mixing with graphite. Langmyhr used this method to measure silver, manganese and zinc.[439]

Since copper was measured directly in finger nail clippings[81] as an early application of electrothermal atomization, the method has been used for copper and zinc determinations.[695] Samples are washed with acetone, dried at 100 °C and 10–20 μg pieces introduced into the atomizer. Alternatively, 1 mg portions are digested with nitric acid at 65 °C and diluted to 1 ml.

Blomfield *et al.*[103] used atomic absorption to estimate patients' uptake of copper and zinc during regular haemodialysis. The total copper and zinc of red cells and free copper and zinc of plasma and dialysis fluids were determined at all points during the process as well as subsequent blood levels. Copper plumbing and zinc oxide plasters were found to be the major source of contamination by these elements.

Lead

There is continuing public concern about the degree of pollution of the environment by lead from various sources and particularly from the combustion of lead anti-knock agents in gasoline. Many thousands of workers in industry are directly exposed to the dangers of lead poisoning (metals, batteries, paints, etc.) as are children, through the paint and metal of their toys and surroundings. The determination of lead as a routine screening test will therefore continue to be of great importance in industrial hygiene, public health, paediatric and clinical laboratories.

The normal blood and excretory levels (Table 20) are low and, when raised, indicate lead poisoning. The action of lead is insidious as it tends to accumulate in the body, causing damage to the brain and central nervous system. Lead also interferes with synthesis of haem for haemoglobin and may thus cause anaemia.

Blood. In blood, lead concentrates in the red cells, and the generally accepted danger level is 80 mg 100 ml^{-1}. If a worker's blood lead concentration approaches this level, he should be removed from the source of ingestion.

Because of the low normal level of lead in whole blood, direct measurement in the flame with a convenient dilution of the sample is still a somewhat marginal procedure. Extraction of the lead from a 1 ml, 2 ml or even larger sample would be necessary. Witness to the difficulties of such methods are the very large number of papers published on atomic absorption determination of lead in blood. For this reason the Delves microsampling cup was evolved and in recent years very much more use has been made of electrothermal atomization.

Kopito et al.[423] have shown that deproteinization can lead to serious losses of lead by sorption on the precipitates. They investigated the effect of different concentrations of trichloracetic acid, alone and with perchloric acid. The best recovery was in the case where an APDC–MIBK extraction followed trichloracetic acid precipitation.

The method of Farrelly and Pybus[223] is still probably the most reliable of the flame methods for lead, but unfortunately this required 5 ml of sample and used formamide, which has been shown to be teratogenic, to break the emulsions. The following modified version of the method has been found to be satisfactory:

> Pipette 1 ml of well mixed heparinized blood into a stoppered centrifuge tube. Add 1 drop of saponin/triton solution (5 ml of Triton-X +5 g of saponin in 25 ml of water), mix thoroughly and add 1 ml of 2% ammonium pyrrolidine dithiocarbamate solution. Shake the tube for 30 s then add by pipette 2 ml of methyl isobutyl ketone and shake for 1 min. Decant the mixture into a small-bore thick glass test-tube and centrifuge for 10 min at 3000 rev. min^{-1}. Aspirate the top layer into a lean air acetylene flame. Prepare standards by taking 1 ml of lead standards containing 0–1 μg of lead per ml through the procedure. Read the lead content of the sample extracts in μg, multiplying the result by 100 to express it in μg of lead per 100 ml of whole blood.

The Delves microsampling method[187] offers a simple and rapid screening procedure for 10 μl aliquots of blood and is thus suitable for use with samples obtained from the prick of a finger or ear-lobe. The use of such small sample volumes incurs the risk of contamination from at least two sources. In an industrial environment, or in these days even in a normal city environment, lead-containing particles deposited on the surface of the skin can give rise to levels of lead comparable with those being measured. Also, similar amounts of lead can readily be picked up from laboratory glassware. This should be soaked overnight in 1 + 1 nitric acid and repeatedly rinsed in deionized water before use. It should be used only for this particular determination.

The accessory is described on p. 52. The procedure is as follows:

> Prepare calibration solutions containing 0, 25, 50, 75 and 100 μg of lead per 100 ml in a slight excess of nitric acid.

If the samples are taken from a finger-prick, a 10 μl aliquot can be transferred direct to the crucible or 'Delves cup'. The crucible containing the sample is dried on a hot plate at 90–100 °C for about 10 min. Standards are treated similarly.

Alternatively, and this is the preferred procedure, the blood sample may be kept with EDTA as anticoagulant and quantitation achieved by the method of standard additions. Transfer 10 μl aliquots of sample to five crucibles, and add to these respectively 10 μl aliquots of the blank and of the four standards. Dry as before. Introduce the dried samples into a lean air acetylene flame in the microsampling accessory, and measure maximum absorbance ('peak height'), if possible using a simultaneous background corrector. If background correction is not applied, two peaks should be observed of which the first is the 'smoke' signal and the second that of the lead absorption.

In a variation of the Delves cup method, instead of introducing the liquid sample into the crucible, the fresh blood sample, with the addition of EDTA anticoagulant, is applied to a piece of filter paper and allowed to spread.[136] It is found that discs of standardized diameter, e.g. 4 mm, punched from the dried paper then contain constant amounts of sample. The discs are transferred to the Delves cup and atomized in the normal way.

Electrothermal atomization undoubtedly offers the simplest solution to the problem of blood lead determinations, but even here several different ways of preparing the sample have been proposed. These mostly involve wet oxidation of the sample either by digestion with nitric acid before introduction into the furnace[87,274] or by addition of nitric acid to the sample in the furnace.[273] Wet oxidation effectively removes much of the organic matrix, but provided adequate sensitivity is available so that the sample can be diluted up to 20 ×, the residual background signal is readily dealt with by simultaneous background correction. The procedure is as follows:

Dilute the whole blood (1+9) or (1+19) with deionized water. Calibrate by method of standard additions, adding 20 μl aliquots of standards containing 0, 1, 2, 4 μg of lead 100 ml^{-1} to 20 μl aliquots of the above sample dilution dried in the graphite tube (these correspond to 0–40 or 0–80 μg 100 ml^{-1} of lead in original sample). Ash for 50 s at 500 °C, atomize for 3 s at 2450 °C.

Plasma lead levels have recently been measured[133] after chelation with APDC and extraction from 1 ml of plasma sample into MIBK, sensitivity being 0.2 μg 100 ml^{-1}. It was found that the ratio of plasma lead/whole blood lead was similar and in the range 2.3–3% for all groups of workers, whether or not they had been exposed to absorption of lead.

Urine. Lead is determined in urine in order to follow the progress of therapy, which usually consists of the administration of a lead-chelating reagent such as EDTA or penicillamine. When the former is used, however, excreted lead is so strongly bound that the standard solvent extraction procedure using ammonium pyrrolidine dithiocarbamate is ineffective.

Selander *et al.*[674] used wet-ashing before extraction, and for lead concentrations below 5 μg ml^{-1}, which could not be measured by direct aspiration of the sample. Roosels and Vanderkeel[646] added calcium to displace lead from its

EDTA complex so that the lead could be extracted with dithizone. This leads to a chloroform solution, which is convenient for electrothermal methods.

The problem has also been solved by co-precipitation methods. Kopito[424] precipitated lead, with other polyvalent ions, from ammoniacal samples with bismuth nitrate.

Where therapy is known to have been by way of penicillamide, extraction of the lead with ammonium pyrrolidine dithiocarbamate is satisfactory:

> Adjust the pH of a 50 ml sample aliquot in a 100 ml separating funnel to 2.5 ± 0.1 with hydrochloric acid, add 5 ml of 1% ammonium pyrrolidine dithiocarbamate solution and 6 ml of methyl isobutyl ketone. Shake for 10 min, allow to stand for 5 min, discard the lower (aqueous) layer and collect the organic layer and any emulsion in a 10 ml centrifuge tube. Centrifuge for 10 min at 3000|rev. min^{-1} and aspirate the organic layer into an air propane or lean air acetylene flame. Compare with standards prepared by taking 50 ml aliquots of lead solution containing 0–1 μg ml^{-1} of lead through the procedure. These standards are equivalent to 0–1 μg of lead per ml of the urine sample.

Alternatively, lead is measured in urine using the Delves cup method[312] or electrothermal atomization. In the method of Ebert[208] the urine is mixed 1:1 with a solution containing 0.1 M EDTA and hydrogen peroxide. After drying at 100 °C, the residue is ashed at 330 °C and atomized at 1100 °C. Calibration is by standard additions.

Other elements

While the foregoing elements are the ones most frequently determined in clinical laboratories, they are by no means the only ones. Not only are some elements administered in therapy or as markers, but as the determination of trace metals becomes easier and more sensitive, 'normal' values become more readily established and abnormal values related to certain conditions or diseases. In principle, any metallic element can be determined in biological material provided that sufficient sample is available from which an adequate quantity of the required metal can be separated.

In some cases sensitivity is good enough in electrothermal atomization for separation or concentration not to be required. This technique is now being used for many analyses (nearly half the clinical atomic absorption methods published during 1976 utilized electrothermal atomization) particularly where low levels or small samples are involved. Consequently, many 'new' metals are being investigated in a clinical context.

Unless a standard or proven method is to be employed for a particular element, it is usually best to rely on the type of sample preparation that effectively removes organic matter, and brings trace elements into a standard form, i.e. wet oxidation. Many workers have used this successfully where a 'short cut' method has not been, or could not be, shown to be successful. Best wet oxidizing reagents for whole blood and tissue appear to be nitric/perchloric acid mixtures and this has also frequently been used for urine. The lipid fraction of liver requires vigorous mineralization, and for this Johnson[363] advocated digestion with aqua regia at 100 °C in preparation for measuring a number of elements in the flame. An alternative treatment for tissue is to dry ash at a comparatively low temperature (in order not to vaporize elements to be measured) and then to take up in 1 M nitric acid. In view of the comments recorded earlier that some trace elements can be retained by the crucible in dry ashing procedures, however, this kind of treatment should always be checked by recovery experiments.

Silver. Silver has been measured in tissue from 0.1–100 μg g^{-1} in an air acetylene flame after wet oxidation with nitric/perchloric acid mixture, and buffering with tartaric acid and ammonia.[638] It has been determined in liver tissue at 0.06 μg g^{-1} by electrothermal atomization after charring at 280 °C, calibration being by standard additions.[586]

Aluminium. Electrothermal atomization is the most successful method for serum[250] in the range 0.02–0.1 μg ml^{-1}. Tissues may be wet ashed with nitric/perchloric acid mixture in a PTFE vessel, or dry ashed and redissolved in nitric acid[502] prior to injection in the furnace using calibration by standard additions. Differences in aluminium results for bone and urine by electrothermal atomization and neutron activation analysis have been shown to be due to the presence of calcium and phosphorus.[272] After digestion of the sample at 60 °C with concentrated nitric acid, calcium chloride and dipotassium hydrogen phosphate, added to overcome effects of varying calcium and phosphorus contents, the residue was diluted to volume and analysed in the electrothermal atomizer. Work on aluminium in whole blood with a crucible type furnace was reported by Langmyhr.[448] 2 μl of heparinized and haemolysed undiluted blood were atomized at 2500 °C after two ashing stages. Aluminium was measured over the range 0.05–0.50 ppm.

Arsenic. The normal level of arsenic in urine is believed to be not more than 0.1 ppm, and a level of 1 ppm is considered to be evidence of harmful exposure. Poisoning is not necessarily inferred from the presence of arsenic in the stomach contents as it may not have been absorbed.

Because of the sensitivity of the hydride generation and electrothermal atomization methods for arsenic and the other similar volatile elements, direct flame atomization is now rarely used.

Samples cannot be prepared by simple dry ashing because arsenic is vaporized at temperatures well below 500 °C, though careful ashing with magnesium nitrate prevents loss. This method was used by Woidich[825] as a preliminary to redissolution and arsenic was then measured by hydride generation with sodium borohydride. Alternatively, biological materials can be wet oxidized with sulfuric/nitric acid mixture and hydrogen peroxide before hydride generation.[567]

In preparation for electrothermal atomization of arsenic, tissues have been wet oxidized with nitric/sulfuric acid mixture in presence of vanadium pentoxide in order to keep arsenic in the most highly oxidized form.[245] Nickel solution was added to the sample solution in the furnace atomizer in order to stabilize the arsenic and allow ashing at 630 °C before atomization. Arsenic was measured in urine[235] by treatment of the sample with hydrochloric acid and potassium iodide and extraction of the arsenic with carbon tetrachloride.

Gold. Gold is not believed to be an essential element though it may be accumulated as an environmental contaminant. Gold complexes, e.g. sodium aurothiomalate (Myochrysine) or sodium aurothioglucose (Solganol), are injected in the treatment of rheumatoid arthritis. The element may therefore have to be determined in the injection material itself[690] and in patients' sera during therapy with the object of avoiding the unpleasant side effects of overdosage. The change in serum gold levels with time following such overdosage was measured by Christian.[145]

Lorber *et al.*[459] used the method of standard additions to measure gold levels in serum, urine and synovial fluid, the diluent for serum being sodium dodecyl-sulfate solution.

For measurement of gold in blood and urine in treatment of rheumatoid arthritis, the following method details have been given by Schattenkirchner:[667]

> For flame atomic absorption, dilute plasma 1:1 and urine 1:3. Prepare standards containing 0.25, 0.5, 1 and 2 μg of gold per ml of 1:1 pooled serum or 1:3 normal urine. A small amount of methanol should be added to samples and standards to prevent foaming.
>
> For measurement by electrothermal atomization, dilute serum 1:4 with 0.1% Triton X solution, and urine 1:10 with 0.01 M hydrochloric acid. Calibrate by method of standard additions using standard gold solutions in the range 0.03–0.1 μg ml^{-1}. Inject 20 μl aliquots, dry at 100 °C for 30 s, ash at 500 °C for 30 s and atomize at 2400 °C for 10 s.

For low levels of gold in body fluids by flame atomic absorption the sample should first be digested with 6 N hydrochloric acid and potassium permanganate[74] at 50 °C, or with nitric/perchloric acid mixture, evaporated to dryness and redissolved in 3 M hydrochloric acid.[203] In both cases the gold is then extracted directly into MIBK and measured in an air acetylene flame. Gold has been determined directly in serum by electrothermal atomization, and Ottaway[379] has measured gold in individual protein fractions of blood serum either separated on cellulose acetate strips, or as solutions resulting from other separation procedures. Portions of the cellulose acetate strip measuring 5 × 8 mm were placed carefully and reproducibly in the graphite atomizer. Ashing could be carried out at 780 °C for 50 s but at higher temperatures appreciable amounts of gold were lost.

Beryllium. Beryllium and its compounds are insidiously toxic, causing berylliosis, affecting bone calcification mechanisms, displacing magnesium from, and inhibiting, magnesium-dependent enzymes. Bokowski[106] detected 0.002 ppm of beryllium in urine with the nitrous oxide acetylene flame, but more recently electrothermal atomization has been used. Direct atomization of urine from a graphite atomizer produces a similar lower limit of determination[340] but the presence of some cations and anions, in particular perchlorate, chloride and fluoride, affect the response of beryllium.[341] Oxidation during the ashing step is ensured in the method of Grewal and Kearns[297] who prepared samples and standards in 0.5% nitric, 4% sulfuric acid. The method of standard additions was advised if interference from calcium was suspected. Drying at 100 °C for 20 s, ashing at 1200 °C for 15 s and atomization at 2700 °C for 8 s is the recommended programme. Beryllium was measured in various biological matrices in the range 0.8–4.6 ng g^{-1} after decomposition with nitric acid in a pressure vessel. The beryllium was then extracted into benzene as acetylacetonate.[704]

Boron. Subtoxic levels of boron were measured in both serum and urine by aspiration into a nitrous oxide acetylene flame.[70] For bone and tissue, or for very low concentrations a wet oxidation procedure must be used as dry ashing leads to serious boron losses.

Bismuth. Bismuth can be measured either by hydride generation or electrothermal atomization. A procedure for the former is given by Rooney[640] in which samples of blood or urine are wet oxidized with nitric/perchloric acid mixture and ashed. The residue is taken up in hydrochloric acid, and an aliquot transferred to sodium borohydride solution contained in the hydride generation apparatus. Nitrogen is the flow-gas.

Bismuth has been extracted from body fluids using APDC/hexane prior to measurement in the furnace atomizer.[47] For 10–300 mg tissue samples, dried at 105 °C and ground, low temperature ashing in an RF dry asher with oxygen passing at 170 ml min^{-1} was followed by dissolution in 0.5 ml of 0.2 M hydrochloric acid.[195] After resumed ashing the residue was dissolved in 10 ml of 2% sulfuric acid. Drying 45 s at 100 °C; ashing 30 s at 470 °C; atomization 10 s at 2650 °C; calibration by method of standard additions.

Cadmium. There is as much interest currently in the measurement of bodily cadmium levels as there has been in lead. Workers exposed to fumes of cadmium may exhibit chronic poisoning and show much increased urinary excretion levels, e.g. 40–400 μg l^{-1} instead of 2–20 μg l^{-1}.[693] Ingestion can also be through eating contaminated shellfish, etc. Urine screening appears to be giving place to whole blood screening now that ultramicro methods have become established. Sensitivity is not lacking either in the flame or the graphite furnace atomizer even though the normal whole blood level of cadmium appears to be about one hundredth or less of that of lead. The main problem lies in deciding which of the very many published methods is most suitable.

Cadmium, like lead, is normally measured in whole blood. With the flame, an extraction method is necessary, and a procedure similar to the one given on p. 268 for lead, but using cadmium standards of suitably lower concentration, would be satisfactory. An alternative method[54] is to treat the sample with citrate, sodium diethyldithiocarbamate and Triton X-100. After 10 min, the cadmium is extracted with MIBK and aspirated into a lean air acetylene flame.

The Delves cup method has also been recommended for blood cadmium determination[137,231] and a sensitivity of 9 nmole l^{-1} with good accuracy has been attained.[766] Delves suggests[191] that if the smoke signal causes serious interference with the cadmium response the addition of phosphate will retard the latter and provide adequate separation.

The punched disc method of sample introduction into the Delves accessory, mentioned in the section on lead, has also been used for cadmium.[138]

For the measurement of cadmium in blood by electrothermal atomization some workers have wet oxidized the sample. Nitric acid and hydrogen peroxide was successfully used by Koirtyohann.[584] Lundgren[462] diluted samples 1:10 adding 20 μl of heparin and 20 μl of 2% Triton X-100 per 1 ml of prepared solution and taking 3 or 4 μl aliquots for atomization. His method relied on fast rise of atomization temperature to separate the cadmium absorption peak, which now appears before the main smoke peak.

The need for these additives is questionable and again, provided the atomization temperature can be rapidly attained, the following very simple dilution procedure will be found suitable:

> Dilute the whole blood $1+9$ or $1+19$ with deionized water. Prepare standards containing 0, 1, 2, 3 μg of cadmium per 100 ml. Use method of standard additions by injecting 20 μl of diluted sample into the atomizer, drying and then adding, for four separate atomizations, 20 μl of each standard. It is essential to use simultaneous background correction. Dry for 20 s at 100 °C, ash for 50 s at 450 °C, atomize for 5 s at 2050 °C.

For measurement of cadmium in urine by electrothermal atomization, the sample should not need pretreatment provided that efficient background correction is available. 10 μl of original sample are placed in the furnace. Alternatively, an APDC–MIBK extraction can be used.

In the analysis of tissue and other solid samples, the dry ashing procedure should be used with great care and ashing should not take place at temperatures more than about 420–450 °C. Wet oxidation procedures with nitric/sulfuric acid mixture are commonly used. Ottaway[569] determined 0.1–100 μg of cadmium per gram of dried liver or kidney tissue after digesting 2–100 mg in this mixture.

Cobalt. Electrothermal atomization allows direct measurement in serum and whole blood[543] in the range 5–10 ng ml^{-1}. Whole blood may be diluted with water, but serum is first dried at 750 °C for 5 min and the residue transferred to the atomizer. For tissue analysis, it may be sufficient to ash at 900 °C in the atomizer provided a rapid temperature rise to the atomize temperature (~ 2800 °C) can be achieved.[463]

For flame analysis of fish tissue, the freeze-dried and homogenized sample was ashed at 480 °C, dissolved in nitric acid and evaporated.[371] The residue was then dissolved in hydrochloric acid and cobalt extracted with APDC–MIBK for aspiration into the air acetylene flame.

Chromium. Chromium is used as a faecal marker and there is concentration with time in the spleen. Plasma chromium concentration has been related to glucose tolerance and may be of some use in diabetes diagnoses.

Williams *et al.*[800] measured chromic oxide in faecal material after ashing, and Feldman[228] dry ashed serum, plasma, whole blood, urine and diet at 550 °C as well as using a wet oxidation method.

Most chromium analyses now appear to be carried out on biological materials down to about 1 ng g^{-1} with electrothermal methods. If oxidizing acids have been used, conditions must be chosen to prevent chromium being lost as chromyl chloride during the ashing step. In a direct method for serum and plasma,[59] nitric acid and hydrogen peroxide were added to the furnace followed by an aliquot of the sample.

For measurement of chromium in urine little sample preparation appears to be necessary. Ross et al.[649] achieved selective volatilization by drying 20 µl of sample at 230 °C, ashing at 1300 °C and atomizing at 2700 °C.

Flame determination of chromium in urine was carried out after the chromium had been coprecipitated with aluminium hydroxide.[131] The sample is saturated with ammonium chloride, and ammonium hydroxide and aluminium chloride solutions added. After filtering, the precipitate is redissolved in sodium hydroxide solution with hydrogen peroxide, and chromium measured in an air acetylene flame.

Manganese. Although manganese levels in most biological samples are very low (ng g^{-1} levels) there is an increasing interest in its measurement. Mahoney et al.[475] used the method of standard additions for flame determination of manganese in serum, but most recent work has been done with an electrothermal atomizer. 50 µl samples of serum and cerebrospinal fluid may be atomized without extraction[175] and it was confirmed that a direct method compares well with the APDC–MIBK extraction procedure.[648]

A simple direct procedure is as follows:

> Dilute the serum 1+9 with deionized water. Prepare calibration standards containing 0, 1, 2, 5 µg of manganese per 100 ml. Use the method of standard additions by injecting 20 µl of the diluted serum into the atomizer, drying, and then adding, for four separate atomizations, 20 µl aliquots of each standard. It is essential to use simultaneous background correction. Recommended atomization programme is: dry for 20 s at 110 °C, ash for 50 s at 1100 °C, atomize for 5 s at 2850 °C.

Small animal tissue samples have been analysed electrothermally after being dried at 110 °C for 48 h;[108] 100 mg are dissolved in 2 ml of concentrated nitric acid. 5 µl of the clear acid solution are then placed in the furnace, dried at 120 °C for 20 s, ashed at 800 °C for 70 s and atomized at 2400 °C for 8 s. Calibration by method of standard additions is recommended as sodium and potassium may interfere.

Mercury. Both environmental pollution studies as well as industrial hygiene prompted considerable interest in this element in the late 1960s and the introduction of the cold vapour atomic absorption method has enabled it to be measured in many types of sample at very low levels.

Workers handling mercury metal or compounds face risk of poisoning and the analysis of urine still provides the most-used screening test. The severity of symptoms, however, is not always in direct relation to the level found. Most atomic absorption methods do not distinguish between different forms of ingested mercury, and therefore diuretics and mercurial antiseptics are possible sources, as well as poisoning. Some work has been done, however, on the speciation of mercury within biological samples, human[519] and otherwise.[58,336]

Prevention of loss of mercury by volatilization is the main aim of methods of sample preparation. Neither dry ashing nor hot digestion in the open can be considered. Lindstedt[457] described the preparation of urine where the organic matter was oxidized by standing 1 ml samples in the cold overnight with 1.5 ml of 6% potassium permanganate solution and 0.2 ml of sulfuric acid. The sulfuric acid was added first and the mixture kept cool before addition of the potassium permanganate. The next day, the excess potassium permanganate was reduced with 0.3 ml of 20% hydroxylamine hydrochloride solution. This solution was then transferred to the cold vapour apparatus and reduced with stannous chloride. Many subsequent methods for urine differ only in detail, e.g. some workers thought it necessary to carry out the wet oxidation at 0 °C,[431] others used cadmium chloride and stannous chloride as reductant[451] or used potassium persulfate in the oxidation process in an automated version of the method.[632] In most such methods the concentration of sulfuric acid was increased.

In other methods the urine was first treated with l-cysteine[205,458] which prevents loss of mercury during the addition of the reductant,[263] e.g. 5 ml of sample were mixed with 2 ml of 1% cysteine, then 0.5 ml of stannous chloride was added. After 3 ml of 30% caustic soda were added, the mercury vapour was swept out with air.

Normally, if only inorganic mercury is to be measured, the analysis of urine requires no oxidation stage.[135] However, sodium borohydride is said to reduce rapidly all forms of mercury including methyl mercury, the sample being simply buffered.[745]

Most methods for determining mercury in tissue employ the sulfuric acid/potassium permanganate oxidation step. Uthe et al.[758] believed a temperature of 50–60 °C to give complete recovery of mercury from fish tissues. A temperature of up to 70 °C is permissible for digestion in the open, and the following procedure is safe and reasonably fast:

Place 0.4–0.8 g of homogenized sample in a weighed 150 ml conical flask with a B24 socket neck. Reweigh and cover with a watchglass. Pipette 5 ml of sulfuric acid into the flask and place in a water bath at 70 °C for 1 h, at the end of which the solution should be homogeneous though highly coloured and perhaps a little cloudy. Cool in an ice bath, add carefully 50 ml of 6% potassium permanganate solution and replace in the water bath for 2 h. Cool to room temperature and add, by pipette, 15 ml of 20% hydroxylamine chloride. Prepare two reagent blanks by taking 2 ml of water through the same procedure.

The measurement is made with the apparatus shown in Fig. 23. Immediately before connecting the socket of the flask to the Drechsel bottle head of the measuring apparatus, introduce 2 ml of 10% stannous chloride solution, then switch on the pump. The absorbance value increases to a constant level in 1–2 min. When this value has been recorded, disconnect the ground glass joint and continue the aeration until the absorbance returns to its minimum value. The equipment is then ready to receive the next sample. Calibration solutions containing 0–3 ml of mercuric sulfate solution, 0.1 mg

Hg l^{-1} ($\equiv$ 0–300 ng Hg) are taken through the measurement sequence after reduction with stannous chloride solution as in the preceding paragraph. The standards should be prepared immediately before analysis. Plot a graph of ng Hg against absorbance, subtracting the absorbance of the zero standard from the other standards. Subtract the reagent blank absorbance from sample absorbance readings and read off ng Hg from calibration graph. The reagent blank should not be more than 30 ng of mercury, and if it is, the reagents and glassware should be checked.

The properly digested mixture may contain suspended hydrated oxides of manganese. These should dissolve completely giving a colourless or slightly opalescent solution when the reductant is added. Digestion with sulfuric acid and potassium permanganate is said to release mercury in all forms, including that present as methylmercury.[564]

Paus recommended that the oxidation of tissue with sulfuric acid and potassium permanganate should be carried out at higher temperatures, e.g. 100 °C in an enclosed PTFE dissolution vessel[579] and measured mercury and a number of other elements in the clear solution thus given. More recently, a temperature of 180 °C has been used in such a vessel.[313]

Nickel. Nickel and cobalt have been measured by flame directly in urine[700,808] and may also be determined after extraction with APDC–MIBK.

Blood and other samples may be analysed for nickel by electrothermal atomization after the sample has been ashed at about 560–600 °C, the residue dissolved in 1 N hydrochloric acid, and extracted as either the dimethylglyoxime or APDC complex into MIBK.[837]

Palladium and Platinum. These two metals are of very similar sensitivity and are conveniently measured in the same sample solution. Interest in the measurement of platinum has arisen since it has been learned that this element, if ingested in appropriately active form, can have long term pathological effects. Electrothermal atomization is the preferred method, and levels down to 0.003 μg g^{-1} may be measured in blood and urine after decomposition of the sample with nitric/perchloric acid mixture.[366] Lead, platinum and palladium were all measured[742] in biological samples, after these had been digested, by extraction with tri-*n*-octylamine into xylene.

Phosphorus. Total inorganic and organically bound phosphorus may be measured in serum with electrothermal atomization and an electrodeless discharge lamp source. Calibration is by the method of standard additions, using four 20 μl aliquots of undiluted serum for each sample with additions appropriate to the range.

Selenium. Some concentration is required, even for electrothermal atomizers, and a non-extractive method was given by Ihnat[347] for 1 g tissue samples.

The hydride evolution method of Clinton[151] also requires 1 g of blood per determination. This is digested in 7 ml of nitric/perchloric acid mixture 2:5, two drops of kerosene also being added to prevent frothing. The temperature is raised to 210 °C and charring is avoided. Fumes of perchloric acid are allowed to continue for 15 min and after addition of 2 ml of hydrochloric acid, the solution is transferred to the hydride generator. Sodium borohydride solution (5 g in 100 ml of 0.1 % m/v caustic soda solution) is the reductant.

Strontium. The similarity in chemical behaviour between strontium and calcium causes strontium to be absorbed in bone, a cause for serious concern because of the increase in radiostrontium from nuclear fallout. This similarity can be utilized in chemical preparation and concentration steps by co-precipitating strontium and calcium and then determining strontium after redissolution. Curnow et al.,[166] analysing serum, redissolved in the lanthanum-containing diluent, while Montford and Cribbs,[529] analysing urine, co-precipitated with lanthanum and redissolved in hydrochloric acid. Tompsett[747] used Dowex 50W cationic exchange resin to concentrate strontium after dry-ashing samples of urine and faeces. Strontium has been measured in serum at the 10^{-9} g level by electrothermal atomization.[88]

Thallium. There is some danger of thallium poisoning as in some countries this metal is used in rat poisons. Its action in the body is similar to that of lead but in atomic absorption it is more sensitive. Curry et al.[167] described methods for determining thallium in biological material. Blood, tissue and stomach contents were digested with sulfuric and nitric acids. The thallium was then converted to bromide at pH 3–4 (adjusted with ammonia) and extracted into water-saturated methyl isobutyl ketone. This specific extraction method for thallium can be used for both atomic absorption and flame emission measurements. It is suggested that when atomic absorption is used, specificity is not required and thallium can be extracted with sodium diethyl dithiocarbamate instead of as bromide. Tissue, blood and urine were treated by this method, digesting as above, but adjusting the pH to 5–6 with sodium hydroxide solution. Absolute detection limits were 40 ng for the extraction method and 200 ng for a direct aqueous method, requiring 1 ml of prepared solution. A detection limit of <2 ng was reported when using the tantalum boat technique, which required 50 μl of urine or prepared sample.

Thallium may be measured in urine by electrothermal atomization by simply diluting 1+9 with 1 % sulfuric acid in order to overcome the effects of presence of chlorides.[258]

Vanadium. Vanadium and aluminium were determined in biological tissue[430] by ashing 30 mg of the dried sample at 650 °C for 18 h and dissolving the residue in 50 μl of 5 % nitric acid.

INDIRECT ATOMIC ABSORPTION METHODS

Indirect atomic absorption methods are those in which the species being determined is not itself measured, and very often is not even measurable. The measurement actually made is that of the absorption of an element which takes part in a stoichiometric reaction with the wanted species. Sometimes the procedures are very ingenious and one is tempted to wonder if atomic absorption is really the most appropriate technique to be involved in such cases. In practice, however, the method may work out well. Consider the determination of sulfate by precipitation of barium sulfate from a solution of known barium content followed by measurement of the barium remaining in solution. If an atomic absorption instrument is available, this procedure is very much quicker than the classical gravimetric finish. The precipitate needs only to settle (aided by centrifugation if necessary) and the result is not subject to the errors caused by sorption of other compounds or by filtration losses.

A number of different types of indirect atomic absorption methods were listed by Pinta.[587] Some anions or other organic species can be measured by their interfering effects upon the absorbance of some metal. This is perhaps the most obvious example. Precipitation of the species to be determined, followed by atomic absorption of the excess precipitant, is another, and 'chemical amplification' such as formation of a heteropoly acid from phosphorus, silicon, vanadium, etc. and determination of the molybdenum which is present in a much greater proportion in the resulting molecule, is a third.

Some examples of indirect methods are given here, but many analytical chemists will be able to apply similar principles to their own specific problems.

Use of interfering effects

Oxyacids. Aluminates, phosphates, sulfates, silicates and other oxyacids exert a considerable depressive effect on the atomic emission and absorption of calcium and strontium in air acetylene or air propane flames. The reasons for this have been discussed earlier in this book. Pinta[587] has shown that it is possible, using 20 ppm of calcium as the basis for measurement, to obtain a graphical relationship for concentration of phosphate (P between 4–20 ppm) and sulfate (S between 4 and 10 ppm). A method for determining phosphate in waters, rocks and steel was described also by Singhal *et al.*[687] making use of the depression of strontium absorption.

Fluorides. In low temperature flames, fluoride depresses the response of magnesium.[110] Fluoride between 2 and 15 ppm can be measured by its effect on the absorbance of 10 ppm of magnesium. Phosphate and sulfate interfere, but the other halides and many metals do not.

Fluoride increases the response of both zirconium and titanium in the nitrous oxide acetylene flame, and this phenomenon has been used by Bond and Willis[111] also to determine fluoride.

Nitrogenous Materials. The same workers showed that ammonia, amines, amino-acids and other organo-nitrogen compounds also increased the response of zirconium. For example, concentrations between 1.7 and 85 ppm of ammonia increase the absorbance of 1350 ppm of zirconium rationally, in the presence of 0.006 M potassium chloride as ionization buffer.

Other Organic Materials. The depressive effect of protein on calcium has already been discussed. Sugars exert a similar effect on calcium in low temperature flames[146] and concentrations below 10^{-6} M may be related to the absorbance of 25 ppm of calcium. Proteins were determined in a similar procedure. As the concentration of these organic materials increases, the depressive effect reverses into an 'enhancing' effect. The value of such methods is therefore limited to certain concentration ranges.

Determination of Traces of Aluminium and Titanium. The increase in absorbance of iron in a stoichiometric air acetylene flame in the presence of very low concentrations of aluminium and titanium was reported by Ottaway,[570] and this observation was made the basis of a method in which the detection limits of these two elements were 0.02 and 0.01 ppm respectively.

General. 'Interference' methods clearly have a limited application as the extent of the effect depends critically on flame temperature and flame chemistry. Flame conditions must therefore be stable and reproducible. Furthermore, the effects are rarely specific. They can usually be employed only if the sample solution contains no variable constituent other than the substance to be measured.

Precipitation and substitution methods

Sulfates are determined by precipitation with a known quantity of barium, followed by determination of the excess by atomic absorption. Ecrement (see Pinta)[587] determined between 5 and 60 ppm of sulfate, adding 0–100 ppm of barium. An alternative approach would be to determine the precipitated barium by separation of the barium sulfate, and redissolving in ammonium EDTA solution.

Rose and Boltz[647] determined sulfur dioxide by formation and precipitation of lead sulfate.

Chlorides (5–100 ppm) may be precipitated with the addition of 300 ppm of silver.[587] Determination is made either of the remaining silver, or of that precipitated after separation and redissolution in ammonia.

Fluorine has been determined in compounds by mixing with silica and potassium sulfate and heating to form silicon tetrafluoride, which may either be passed into a hot flame or decomposed in a heated cell. In either case, the silicon absorption is measured.[300]

In a more complex procedure fluorine has been determined after decomposition of its compounds with sodium biphenyl. The sodium fluoride formed has the effect of depressing the extraction of iron from aqueous ammonium thiocyanate solution into MIBK[391] and the species finally measured is iron.

A procedure for determining low levels of cyanide is based on the formation of the complex ion $Cu(CN)_3^-$ with basic copper carbonate in alkaline solution.[478] The copper thus solubilized is measured by atomic absorption, the sensitivity being claimed to be 2×10^{-5} M CN^-. In another method, similar in principle, the neutralized sample solution is passed over insoluble cuprous cyanide. Copper is measured in the soluble complex formed.[372]

Oxalic acid can be determined by precipitation with calcium, separation, redissolution and measurement of calcium. This is the basis of its determination in urine.[509] Thiol groups have also been measured in protein by their ability to combine with mercury.[130] After separation of the mercury-containing fraction by gel filtration, mercury is determined by a standard cold vapour method.

Reducing agents may be determined by precipitation of copper (II). Between 0.1 and 5 ppm of dextrose were determined by Potter *et al.*[597] using this principle.

Chemical amplification with heteropoly acids

Ammonium salts of heteropoly acids are formed in solution between ammonium molybdate and phosphate, silicate, titanate, vanadate, arsenate, niobate and germanate ions. The compounds formed are stoichiometric under given conditions of temperature, pH, etc. and are extractable, often selectively, into organic solvents. This principle is, of course, well known in spectrophotometry. The fact

TABLE 23. Indirect methods using heteropoly acids

	Complex extracted	Mo/el. ratio	Extraction		Ref.
			pH	extractant	
Phosphorus	phosphomolybd.		pH 0.7	isobutyl acetate	
Arsenic	arsenomolybd.		pH 0.7	(1+1) ethyl acetate+ butanol	618
Silicon	silicomolybd.	41	pH 3.2	MIBK	
Phosphorus	phosphomolybd.		0.15 M HCl	chloroform/butanol	412
Titanium	P–Ti–molybd.		0.15 M HCl	butanol	
Titanium	P–Ti–molybd.	24	pH 1	isobutyl acetate	587
Vanadium	P–V–molybd.	20–6	ammonia	(4+1) diethyl ether+ pentanol	356
Niobium	P–Nb–molybd.	11.3	0.5 M HCl	butanol	411

of particular interest in atomic absorption is that the ratio of molybdenum to combining element is very high. Silicon forms the complex $NH_4SiO_4(MoO_3)_{12}$ and the weight ratio of molybdenum to silicon is 41:1. If the molybdenum absorbance is measured there is thus a considerable 'amplification' factor over the silicon and this is further increased by the extraction factor and the improvement expected from the use of an organic solvent.

Zaugg and Knox[838] used this principle to determine phosphate, and Kirkbright *et al.*[410] extended it first to the sequential determination of phosphorus and silicon, and later to niobium[411] and titanium.[412]

Hurfurd and Boltz[339] also discussed the determination of phosphate and silicate and Ramakrishna *et al.*[618] gave procedures for these elements and arsenic.

Some details and references are summarized in Table 23.

Determination of compounds forming extractable metal chelates

In the same way as metals are extracted quantitatively after addition of excess chelating agent, so may small quantities of the chelating agent be extracted after addition of excess of a suitable metal. The extracted metal chelate may then be aspirated to obtain the chelating agent concentration by measuring the metal in atomic absorption, or alternatively the excess metal remaining in the aqueous solution can be measured, giving the chelate by difference.

Dithiocarbamates, 8-hydroxyquinoline and EDTA have all been determined in this way.[587] Extractable complexes with metal chelates are also given by some compounds. Kumamaru[432] determined phthalic and nitric acids by extracting the complexes which they form with the copper (I)–neocuproin chelate. Sensitivity for phthalic acid was between 4×10^{-6} and 4×10^{-5} M. Collinson and Boltz[154] determined perchloric acid (0.2–6 ppm) with the same chelate. Thiocyanates may be determined by the chloroform-extractable $Cu(py)_2(SCN)_2$ complex with pyridine and iodates[833] by their nitrobenzene-extractable cadmium–1 : 10 phenanthroline complex. An indirect determination of mercury (II) by extraction with 1,2'-bipyridyl zinc chelate followed by determination of the zinc was also described.[834] The sensitivity of mercury thus becomes similar to that of zinc, but a number of metals may interfere.

Organic bases such as alkaloids may be extracted into MIBK as their phosphomolybdate complex and measured as molybdenum.[683]

A number of indirect methods for drugs have already been mentioned on p. 247.

Oxidizing and reducing agents

Making use of the fact that chromium (VI) can be extracted from acid solution into methyl isobutyl ketone, if part of the chromium is first reduced to Cr(III) by another compound, for example iodide, the latter can be determined in relation to the chromium actually extracted.[146] Conversely, iron is converted from Fe(II) to Fe(III) by iodates and the latter can be determined by extracting the resulting Fe(III) in presence of hydrochloric acid into ether or another more suitable solvent.

APPLICATIONS OF ATOMIC FLUORESCENCE

The number of known genuine analytical applications of atomic fluorescence spectroscopy remains small though much valuable information has been published by research workers on the sensitivity of various elements with different sources of excitation and in different atomizing flames. Much of this was initiated by West, who discussed[788] the practicality of atomic fluorescence as an analytical technique. He suggested that scatter and quenching effects should not generally limit the sensitivity that can be obtained. The selection of atom reservoir is more significant as a choice may have to be made between one which gives best sensitivity for the test element and one which achieves complete vaporization of the sample. West reviewed the method again in 1976.[789]

A comprehensive textbook devoted to atomic fluorescence spectrometry has also been published.[17] This gives fluorescence characteristics of thirty-nine elements but also confirms that relatively few exclusive applications of atomic fluorescence to the analysis of 'real' samples, i.e. where atomic fluorescence solves the problem better than any other technique, have been reported. The tendency in some quarters to blame lack of progress on the instrument manufacturers who do not provide atomic fluorescence instrumentation, is unfortunate. If the method had shown the same potential versatility, appeal and ease of application as atomic absorption, with sufficient advantages, the commercial suppliers would certainly have responded to the situation.

Many papers published on atomic fluorescence methods have either described apparatus of a specialized or unconventional nature (e.g. cathodic sputtering as an atom source for fluorescence measurements,[292] use of a non-dispersive spectrometer)[554] or relate to very specialized applications, which, though undoubtedly interesting, are not particularly helpful to readers of a practical text book. Most probably, during the foreseeable future, worthwhile applications of atomic fluorescence will be limited to those elements, particularly zinc, cadmium and mercury, which offer especially good sensitivity without incurring the cost of electrothermal atomization.

Choice of fluorescence line

The most useful analytical fluorescence line is usually the basal resonance line,[789] i.e. the one of lowest frequency. This has the added virtues of being the easiest to excite, has maximum stepwise contribution and would be affected less by scatter than resonance lines of higher frequency. The actual sensitivity, however, depends on a number of factors both spectroscopic and instrumental. The best lines, as recorded or recommended by Sychra, Svoboda and Rubeska,[17] are included in the element tables in Chapter 9.

Flame atomization

An ideal flame atomizer for atomic fluorescence should give the highest degree of atomization, lowest emission background and flame noise and best freedom from quenching effects. These requirements are, unfortunately, usually incompatible. Source modulation allows the emission signal to be rejected but does not overcome the problems of scatter or associated noise. Much use has therefore been made of hydrogen-based diffusion flames, though because they allow physical and chemical matrix interferences and have low atomization efficiencies, they are effective only for a limited range of elements. The premixed air hydrogen flame is much more useful, but probably best of all are premixed laminar flow hydrocarbon flames where separation of the chemiluminescent outer diffusion mantle with a concentric stream of nitrogen or argon allows the fluorescence to be generated in a virtually non-radiative zone and also protects the atoms from combination with atmospheric oxygen.

Using a continuum source, the 450 W xenon arc, Cotton and Jenkins[157] analysed kerosene for traces of copper, iron and lead. A specially made burner gave premixed flames with hydrogen or with the liquid hydrocarbon itself as fuel. Non-luminous low background flames were produced both with kerosene and with benzene, resulting in detection limits for copper 0.004, iron 0.04 and lead 0.06 ppm. Matousek and Sychra[495] reported detection limits for iron 0.02, cobalt 0.01 and nickel 0.003 ppm, in an oxygen argon hydrogen flame, and for gold[496] 0.005 ppm in a similar separated flame. A detection limit of 0.04 ppm for palladium was obtained[724] with the same type of flame, and a nitrogen separated air acetylene flame gave a detection limit of 0.007 ppm for nickel in petroleum products and residues, good calibration linearity being shown from 0.01 to about 5 ppm of nickel in solution with the line 232.0 nm.

Zinc and cadmium analyses have been investigated by several groups of workers. In particular an automatic system for correction of scatter in atomic fluorescence[617] was applied by Rains et al. to the analysis of various reference materials[216] including minerals and plant material. Samples were digested with nitric/hydrofluoric/perchloric acid mixture in a teflon vessel evaporated to fuming and redissolved in water prior to aspiration into an air acetylene or argon hydrogen flame. Fell et al.[230] determined cadmium and zinc in plasma, blood and urine by a simple dilution procedure and aspiration into an air hydrogen flame.

In the determination of cadmium in sea and surface waters the addition of perchloric acid was shown to reduce the degree of scatter.[301]

An argon hydrogen flame was used by Norris and West[556] for the determination of antimony in metallurgical samples. Copper samples were dissolved in hydrochloric/nitric acid mixture, and after addition of sodium nitrite, the antimony was extracted into MIBK.

In a continuation of the atomic fluorescence studies already mentioned (p. 107) Ebdon et al. measured tin in steel (after removing the silicon) with a detection limit of 0.05 $\mu g\ g^{-1}$ in an argon separated air acetylene flame and 0.01 $\mu g\ g^{-1}$ in an argon oxygen hydrogen flame.[206]

Electrothermal atomizers

The relative improvements in sensitivity which are a feature of the 'total sample' atomization methods described on p. 54 et seq. can also be attained in atomic fluorescence. The use of non-flame atom reservoirs has been discussed at some length by Winefordner.[820] Though there is little evidence so far of their application outside the research laboratory, there is little doubt that these enable a very great degree of sensitivity to be achieved.

Both the Massmann furnace[491] and West's carbon filament[792] have been employed for this purpose. The accessibility of the atom cloud produced by the latter device would seem to make it ideal for fluorescence measurements, and detection limits of about three orders of magnitude better than for flame fluorescence have been reported. Actual limits of detection quoted[61] were magnesium and silver 10^{-12} g, lead and bismuth 10^{-11} g, thallium and gallium 5×10^{-11} g, and zinc 2×10^{-14} g. This last result then represented one of the best sensitivities ever achieved with an optical spectroscopic technique. With the filament unenclosed but shielded with argon Alder and West[43] found a detection limit of 1.5×10^{-13} g for cadmium.

It must be remembered that the time span of an atomic fluorescent event will certainly be less than the response time of the amplifier-readout system of an instrument designed solely for flame work. Special fast response electronics, and digital peak seeking or peak area measuring facilities are needed instead of the traditional recorder output. Atom lifetimes with the 'open' rods, ribbons etc. are shorter than with tubular furnaces because after generation at the heated surface of the atomizer, the atoms come into immediate contact with a cold environment where they rapidly condense into polyatomic aggregates or form molecules. The presence of other species makes the latter process even faster.

Mercury by atomic fluorescence

The analysis for which atomic fluorescence appears to offer the most advantages is the determination of mercury. The total amount of mercury present in a particular sample can be brought into the fluorescence cell at one time without the expense of an electrothermal atomizer and without the quenching effects present in a flame. With the cold vapour method there is minimal background interference and scatter. Argon must be used as the carrier gas as air quenches the fluorescence of mercury. Very simple instruments have been constructed for this analysis, needing only a linear amplifier. Muscat et al.[541] used, as source, a mercury pen lamp (Ultraviolet Products Inc.), an f.7 scanning monochromator, and a solar blind photomultiplier. A small volume fluorescence cell was constructed with Vycor windows. Samples were either oxidized with sulfuric acid and potassium permanganate, then reduced with stannous chloride, or oxidized in a stream of argon and oxygen at a temperature of 800 °C. In both cases, the elemental mercury formed was collected on a silver amalgamator. After collection, the amalgamator was heated to liberate the mercury which was passed in a single 'slug' through the fluorescence cell.

Thompson and Godden[738] used a commercial atomic absorption instrument, but described a mercury fluorescence unit in which they decreased the cell volume to improve sensitivity and reduce measurement time so that a collector was not necessary. A Philips OZ4W low pressure mercury lamp, operated at an optimum current of 0.38 A, was cooled at the rear surface, also to improve sensitivity and baseline stability and to minimize the formation of ozone which has a strong absorption at 253.7 nm. Instead of a formal fluorescence cell, the mercury vapour was emitted into the open but contained within a shield of argon. This arrangement proved more transparent, and was not subject to deposits on cell windows. The absolute detection limit claimed for this system is 0.02 ng.

A reduction vessel which permits the mercury to be swept out of solution with greater efficiency with a smaller volume of gas[309] had the sample solution placed above a frit through which the carrier gas (argon) was bubbled. This type of reduction cell was also used by Caupeil et al.[132] whose non-dispersive atomic fluorescence instrument consisted simply of a mercury electrodeless discharge lamp and a solar blind photomultiplier, which detected both 253.7 and 184.9 nm mercury lines. The apparatus was operated in a room illuminated only by sodium lamps. There was no formal optical cell, the mercury vapour being emitted vertically downwards, from a tube connected to the reduction vessel, immediately before the source lamp and detector aperture. A rapid preparation procedure for fish tissues was also described. One gram of sample was mineralized in twenty minutes with nitric/sulfuric acid mixture and hydrogen peroxide in the presence of vanadium pentoxide. This was heated under a very efficient condenser so that no mercury was allowed to escape. Detection limits are claimed to be in the μg kg^{-1} range.

Cavalli and Rossi[134] used a gold collector for the mercury vapour and designed optics of high light-gathering power to operate with a Philips 90W low pressure mercury lamp and a gas-shielded windowless cell. Their claimed absolute detection limit was 0.03 ng.

Chapter 9

Analytical data for the individual elements

The following information is tabulated, when known, by element in alphabetical order:

Atomic number; atomic weight.

D_o, the dissociation energy of the oxide in eV.*

E_i, the ionization energy in eV.*

f, the oscillator strength.*

Wavelength of strongest absorption line, with a guide to typical reciprocal sensitivity and detection limit in flame and electrothermal atomization[795] as obtained on a Pye Unicam SP 1900 or SP 192 Atomic Absorption Spectrometer, unless stated otherwise (electrothermal figures quoted in brackets are taken from Fuller).[19]

Wavelengths of other, less sensitive, lines suitable for use in atomic absorption with a factor showing their sensitivity relative to the strongest line.

Best flame type.

Guide to ashing and atomizing temperatures in electrothermal atomization from author's laboratory.

Weights of reagents and suitable solvents for making up 1 litre of 1000 mg l^{-1} stock solutions; dissolve the amount of material in the solvent given, dilute to 1 litre in a volumetric flask with deionized water. Store in polythene bottles.

Organometallic reagents for making up non-aqueous solutions with source of supply.

B = British Drug Houses Chemicals Ltd., Poole, Dorset, UK. E = Eastman Organic Chemicals, Rochester, NY, USA and Kodak Ltd., Kirby, Liverpool, UK. M = E. Merck, Darmstadt, West Germany.

Useful solvent extraction systems.

Some of the more widely experienced interferences.

Suitable flame emission lines.

Best atomic fluorescence lines.[17]

* Taken mainly from Rubeska and Moldan.[4]

ALUMINIUM   Al

PHYSICAL CONSTANTS

Atomic number	Atomic weight	D_0 eV	E_i eV
13	26.98	5.0	5.98

CHARACTERISTIC LINES

λ	f		Flame AAS			Electrothermal atomization			
nm		Flame type	Rec. sens. ppm	Det. lim. ppm	Ashing °C	Atom'n °C	Rec. sens. pg	Det. lim. pg
309.3	0.23	N_2O C_2H_2 rich	0.34	0.01	1100	2850	44	(5)
308.2	0.22		(1/1.5)					
396.15	0.15		—					
394.4	0.15		(1/2)					

STANDARD AQUEOUS SOLUTIONS

1000 mg l^{-1}	Wt (g)	Reagent	Dissolution
Al	1.000	Metal foil	25 ml conc. HCl+few drops HNO_3
Al_2O_3	0.5294	Metal foil	20 ml conc. HCl+few drops HNO_3

REAGENTS FOR NON-AQUEOUS SOLUTIONS

Aluminium cyclohexanebutyrate, aluminium 2-ethylhexanoate.

EXTRACTIONS

(i) Cupferron–MIBK; (ii) 8 Quinolinol–chloroform or MIBK (pH 8).

INTERFERENCES

Aluminium can be as much as 15% ionized in a nitrous oxide acetylene flame. This is considerably reduced in presence of 0.1% potassium or lanthanum, which must be added to both sample and standard solutions. 'Enhancements' of aluminium response have been reported in presence of iron and titanium, and in fluoroborate media.

In electrothermal atomization aluminium sublimes readily as $AlCl_3$, hence chloride should be absent. Aluminium also forms nitride when nitrogen is used as the purge gas.

FLAME EMISSION

396.2 nm, stoichiometric nitrous oxide acetylene flame.

ATOMIC FLUORESCENCE

396.2 nm, nitrogen shielded nitrous oxide acetylene flame.

Sb ANTIMONY

PHYSICAL CONSTANTS

Atomic number	Atomic weight	D_0 eV	E_i eV
51	121.75	3.2	8.64

CHARACTERISTIC LINES

λ	f	Flame AAS			Electrothermal atomization			
		Flame type	Rec. sens.	Det. lim.	Ashing	Atom'n	Rec. sens.	Det. lim.
nm			ppm	ppm	°C	°C	pg	pg
206.8	0.1	Air C_2H_2 stoichiometric	0.25	0.02	500	2800	28	(5)
217.6	0.045		(1/1.2)					
231.2	0.03		(1/2)					

STANDARD AQUEOUS SOLUTIONS

1000 mg l^{-1}	Wt (g)	Reagent	Dissolution
Sb	1.000	Metal shot	10 ml conc. HCl + few drops conc. HNO_3 + 10 g tartaric acid

EXTRACTIONS

APDC–MIBK pH 3–5.

INTERFERENCES

High concentrations of both copper[536] and lead[813] cause spectral interference by line overlap at 217.6 nm, hence 206.8 nm or 231.2 nm are preferred for samples containing these as major elements. A remarkable freedom from interferences was noted for antimony in both air acetylene and argon hydrogen flames (Johns).[359]

Antimony is nevertheless best determined using a hydrogen generation system, using borohydride as the reductant. The detection limit should be about 1 ng. In electrothermal atomization, antimony is readily volatilized and should be stabilized during the ashing step by addition of molybdenum salts.

ATOMIC FLUORESCENCE

217.6 nm, argon hydrogen or argon oxygen hydrogen flame.

ARSENIC   As

PHYSICAL CONSTANTS

Atomic number	Atomic weight	D_0 eV	E_i eV
33	74.922	4.9	9.81

CHARACTERISTIC LINES

λ	f	Flame AAS			Electrothermal atomization			
		Flame type	Rec. sens.	Det. lim.	Ashing	Atom'n	Rec. sens.	Det. lim.
nm			ppm	ppm	°C	°C	pg	pg
193.7	0.095	Air C$_2$H$_2$ just luminous	0.50	0.20	350	2200	38	(20)
193.7	0.095	Argon hydrogen*	0.22	0.10				
193.7	0.095	N$_2$O C$_2$H$_2$ lean						
197.2	0.07	Air C$_2$H$_2$ just luminous	(1/2)					
197.2	0.07	Argon hydrogen						
189.0		Argon hydrogen						

* Argon hydrogen flame: 1500 cm^3 min^{-1} hydrogen, 6 l min^{-1} argon.

STANDARD AQUEOUS SOLUTIONS

1000 mg l^{-1}	Wt (g)	Reagent	Dissolution
As	1.0000	Metal powder	50 ml conc. HCl or 50 ml conc. HNO$_3$
As	1.3200	As$_2$O$_3$	50 ml conc. HCl or 50 ml 0.880 NH$_4$OH or 20 ml 5% w/v NaOH

EXTRACTIONS. APDC–MIBK [As(III), pH ≯ 4].

INTERFERENCES

Many minor interferences in air acetylene flame, but major background problem as this flame absorbs/scatters at least 60% of the incident radiation at 193.7 nm, and much more at 189.0 nm.

The nitrous oxide acetylene flame has lower sensitivity but is much more transparent and has fewer elemental interferences.

The argon hydrogen flame shows least background, but there are major depressive interferences from many cations. This flame is therefore used only for virtually 'pure' waters, etc. and as atomizer for hydride method. Background correction must be used with all these flames.

Arsenic is best determined by hydride generation, using sodium borohydride as reductant. The detection limit should be about 1 ng.

In electrothermal atomization, arsenic should be measured in presence of oxyacids and stabilized during ashing step by addition of nickel salts.

The use of a radiofrequency electrodeless discharge lamp as radiation source is recommended.

FLAME EMISSION. 235.0 nm, nitrous oxide acetylene flame.

ATOMIC FLUORESCENCE. 235.0 nm, separated air acetylene flame.

Ba   **BARIUM**

PHYSICAL CONSTANTS

Atomic number	Atomic weight	D_0 eV	E_i eV
56	137.34	5.95	8.29

CHARACTERISTIC LINES

λ	f	Flame AAS			Electrothermal atomization			
		Flame type	Rec. sens.	Det. lim.	Ashing	Atom'n	Rec. sens.	Det. lim.
nm			ppm	ppm	°C	°C	pg	pg
553.6	1.4	N_2O C_2H_2 rich	0.16	0.008	900	2800	(150)	(50)
553.6	1.4	Air C_2H_2 rich						
455.4	(ion)	N_2O C_2H_2 rich	(1/5)					
350.1	—	N_2O C_2H_2 rich	(1/12)					

STANDARD AQUEOUS SOLUTIONS

1000 mg l^{-1}	Wt (g)	Reagent	Dissolution
Ba	1.4380	$BaCO_3$	20 ml M HCl

REAGENTS FOR NON-AQUEOUS SOLUTIONS

Barium cyclohexanebutyrate, E, M.

INTERFERENCES

Ionization of barium in nitrous oxide acetylene flame can be as high as 90%. If 553.6 nm is used, this is minimized by addition of 0.1% potassium chloride. Considerable interference at 553.6 nm in presence of calcium because of molecular absorption by CaOH bands. This is avoided at 455.4 nm which is an *ion* line, and therefore potassium and other easily ionized elements should be *absent*.

Stable compound interference from phosphate, silicate, aluminate, etc. is minimized in nitrous oxide acetylene flame and/or by use of lanthanum chloride.

Nitrous oxide acetylene flame gives strong emissions in the region of 553.6 nm contributing noise to the absorption signal. This effect is only reduced by use of intense sources.

In electrothermal atomization, barium sensitivity is better with argon as purge gas, and also in presence of methane (Fuller[19]).

Formation of barium carbide has been avoided by use of a thin tantalum liner.[631]

FLAME EMISSION

553.6 nm or 455.4 nm, nitrous oxide acetylene flame, lean.

BERYLLIUM   Be

PHYSICAL CONSTANTS

Atomic number	Atomic weight	D_0 eV	E_i eV
4	9.0122	4.6	9.32

CHARACTERISTIC LINES

| λ | f | Flame AAS | | | Electrothermal atomization | | | |
| | | Flame type | Rec. sens. | Det. lim. | Ashing | Atom'n | Rec. sens. | Det. lim. |
nm			ppm	ppm	°C	°C	pg	pg
234.9	0.24	N_2O C_2H_2 rich	0.02	0.002	1200	2850	(2)	(0.5)

STANDARD AQUEOUS SOLUTIONS

1000 mg l^{-1}	Wt (g)	Reagent	Dissolution
Be	19.6390	$BeSO_4$	Water

EXTRACTIONS

(i) Acetylacetone–MIBK (from aqueous solution containing EDTA and sodium chloride, buffered at pH 5–7 with sodium acetate).[499]
(ii) Alternatively, add NaEDTA and acetylacetone at pH 1.5, adjust to pH 7 with ammonia and extract with MIBK.

INTERFERENCES

Few interferences and little ionization in nitrous oxide acetylene flame. Silicon and magnesium depress the response. Interference by aluminium is reduced by addition of hydrofluoric acid to samples and standards.

In electrothermal atomization with a graphite furnace, the presence of hydrogen suppresses the beryllium signal.[635]

FLAME EMISSION

234.9 nm, nitrous oxide acetylene flame, stoichiometric.

ATOMIC FLUORESCENCE

234.9 nm, nitrous oxide acetylene flame.

Bi   BISMUTH

PHYSICAL CONSTANTS

Atomic number	Atomic weight	D_0 eV	E_i eV
83	208.9806	4.0	7.29

CHARACTERISTIC LINES

λ	f	Flame AAS			Electrothermal atomization			
		Flame type	Rec. sens.	Det. lim.	Ashing	Atom'n	Rec. sens.	Det. lim.
nm			ppm	ppm	°C	°C	pg	pg
223.1	0.012	Air C_2H_2 lean	0.16	0.015	400	2100	37	(20)
222.8	0.0025		(1/2.2)					
306.8	0.25		(1/3.5)					
206.2	—		(1/8)					
227.7	—		(1/14)					
202.1	—		(1/70)					

STANDARD AQUEOUS SOLUTIONS

1000 mg l^{-1}	Wt (g)	Reagent	Dissolution
Bi	1.0000	Metal chips	50 ml conc. HNO_3

EXTRACTIONS

(i) APDC–MIBK or hexane pH 1–10; (ii) Dithizone–carbon tetrachloride, from solution buffered at pH 9.2 with ammonium citrate and potassium cyanide.

INTERFERENCES

No major interferences in air acetylene flame.

Low concentrations are measured using the hydride generation method with sodium borohydride as reductant. The detection limit should be about 0.5 ng.

In electrothermal atomization the ashing temperature must be kept low to avoid losses by vaporization.

A radiofrequency electrodeless discharge lamp is the recommended source.

The intensity of the 306.8 nm line from the hollow cathode lamp is usually much greater than that of the 223.1 nm line and it may well therefore give an equally good detection limit.

FLAME EMISSION

223.1 nm, nitrous oxide acetylene flame, stoichiometric.

ATOMIC FLUORESCENCE

302.5 nm or 306.8 nm, argon hydrogen, air hydrogen flame.

BORON   B

PHYSICAL CONSTANTS

Atomic number	Atomic weight	D_0 eV	E_i eV
5	10.81	7.95	8.29

CHARACTERISTIC LINES

λ	f	Flame AAS			Electrothermal atomization			
		Flame type	Rec. sens.	Det. lim.	Ashing	Atom'n	Rec. sens.	Det. lim.
nm			ppm	ppm	°C	°C	pg	pg
249.8	0.33	N_2O C_2H_2 rich	7.0	2.0	500	2850	—	—
249.8	0.33	N_2O hydrogen						
249.7	0.32	N_2O C_2H_2 rich	(1/2)					

STANDARD AQUEOUS SOLUTIONS

1000 mg l^{-1}	Wt (g)	Reagent	Dissolution
B	5.7144	Boric acid	Water
B_2O_3	1.7760	Boric acid	Water

REAGENTS FOR NON-AQUEOUS SOLUTIONS

Menthyl borate, E

EXTRACTIONS

2-ethylhexane-1,3-diol–chloroform or MIBK.

INTERFERENCES

No major interferences reported in nitrous oxide acetylene flame.
   Sensitivity in electrothermal atomization poor because of tendency to form carbide with graphite tube.

FLAME EMISSION

249.8 nm, nitrous oxide hydrogen flame; 518 nm (oxide band), nitrous oxide hydrogen or nitrous oxide acetylene (lean flames).

Cd   CADMIUM

PHYSICAL CONSTANTS

Atomic number	Atomic weight	D_0 eV	E_i eV
48	112.40	3.8	6.11

CHARACTERISTIC LINES

λ	f		Flame AAS			Electrothermal atomization			
		Flame type	Rec. sens.	Det. lim.	Ashing	Atom'n	Rec. sens.	Det. lim.
nm			ppm	ppm	°C	°C	pg	pg
228.8	1.2	Air C_2H_2 lean	0.009	0.002	500	2050	1	(0.1)
326.1	0.0018		(1/1000)					

STANDARD AQUEOUS SOLUTIONS

1000 mg l^{-1}	Wt (g)	Reagent	Dissolution
Cd	1.1423	CdO	20 ml 5 M HCl
Cd	1.0000	Metal rod	20 ml 5 M HCl + 2 drops conc. HNO_3

REAGENTS FOR NON-AQUEOUS SOLUTIONS

Cadmium cyclohexanebutyrate, B, E, M.
Cadmium 2-ethylhexanoate, E.

EXTRACTIONS

(i) APDC–MIBK pH 0–11; (ii) NaDDC–MIBK (from ammonium citrate, ammonium sulfate buffered solution pH 9.5); (iii) Dithizone–carbon tetrachloride or chloroform, pH 6–9.

INTERFERENCES

Few interferences reported except that of high concentrations of silicate.

Cadmium is also determined in biological matrices in the Delves microsampling accessory.

Presence of chloride causes serious loss of cadmium in electrothermal atomization. This is reduced in presence of oxyacids, sulfate, phosphates, etc., but high ashing temperatures cannot be used. Yellow micropipette tips may contain cadmium which can be leached out to cause contamination.

Background correction is normally recommended in flame and electrothermal atomization because of the low resonance wavelength.

FLAME EMISSION

326.1 nm, nitrous oxide acetylene flame, lean.

ATOMIC FLUORESCENCE

228.8 nm, separated air acetylene flame.

CAESIUM   Cs

PHYSICAL CONSTANTS

Atomic number	Atomic weight	E_i eV
55	132.906	3.87

CHARACTERISTIC LINES

λ	f	Flame AAS				Electrothermal atomization			
		Flame type	Rec. sens.	Det. lim.	Ashing	Atom'n	Rec. sens.	Det. lim.
nm			ppm	ppm	°C	°C	pg	pg
852.1	0.8	Air C$_2$H$_2$ lean	0.10 (1/50)	0.002	450	2350	—	—
455.5	—							

STANDARD AQUEOUS SOLUTIONS

1000 mg l^{-1}	Wt (g)	Reagent	Dissolution
Cs	1.2670	CsCl	Water

EXTRACTIONS

Na–tetraphenyl boron/hexane–cyclohexane.[239]

INTERFERENCES

Ionization naturally depresses the absorption and emission response. Presence of another alkali element therefore readily enhances it. 0.5% potassium should therefore always be used as ionization buffer. Strong mineral acids also depress the response, and standards should be matched to samples in this respect. Higher sensitivity is found in cooler flames, e.g. air propane, but interferences are more liable to occur.

There is little experience with electrothermal atomization. Caesium hollow cathode lamps may not prove entirely satisfactory. Vapour discharge lamps may be used instead.

FLAME EMISSION

852.1 nm, air acetylene flame, lean. An ionization buffer should be used (0.5% potassium as chloride).

Ca   CALCIUM

PHYSICAL CONSTANTS

Atomic number	Atomic weight	D_0 eV	E_1 eV
20	40.08	5.0	6.11

CHARACTERISTIC LINES

λ	f	Flame AAS			Electrothermal atomization			
		Flame type	Rec. sens.	Det. lim.	Ashing	Atom'n	Rec. sens.	Det. lim.
nm			ppm	ppm	°C	°C	pg	pg
422.7	1.49	Air C_2H_2 stoichio-metric	0.05	0.001	1100	2850	13	—
422.7	1.49	N_2O C_2H_2 lean	0.02	< 0.0005				
239.9	0.037		(1/300)					

STANDARD AQUEOUS SOLUTIONS

1000 mg l^{-1}	Wt (g)	Reagent	Dissolution
Ca	2.4970	$CaCO_3$	25 ml M HCl
CaO	1.7850	$CaCO_3$	25 ml M HCl

REAGENTS FOR NON-AQUEOUS SOLUTIONS. Calcium 2-ethylhexanoate, B, E.

EXTRACTIONS. 8-Quinolinol–MIBK pH 11.

INTERFERENCES

Depression of calcium response in air acetylene flame is caused by elements which give rise to stable oxysalts: aluminium, beryllium, phosphorus, silicon, titanium, vanadium, zirconium, etc. This effect is much reduced in presence of a releasing agent, e.g. 0.2% lanthanum or strontium in all solutions. Protein in biological specimens may reduce the response and this can be avoided by addition of EDTA. These effects are virtually non-existent in the nitrous oxide acetylene flame, but as calcium is then appreciably ionized, 0.1% of potassium as chloride should be added to both sample and standard solutions.

A cyanogen emission bandhead occurs at about 421.5 nm, and this can cause an increase in background noise if not completely separated by the monochromator.

In electrothermal atomization, the same bandhead can interfere if nitrogen is present. Carbide formation in the furnace and a strong liability for calcium determinations to be affected by contamination have militated against use of electrothermal methods for calcium, though actual sensitivity is excellent.

FLAME EMISSION. 422.7 nm, nitrous oxide acetylene flame, lean.

ATOMIC FLUORESCENCE. 422.7 nm, air hydrogen, air acetylene flames.

CERIUM   Ce

PHYSICAL CONSTANTS

Atomic number	Atomic weight
58	140.12

CHARACTERISTIC LINES

λ	f	Flame AAS	
		Flame type	Rec. sens.
nm			ppm
520.0	—	N_2O C_2H_2	30
569.7	—		39

STANDARD AQUEOUS SOLUTIONS

1000 mg l^{-1}	Wt (g)	Reagent	Dissolution
Ce	2.6957	$CeCl_3 . 7H_2O$	Water

This is one of the least sensitive elements in atomic absorption and little work on it has ever been done. The cerium emission spectrum is exceedingly complex and the likelihood of band pass effects in a normal monochromator is therefore high.

Cr   CHROMIUM

PHYSICAL CONSTANTS

Atomic number	Atomic weight	D_0 eV	E_i eV
24	51.996	4.2	6.76

CHARACTERISTIC LINES

λ	f	Flame AAS			Electrothermal atomization			
		Flame type	Rec. sens.	Det. lim.	Ashing	Atom'n	Rec. sens.	Det. lim.
nm			ppm	ppm	°C	°C	pg	pg
357.9	0.34	Air C_2H_2 rich	0.04	0.004	1350	2850	13	(10)
357.9	0.34	N_2O C_2H_2						
359.3	0.27		(1/1.5)					
360.5	0.19		(1/3)					
425.4	0.10		(1/3.5)					
427.5	—		(1/4)					
428.97	—		(1/6)					

STANDARD AQUEOUS SOLUTIONS

1000 mg l^{-1}	Wt (g)	Reagent	Dissolution
Cr	1.0000	Metal rod	50 ml conc. HCl

REAGENTS FOR NON-AQUEOUS SOLUTIONS

Tris(1-phenyl-1,3-butanediono)chromium(III), E.

EXTRACTIONS

(i) Cr(VI) (cold, in presence of hydrochloric acid)–MIBK;[229] (ii) Cr(VI) Diphenyl carbazone–MIBK (pH 0.5 sulfuric acid);[186] (iii) APDC–MIBK pH 3–7.

To determine both Cr(VI) and total Cr (or Cr(III) by difference) boil with 0.2 M EDTA, adjust pH to 4–5, extract with DDC–MIBK giving Cr(VI). Oxidize another part of sample with sulfuric acid and permanganate, extract as above, giving total Cr.[498]

INTERFERENCES

Cr(III) and Cr(VI) give different responses and samples are best oxidized with perchloric acid to give Cr(VI), or reduced with hydrogen peroxide to give Cr(III).[295] There is a major interference by iron in an air acetylene flame, which is reduced by presence of ammonium chloride, and which is removed in nitrous oxide acetylene flame. Depression of response by excess of phosphate is overcome by addition of calcium.

In electrothermal atomization there are a number of minor interferences and the response is said to be improved with methane in the purge gas.

FLAME EMISSION. 425.4 nm, nitrous oxide acetylene flame.

ATOMIC FLUORESCENCE. 357.9, 359.3, 360.5 nm triplet, air acetylene flame.

COBALT   Co

PHYSICAL CONSTANTS

Atomic number	Atomic weight	E_i eV
27	58.9332	7.86

CHARACTERISTIC LINES

λ	f	Flame AAS			Electrothermal atomization			
		Flame type	Rec. sens.	Det. lim.	Ashing	Atom'n	Rec. sens.	Det. lim.
nm			ppm	ppm	°C	°C	pg	pg
240.7	0.22	Air C$_2$H$_2$ lean	0.04	0.004	1100	2850	38	(5)
242.5	0.19		(1/1.2)					
252.1	0.19		(1/3)					
241.2	—		(1/2)					
243.6	—		(1/3)					
304.4	—		(1/12)					
352.7	—		(1/15)					
346.6	—		(1/40)					
341.3	—		(1/40)					

STANDARD AQUEOUS SOLUTIONS

1000 mg l^{-1}	Wt (g)	Reagent	Dissolution
Co	1.0000	Metal rod	50 ml 6 M HNO$_3$
Co	4.9379	Co(NO$_3$)$_2$.6H$_2$O	Water

REAGENTS FOR NON-AQUEOUS SOLUTIONS

Cobalt cyclohexanebutyrate, E, M.

EXTRACTIONS

(i) APDC–MIBK (pH 1–10) or dichloromethane; (ii) K-ethyl xanthate–MIBK (acetate buffer pH 8); (iii) 2-nitroso-1 naphthol–chloroform.

INTERFERENCES

Excesses of some transition and heavy metals depress the response of cobalt in an air acetylene flame, and matrix matching of standards is important.

Suppression of response in electrothermal atomization in presence of chloride.

FLAME EMISSION

345.4 nm, nitrous oxide acetylene flame, lean.

ATOMIC FLUORESCENCE

240.7 nm, air hydrogen or separated air acetylene flames.

Cu **COPPER**

PHYSICAL CONSTANTS

Atomic number	Atomic weight	D_0 eV	E_i eV
29	63.54	4.9	7.72

CHARACTERISTIC LINES

λ	f	Flame AAS			Electrothermal atomization			
		Flame type	Rec. sens.	Det. lim.	Ashing	Atom'n	Rec. sens.	Det. lim.
nm			ppm	ppm	°C	°C	pg	pg
324.8	0.74	Air C_2H_2 lean	0.025	0.0015	800	2750	27	(2)
327.4	0.38		(1/2)					
217.9	0.011		(1/4)					
216.5	—		(1/6)					
222.6	0.004		(1/20)					
249.2	—		(1/100)					
244.2	—		(1/300)					

STANDARD AQUEOUS SOLUTIONS

1000 mg l^{-1}	Wt (g)	Reagent	Dissolution
Cu	1.0000	Metal foil	50 ml 5 M HNO_3

REAGENTS FOR NON-AQUEOUS SOLUTIONS

Copper cyclohexanebutyrate, B, E, M
Bis(1-phenyl-1,3-butanediono) copper(II), E

EXTRACTIONS

APDC–MIBK, chloroform or ethyl acetate.

INTERFERENCES

Few interferences in air acetylene flame, though large excesses of some transition elements in presence of mineral acids depress the response, e.g. 10 000 ppm of iron depress 10 ppm of copper by 10%. Matrix matching of standards is important.

Chloride should be removed from sample before electrothermal atomization, or a small excess of an oxyacid (sulfuric or nitric) added before the drying or ashing stages.

FLAME EMISSION

324.8 nm, nitrous oxide acetylene flame, lean.

ATOMIC FLUORESCENCE

324.8 nm, argon hydrogen or oxygen argon hydrogen.

DYSPROSIUM   Dy

PHYSICAL CONSTANTS

Atomic number	Atomic weight	E_i eV
66	162.50	6.2

CHARACTERISTIC LINES

λ	f		Flame AAS			Electrothermal atomization			
		Flame type	Rec. sens.	Det. lim.	Ashing	Atom'n	Rec. sens.	Det. lim.
nm			ppm	ppm	°C	°C	pg	pg
421.2	—	N_2O C_2H_2 rich	0.2	0.06	1000	2750	—	—
404.6	—		(1/1.2)					
418.7	—		(1/1.4)					
419.5	—		(1/1.6)					
416.8	—		(1/7)					

STANDARD AQUEOUS SOLUTIONS

1000 mg l^{-1}	Wt (g)	Reagent	Dissolution
Dy	1.1476	Dy_2O_3	20 ml 5 M HCl

INTERFERENCES

Ionization in high temperature flame causing low response is reduced by alkali metals (0.1% potassium as chloride should be added as buffer). Response enhanced by sulfuric and phosphoric acids, and matrix matching of standards is important.

FLAME EMISSION

404.6 nm, nitrous oxide acetylene flame, lean.

Er   **ERBIUM**

PHYSICAL CONSTANTS

Atomic number	Atomic weight	E_i eV
68	167.26	6.2

CHARACTERISTIC LINES

λ	f	Flame AAS			Electrothermal atomization			
		Flame type	Rec. sens.	Det. lim.	Ashing	Atom'n	Rec. sens.	Det. lim.
nm			ppm	ppm	°C	°C	pg	pg
400.8	—	N_2O C_2H_2 rich	0.7	0.05	800	2750	—	—
386.3	—		(1/1.5)					
415.1	—		(1/1.7)					
389.3	—		(1/4)					
408.8	—		(1/7)					
390.5	—		(1/20)					
394.4	—		(1/25)					

STANDARD AQUEOUS SOLUTIONS

1000 mg l^{-1}	Wt (g)	Reagent	Dissolution
Er	1.0000	Metal	20 ml 5 M HCl

INTERFERENCES

Ionization causing low response is reduced by addition of 0.1% potassium as chloride, or other easily ionizable metal.

FLAME EMISSION

400.8 nm, nitrous oxide acetylene flame, lean.

EUROPIUM   Eu

PHYSICAL CONSTANTS

Atomic number	Atomic weight	E_i eV
63	151.96	5.64

CHARACTERISTIC LINES

λ	f	Flame AAS			Electrothermal atomization			
		Flame type	Rec. sens.	Det. lim.	Ashing	Atom'n	Rec. sens.	Det. lim.
nm			ppm	ppm	°C	°C	pg	pg
459.4	0.22	N$_2$O C$_2$H$_2$ rich	0.3	0.04	850	2600	—	—
462.7	0.20		(1/1.5)					
466.2	0.17		(1/2)					
321.1	—		(1/12)					
321.3	—		(1/15)					
333.4	—		(1/20)					
337.3 (II)	—	N$_2$O C$_2$H$_2$ rich	1.2					
326.5 (II)	—		2.3					
349.9 (II)	—		3.6					

STANDARD AQUEOUS SOLUTIONS

1000 mg l^{-1}	Wt (g)	Reagent	Dissolution
Eu	1.1597	Eu$_2$O$_3$	10 ml 5 M HNO$_3$

INTERFERENCES

Ionization causing low response is reduced by addition of 0.1% of potassium as chloride or other easily ionizable metal. If one of the *ion* lines (II) in the above table is used, potassium and other easily ionized elements must be *absent*.

FLAME EMISSION

459.4 nm, nitrous oxide acetylene flame, lean.

Gd   GADOLINIUM

PHYSICAL CONSTANTS

Atomic number	Atomic weight	D_0 eV	E_i eV
64	157.25	6.0	6.16

CHARACTERISTIC LINES

λ	f	Flame AAS		
nm		Flame type	Rec. sens. ppm	Det. lim. ppm
407.9	—	N_2O C_2H_2 rich	15.0	3.0
368.4	—		(1/1.2)	
405.5	—		(1/1.5)	
419.1	—		(1/2.5)	
367.4	—		(1/3)	

STANDARD AQUEOUS SOLUTIONS

1000 mg l^{-1}	Wt (g)	Reagent	Dissolution
Gd	1.0000	Metal	20 ml 5 M HCl

INTERFERENCES

Ionization causing low response is reduced by addition of 0.1% of potassium as chloride or other easily ionizable metal.

FLAME EMISSION

440.2 nm, nitrous oxide acetylene flame, lean.

GALLIUM   Ga

PHYSICAL CONSTANTS

Atomic number	Atomic weight	D_o eV	E_i eV
31	69.72	2.5	6.00

CHARACTERISTIC LINES

λ	f	Flame AAS			Electrothermal atomization			
		Flame type	Rec. sens.	Det. lim.	Ashing	Atom'n	Rec. sens.	Det. lim.
nm			ppm	ppm	°C	°C	pg	pg
287.4	0.32	Air C_2H_2 lean	1.5	0.05	800	2500	(400)	(200)
294.4	0.29		(1/1.1)					
417.2	0.14		(1/2)					
403.3	0.13		(1/3)					
250.0	—		(1/9)					
272.0	—		(1/20)					

STANDARD AQUEOUS SOLUTIONS

1000 mg l^{-1}	Wt (g)	Reagent	Dissolution
Ga	6.4384	$Ga(NO_3)_2$	Water

EXTRACTIONS

(i) APDC–MIBK pH 3–8; (ii) Cupferron–butanol or chloroform (from solution buffered at pH 4.7 with ammonium acetate); (iii) 6 M Hydrochloric acid–isopropyl ether or MIBK.

INTERFERENCES

Hollow cathode lamp performance should be carefully checked as gallium metal melts at 30 °C.

Sensitivity of gallium falls if the acetylene cylinder pressure falls below 7 kg cm^{-2} (100 p.s.i.). No serious elemental interferences.

For electrothermal atomization standards and samples should be prepared in nitric acid.

FLAME EMISSION

403.3 nm, nitrous oxide acetylene flame, lean.

ATOMIC FLUORESCENCE

403.3 nm and 417.2 nm, argon hydrogen or air hydrogen flames.

Ge   GERMANIUM

PHYSICAL CONSTANTS

Atomic number	Atomic weight	D_0 eV	E_i eV
32	72.59	6.5	7.88

CHARACTERISTIC LINES

λ	f	Flame AAS			Electrothermal atomization			
		Flame type	Rec. sens.	Det. lim.	Ashing	Atom'n	Rec. sens.	Det. lim.
nm			ppm	ppm	°C	°C	pg	pg
265.2	0.84	N_2O C_2H_2 rich	1.5	0.2	700	2650	(100)	—
265.2	0.84	Air C_2H_2 luminous						
259.3	0.37		(1/2)					
271.0	0.43		(1/2.5)					
275.4	0.22		(1/2.5)					
269.1	—		(1/4)					
204.2	—		(1/5)					
303.9	—		(1/20)					

STANDARD AQUEOUS SOLUTIONS

1000 mg l^{-1}	Wt (g)	Reagent	Dissolution
Ge	1.0000	Metal chips	10 ml conc. HCl + 10 ml conc. HNO_3

INTERFERENCES

Few interferences reported.

Germanium can be measured using the hydride generation method with sodium borohydride as reductant, but sensitivity is not as good as with the flame.

In electrothermal atomization standards and samples should be made up in oxyacid medium, e.g. nitric, sulfuric or perchloric.

FLAME EMISSION

265.2 nm, nitrous oxide acetylene flame, stoichiometric.

ATOMIC FLUORESCENCE

265.2 nm, separated nitrous oxide acetylene flame.

GOLD   Au

PHYSICAL CONSTANTS

Atomic number	Atomic weight	E_1 eV
79	196.97	9.22

CHARACTERISTIC LINES

λ	f	Flame AAS			Electrothermal atomization			
		Flame type	Rec. sens.	Det. lim.	Ashing	Atom'n	Rec. sens.	Det. lim.
nm			ppm	ppm	°C	°C	pg	pg
242.8	0.3	Air C$_2$H$_2$ lean	0.08	0.015	500	2750	10	(10)
267.6	0.19		(1/2)	—				
274.8	—		(1/1000)	—				
312.3	—		(1/900)	—				

STANDARD AQUEOUS SOLUTIONS

1000 mg l^{-1}	Wt (g)	Reagent	Dissolution
Au	1.000	Metal foil	15 ml conc. HCl + 5 ml conc. HNO$_3$
Au	1.8110	NH$_4$AuCl$_4$	Water

EXTRACTIONS

(i) Concentrated hydrochloric acid and bromine or hydrobromic acid–MIBK; (ii) 0.2 M dibenzyl sulfoxide–toluene; (iii) APDC–MIBK.

INTERFERENCES

Gold is readily ionized, even in an air acetylene flame, and the resulting reversed curvature of calibration graphs is corrected by addition of 0.1% potassium as chloride to samples and standards.

Major interference from some other noble metals in air acetylene flame and addition of copper may reduce this effect. When unknown samples are to be analysed for gold it is advisable to use the method of standard additions. Otherwise standards should be carefully matrix matched.

Strong acids may also cause depressive interferences. Cyanide complexes depress gold response and should be destroyed by fuming with sulfuric/perchloric acid mixture. Precipitation of gold takes place readily in reducing conditions and these must be avoided in sample and standard solutions.

In electrothermal atomization, high concentrations of sulfuric and nitric acids reduce the sensitivity.

The 267.6 nm line is usually of higher intensity than 242.8 nm and may therefore give equal or better detection limits.

FLAME EMISSION.   268.6 nm, nitrous oxide acetylene flame, lean.

ATOMIC FLUORESCENCE.   242.8 nm or 267.6 nm, air acetylene or argon oxygen hydrogen flames.

Hf   HAFNIUM

PHYSICAL CONSTANTS

Atomic number	Atomic weight	E_i eV
72	178.49	6.8

CHARACTERISTIC LINES

λ	f	Flame AAS			Electrothermal atomization			
		Flame type	Rec. sens.	Det. lim.	Ashing	Atom'n	Rec. sens.	Det. lim.
nm			ppm	ppm	°C	°C	pg	pg
307.3	0.02	N_2O C_2H_2 rich	10.0	2.0	1000	2750	—	—
286.6	—		(1/1.1)					
289.8	—		(1/5)					
296.5	—		(1/6)					
368.2	—		(1/6)					
295.1	—		(1/8)					
377.8	—		(1/10)					

STANDARD AQUEOUS SOLUTIONS

1000 mg l^{-1}	Wt (g)	Reagent	Dissolution
Hf	1.0000	Metal	10 ml HF (40%)

INTERFERENCES

Large excesses of iron depress response. Presence of fluoride improves sensitivity. Calibration solutions and samples should contain 0.1% of hydrofluoric acid or 0.1 M ammonium fluoride.

FLAME EMISSION

368.2 nm, nitrous oxide acetylene flame, lean.

HOLMIUM   Ho

PHYSICAL CONSTANTS

Atomic number	Atomic weight
67	164.9303

CHARACTERISTIC LINES

λ	f		Flame AAS			Electrothermal atomization			
		Flame type	Rec. sens.	Det. lim.	Ashing	Atom'n	Rec. sens.	Det. lim.
nm			ppm	ppm	°C	°C	pg	pg
410.4	—	N_2O C_2H_2 rich	0.8	0.4	600	2900	—	—
405.4	—		(1/1.3)					
416.3	—		(1/2)					
417.3	—		(1/4)					
404.1	—		(1/5)					
410.9	—		(1/10)					
412.7	—		(1/13)					

STANDARD AQUEOUS SOLUTIONS

1000 mg l^{-1}	Wt (g)	Reagent	Dissolution
Ho	1.6449	$HoCl_3$	20 ml 5 M HCl

INTERFERENCES

Ionization causing low response is reduced by addition of 0.1% of potassium as chloride.

FLAME EMISSION

405.4 nm, nitrous oxide acetylene flame, lean.

In **INDIUM**

PHYSICAL CONSTANTS

Atomic number	Atomic weight	D_0 eV	E_i eV
49	114.82	1.1	5.78

CHARACTERISTIC LINES

λ	f	Flame AAS			Electrothermal atomization			
		Flame type	Rec. sens.	Det. lim.	Ashing	Atom'n	Rec. sens.	Det. lim.
nm			ppm	ppm	°C	°C	pg	pg
303.9	0.36	Air C_2H_2 lean	0.5	0.02	500	2400	(40)	(20)
325.6	0.37		(1/1.1)					
410.5	0.14		(1/3)					
256.0	—		(1/12)					
271.0	—		(1/20)					

STANDARD AQUEOUS SOLUTIONS

1000 mg l^{-1}	Wt (g)	Reagent	Dissolution
In	1.0000	Metal	10 ml conc. HCl + 5 ml conc. HNO_3
In	2.2550	$In_2(SO_4)_3$	Water

EXTRACTIONS

(i) APDC–MIBK pH 2–9; (ii) 8-Quinolinol–MIBK or chloroform.

INTERFERENCES

Response reduced by excesses of aluminium, copper, magnesium, zinc and phosphate. Standards should be matrix matched. Narrow monochromator band width, e.g. 0.2 nm, should be used at 325.6 to avoid a weakly absorbing line at 325.86 nm. Sensitivity falls if acetylene cylinder pressure falls below 7 kg cm^{-2} (100 p.s.i.).

In electrothermal atomization chloride must be absent to avoid premature volatilization.

FLAME EMISSION

451.1 nm, nitrous oxide acetylene flame, lean.

ATOMIC FLUORESCENCE

410.18 nm, air hydrogen flame.

IODINE   I

PHYSICAL CONSTANTS

Atomic number	Atomic weight	E_i eV
53	126.90	10.45

CHARACTERISTIC LINES

λ	f	Flame AAS			Electrothermal atomization			
		Flame type	Rec. sens.	Det. lim.	Ashing	Atom'n	Rec. sens.	Det. lim.
nm			ppm	ppm	°C	°C	pg	pg
183.0	0.016	N_2O C_2H_2 (N$_2$ separated)	12	25			30[b]	
206.16	—	Air C_2H_2	1300	600		2400	—	2000[b]
178.3[a]	—						200	
183.0[a]	—						400	

[a] With vacuum monochromator (Ref. 36).
[b] From L'vov (Ref. 469).

STANDARD AQUEOUS SOLUTIONS

1000 mg l^{-1}	Wt (g)	Reagent	Dissolution
I	1.3081	KI	Water

INTERFERENCES

At 183.0 nm an air-free light path (vacuum or nitrogen-flushed monochromator) is required. The nitrous oxide flame is still transparent at this wavelength. This condition is also necessary and perhaps easier to achieve in electrothermal atomization. Few interferences have been reported in limited work on iodine.

Ir   IRIDIUM

PHYSICAL CONSTANTS

Atomic number	Atomic weight	E_i eV
77	192.22	9

CHARACTERISTIC LINES

λ	f	Flame AAS			Electrothermal atomization			
		Flame type	Rec. sens.	Det. lim.	Ashing	Atom'n	Rec. sens.	Det. lim.
nm			ppm	ppm	$^{\circ}$C	$^{\circ}$C	pg	pg
208.9	—	Air C_2H_2 stoichio-metric	8.0	0.8	1000	2750	(500)	—
263.9	—		(1/2)					
264.0	—		(1/2)					
285.0	—		(1/3)					
254.4	—		(1/5)					
250.3	—		(1/6)					

STANDARD AQUEOUS SOLUTIONS

1000 mg l^{-1}	Wt (g)	Reagent	Dissolution
Ir	1.8260	$(NH_4)_2IrCl_6$	Water

EXTRACTIONS

(i) APDC–MIBK pH 1–14; (ii) Methyl triphenyl phosphonium chloride–chloroform.

INTERFERENCES

Because of background interference and poor signal to noise ratio at 208.9 nm, 263.9 nm may be the better line to use. Many common elements decrease the response of iridium, though addition of 1% of sodium and of copper may reduce these effects.[357] Both potassium and sodium also decrease the response of iridium, and this effect is overcome and sensitivity actually improved by addition of lanthanum.[251] Clearly it is necessary to matrix match standards under these circumstances.

In electrothermal atomization, the response of iridium is reduced in presence of gold, palladium, platinum, rhodium and ruthenium.[219]

FLAME EMISSION

380.0 nm, nitrous oxide acetylene flame, lean.

ATOMIC FLUORESCENCE

254.4 nm, air hydrogen flame.

IRON   Fe

PHYSICAL CONSTANTS

Atomic number	Atomic weight	D_0 eV	E_i eV
26	55.847	4.0	7.87

CHARACTERISTIC LINES

λ nm	f	Flame AAS			Electrothermal atomization			
		Flame type	Rec. sens. ppm	Det. lim. ppm	Ashing °C	Atom'n °C	Rec. sens. pg	Det. lim. pg
248.3	0.34	Air C$_2$H$_2$ lean	0.05	0.006	1100	2450	20	(5)
248.8	—		(1/2)					
252.3	0.30		(1/2)					
271.9	0.15		(1/3)					
302.1	0.08		(1/4)					
372.0	0.04		(1/10)					
296.7	0.06		(1/11)					
386.0	0.034		(1/25)					
344.1	0.055		(1/30)					
382.4	—		(1/90)					

STANDARD AQUEOUS SOLUTIONS

1000 mg l^{-1}	Wt (g)	Reagent	Dissolution
Fe	1.0000	Metal powder	20 ml 5 M HCl + 5 ml conc. HNO$_3$
Fe$_2$O$_3$	0.6990	Metal powder	20 ml 5 M HCl + 5 ml conc. HNO$_3$

REAGENTS FOR NON-AQUEOUS SOLUTIONS. Iron cyclohexanebutyrate, B.

EXTRACTIONS

(i) APDC–MIBK pH 1–10; (ii) acetylacetone–chloroform pH 3; (iii) 6 M hydrochloric acid–MIBK or isobutyl acetate removes major quantities of iron as Fe(III).

INTERFERENCES

Slight depression by nickel cobalt and mineral acids. Depression by silicon[592] reduced by addition of 0.2% of calcium chloride. Depression by organic acids, particularly citric,[644] overcome by addition of 0.5% phosphoric acid. Matrix matching of standard solutions is recommended, though most of these effects disappear in a nitrous oxide acetylene flame.

Some minor depressions of iron response have been noted in electrothermal atomization.

FLAME EMISSION. 371.99 nm, nitrous oxide acetylene flame, lean.

ATOMIC FLUORESCENCE. 248.3 nm and 371.99 nm, air acetylene flame.

La   **LANTHANUM**

PHYSICAL CONSTANTS

Atomic number	Atomic weight	D_0 eV	E_i eV
57	138.9055	7.0	5.61

CHARACTERISTIC LINES

λ	f	Flame AAS		
		Flame type	Rec. sens.	Det. lim.
nm			ppm	ppm
550.1	0.15	N_2O C_2H_2 rich	35.0	5.0
418.7	—		(1/1.5)	
357.4	0.12		(1/2.5)	
364.9	—		(1/4)	
392.8	0.18		(1/4)	
403.7	—		(1/7)	

STANDARD AQUEOUS SOLUTIONS

1000 mg l^{-1}	Wt (g)	Reagent	Dissolution
La	1.1782	La_2O_3	20 ml 5 M HCl

INTERFERENCES

Depression of response caused by ionization is reduced by addition of 0.1% of potassium as chloride.

Absorption lines at 357.4 nm and 364.95 nm are seriously affected by emission from the flame in this wavelength region.

There are few other reported interferences, though the lanthanide elements as a group manifest a rather complex pattern of mutual interferences.[734]

It is also apparent that best instrumental conditions are different from one lanthanide element to another. Little has yet been reported on their performance in electrothermal atomization.

LEAD   Pb

PHYSICAL CONSTANTS

Atomic number	Atomic weight	D_0 eV	E_i eV
82	207.2	4.1	7.42

CHARACTERISTIC LINES

λ	f	Flame AAS			Electrothermal atomization			
nm		Flame type	Rec. sens. ppm	Det. lim. ppm	Ashing °C	Atom'n °C	Rec. sens. pg	Det. lim. pg
217.0	0.39	Air C$_2$H$_2$ lean	0.08	0.01	700	2450	10	(2)
283.3	0.21		(1/2.5)					
261.4	—		(1/40)					
368.4	—		(1/100)					

STANDARD AQUEOUS SOLUTIONS

1000 mg l^{-1}	Wt (g)	Reagent	Dissolution
Pb	1.0000	Metal sheet	50 ml 2 M HNO$_3$

REAGENTS FOR NON-AQUEOUS SOLUTIONS.   Lead cyclohexane butyrate, B, E, M.

EXTRACTIONS

(i) APDC–MIBK pH 0–8 (pH 2–2.5 usually recommended); (ii) DEDC–xylene; (iii) TOPO–MIBK (from solution 1.2 M in hydrochloric acid containing potassium iodide).

INTERFERENCES

Few serious interferences reported in air acetylene flame, though the effects of large excesses of other ions should be checked, e.g. 10 000 ppm of iron enhance 5 ppm of lead by 35% in a lean air acetylene flame. The 283.3 nm line of lead is often preferred for routine analysis because of the better signal to noise ratio and lower background interferences than at the 217.0 nm line.

Traces of lead are readily measured in small biological specimens using the Delves micro-sampling cup technique. It has also been measured using hydride generation methods with sodium borohydride as reductant, though its sensitivity in this mode is no better than the flame.

Fuller states that more work has been published on the determination of lead by electro-thermal methods than on any other element.[19] A complex pattern of interference effects is therefore apparent from which it emerges that the most severe of these effects arise from the presence of the chlorides of alkali and alkaline earth metals. However, one thousand fold excesses of many elements as nitrates do not cause interference. Improvements in sensitivity have been reported in presence of EDTA[196] and of organic acids and sucrose.[626] Ashing of samples for lead determination should be carried out below 700 °C to avoid vaporization losses.

FLAME EMISSION.   405.8 nm, nitrous oxide acetylene flame, lean.

ATOMIC FLUORESCENCE.   405.8 nm, air hydrogen flame.

Li   LITHIUM

PHYSICAL CONSTANTS

Atomic number	Atomic weight	E_i eV
3	6.941	5.39

CHARACTERISTIC LINES

λ	f	Flame AAS			Electrothermal atomization			
		Flame type	Rec. sens.	Det. lim.	Ashing	Atom'n	Rec. sens.	Det. lim.
nm			ppm	ppm	°C	°C	pg	pg
670.8	0.71	Air C$_2$H$_2$ lean	0.02	< 0.0005	700	2400	(10)	(5)
323.3	0.026		(1/400)					
610.4	—		(1/3400)					

STANDARD AQUEOUS SOLUTIONS

1000 mg l^{-1}	Wt (g)	Reagent	Dissolution
Li	9.2150	Li$_2$SO$_4$.H$_2$O	Water

REAGENTS FOR NON-AQUEOUS SOLUTIONS

Lithium cyclohexanebutyrate, E, M

INTERFERENCES

Ionization, even in air acetylene flame, overcome by addition of 0.1% of potassium as chloride. No reported elemental interferences, except strontium which causes an increase in response at 670.8 nm because of absorption by SrOH species.

In electrothermal atomization, the main interference is by emission from the heated atomizer at the resonance wavelength, and good optical design is therefore important.

FLAME EMISSION

670.8 nm, nitrous oxide acetylene flame, lean; atomic emission measurements have also been made from a graphite furnace at 2700 °C.[571]

ISOTOPE RATIO DETERMINATION

Lithium isotopes can be measured individually by atomic absorption, even though one line of the doublet of ^{6}Li is coincident with the other line of ^{7}Li.[483,722] Manning and Slavin[483] used an oxy-hydrogen flame as the radiation source, aspirating pure isotope solutions into it to measure the natural isotope ratio of the sample sprayed into an air acetylene flame. Goleb and Yoko-yama[286] used an electrothermal atomizer and cooled hollow cathode lamps as radiation source.

More recently it has been shown by Chapman and Dale[141] that if the 'resonance' and 'continuum' beams of a background correcting atomic absorption spectrometer are equipped with a natural lithium lamp and an enriched ^{6}Li lamp respectively, the difference readout is proportional to the isotopic abundance of ^{6}Li.

LUTETIUM   Lu

PHYSICAL CONSTANTS

Atomic number	Atomic weight	D_0 eV	E_i eV
71	174.97	4.3	6.15

CHARACTERISTIC LINES

λ	f	Flame AAS			Electrothermal atomization			
		Flame type	Rec. sens.	Det. lim.	Ashing	Atom'n	Rec. sens.	Det. lim.
nm			ppm	ppm	°C	°C	pg	pg
336.0	0.076	N$_2$O C$_2$H$_2$ rich	10.0	2.0	1100	2900	—	—
331.2	—		(1/2)					
337.7	—		(1/2.2)					
298.9	—		(1/10)					
451.9	—		(1/15)					

STANDARD AQUEOUS SOLUTIONS

1000 mg l^{-1}	Wt (g)	Reagent	Dissolution
Lu	1.0000	Metal	20 ml 5 M HCl

INTERFERENCES

Ionization causing low response is reduced by addition of 0.1% of potassium as chloride.

FLAME EMISSION

451.9 nm, nitrous oxide acetylene flame.

Mg   MAGNESIUM

PHYSICAL CONSTANTS

Atomic number	Atomic weight	D_0 eV	E_1 eV
12	24.305	4.3	7.64

CHARACTERISTIC LINES

λ	f	Flame AAS			Electrothermal atomization			
		Flame type	Rec. sens.	Det. lim.	Ashing	Atom'n	Rec. sens..	Det. lim.
nm			ppm	ppm	°C	°C	pg	pg
285.2	1.2	Air C_2H_2 stoichio-metric	0.003	0.0001	1100	2800	(0.2)	(0.02)
285.2	1.2	N_2O C_2H_2 stoichio-metric						
279.6^a (II)	1.65	N_2O C_2H_2 stoichio-metric	(1/100)					
202.6			(1/40)					

^{a} Magnesium ion line. Do not use ionization buffer.

STANDARD AQUEOUS SOLUTIONS

1000 mg l^{-1}	Wt (g)	Reagent	Dissolution
Mg	1.0000	Metal ribbon	50 ml 5 M HCl
MgO	0.6040	Metal ribbon	50 ml 5 M HCl

REAGENTS FOR NON-AQEUOUS SOLUTIONS

Magnesium cyclohexanebutyrate, B, E, M

EXTRACTIONS. 8-Quinolinol or 8-hydroxyquinaldine–MIBK, pH 11.

INTERFERENCES

Oxyacids of aluminium, phosphorus, silicon, titanium, etc., interfere in amounts greater than equivalence in air acetylene flame. The effect is overcome by addition of 0.1% of lanthanum or strontium as chloride. These and other reported effects are largely eliminated in the nitrous oxide acetylene flame, but here, to reduce ionization, 0.1% of potassium as chloride should be added. If 279.6 nm ion line used, potassium and other easily ionized elements should be absent.

Although magnesium is reported to be very sensitive in electrothermal atomization the main problem likely to be encountered is that of contamination and of residual 'blank' concentrations in water and reagents.

FLAME EMISSION. 285.2 nm, nitrous oxide acetylene flame, lean.

ATOMIC FLUORESCENCE. 285.2 nm, air hydrogen, air acetylene or argon oxygen hydrogen.

MANGANESE   Mn

PHYSICAL CONSTANTS

Atomic number	Atomic weight	D_0 eV	E_i eV
25	54.9380	4	7.43

CHARACTERISTIC LINES

λ	f	Flame AAS					Electrothermal atomization			
nm		Flame type	Rec. sens. ppm	Det. lim. ppm	Ashing °C	Atom'n °C	Rec. sens. pg	Det. lim. pg
279.5	0.58	Air C_2H_2 stoichiometric	0.02	0.0015	1100	2800	4	(0.2)
279.8			(1/1.3)					
280.1			(1/2)					
403.1			(1/10)					
321.7			(1/2000)					

STANDARD AQUEOUS SOLUTIONS

1000 mg l^{-1}	Wt (g)	Reagent	Dissolution
Mn	1.0000	Metal rod	50 ml conc. HCl
Mn_2O_3	0.6960	Metal rod	50 ml conc. HCl
MnO	0.7744	Metal rod	50 ml conc. HCl
Mn	3.6077	$MnCl_2 . 4H_2O$	50 ml conc. HCl

REAGENTS FOR NON-AQUEOUS SOLUTIONS.   Manganese cyclohexanebutyrate

EXTRACTIONS

(i) APDC–MIBK, pH 4–6 (pH 3 sometimes recommended); (ii) Cupferron–MIBK, pH 7.5; (iii) Diphenyl carbazone–MIBK (pH 0.5 sulfuric acid).

INTERFERENCES

Few major interferences in the air acetylene flame. Silicon depresses the signal and this effect is overcome by incorporation of 0.2% of calcium chloride in samples and standards. The effect of large excesses of other elements should also be checked, e.g. 10 000 ppm of iron increase the signal of 20 ppm of manganese by 10%.

The three strongest absorption lines of manganese form a closely spaced triplet which is not resolved in some monochromators, but which, as a triplet, gives a sensitivity of about 0.05 ppm.

In electrothermal atomization, the chlorides of many elements have been found to depress the manganese absorption signal. In presence of oxyacids, ashing temperatures up to 1100 °C may be used, effectively removing many potentially interfering species.

FLAME EMISSION.   403.1 nm, nitrous oxide acetylene flame, lean.

ATOMIC FLUORESCENCE.   279.5 nm, air hydrogen flame.

Hg   MERCURY

PHYSICAL CONSTANTS

Atomic number	Atomic weight	E_i eV
80	200.59	10.43

CHARACTERISTIC LINES

λ	f	Flame AAS			Electrothermal atomization			
		Flame type	Rec. sens.	Det. lim.	Ashing	Atom'n	Rec. sens.	Det. lim.
nm			ppm	ppm	°C	°C	pg	pg
184.9	1	N_2O C_2H_2 (N_2 separated) (Ref. 414)	0.05	0.02	—	—	—	—
253.6	0.03	Air C_2H_2 lean	10	1	< 100	1500	(200)	(100)
253.6	0.03	Argon hydrogen	4	0.5				
253.6	0.03	Non-flame (cold vapour)	~1 ng	~1 ng				

STANDARD AQUEOUS SOLUTIONS

1000 mg l^{-1}	Wt (g)	Reagent	Dissolution
Hg	1.000	Metal	20 ml 5 M HCl

REAGENTS FOR NON-AQUEOUS SOLUTIONS

Mercury cyclohexanebutyrate, E, M

EXTRACTIONS

APDC–MIBK pH 0–10 (not frequently used because mercury is not now determined directly in the flame, see below).

INTERFERENCES

To reach the most sensitive line, 184.9 nm, the optical path of the instrument must be purged with nitrogen, but the nitrous oxide acetylene flame is quite transparent at this wavelength.

Low sensitivity in most types of flame atomizer causes conventional flame methods for mercury to be abandoned in favour of the cold vapour method of Hatch and Ott (see p. 56).[308]

The separation of the analyte element in this way usually renders unnecessary any other form of separation or extraction procedure and also isolates it from other possibly interfering species.

FLAME EMISSION

253.6 nm, nitrous oxide acetylene flame.

ATOMIC FLUORESCENCE

253.6 nm, cold vapour (see section on p. 285).

MOLYBDENUM   Mo

PHYSICAL CONSTANTS

Atomic number	Atomic weight	E_i eV
42	95.94	7.10

CHARACTERISTIC LINES

λ nm	f	Flame AAS			Electrothermal atomization			
		Flame type	Rec. sens. ppm	Det. lim. ppm	Ashing °C	Atom'n °C	Rec. sens. pg	Det. lim. pg
313.3	0.2	Air C_2H_2 rich			850	2900	(20)	(5)
313.3	0.2	N_2O C_2H_2 rich	0.18	0.025				
317.0	0.12		(1/1.5)					
379.8	0.13		(1/1.8)					
319.4	—		(1/2.0)					
386.4	—		(1/2.2)					
390.3	—		(1/3.8)					
315.8	—		(1/4)					
320.8	—		(1/10)					

STANDARD AQUEOUS SOLUTIONS

1000 mg l^{-1}	Wt (g)	Reagent	Dissolution
Mo	1.5003	MoO_3	10 ml conc. HCl
Mo	1.8290	$(NH_4)_6Mo_7O_{24}.4H_2O$	Water

EXTRACTIONS

(i) APDC–MIBK, pH 2; (ii) 8-Quinolinol–MIBK, pH 1.5; (iii) Thiocyanate–MIBK (extracted from solution containing hydrochloric acid, ammonium thiocyanate, ascorbic acid and stannous chloride).

INTERFERENCES

Depression of signal in an acetylene flame by calcium, strontium, sulfate and iron, the last being serious. Extent of the depression depends on observation height but is much reduced in the presence of 0.5% of aluminium chloride or 2% of ammonium chloride.[178,534]

In the nitrous oxide flame depressive effects are also very much reduced.

Molybdenum sensitivity may also be found to fall if the acetylene cylinder pressure drops below 7 kg cm^{-2} (100 p.s.i.).

In electrothermal atomization, the molybdenum signal can be severely reduced because of carbide formation, either with the carbon of the atomizer or with the carbon of an organic matrix. Saturation of the graphite tubes with solutions of salts of other carbide-forming elements and heating to form a carbide layer, prior to their use for analysis, gives a very marked improvement in sensitivity. Complete removal of molybdenum from the graphite atomizer requires long atomization periods and high temperatures.

FLAME EMISSION. 390.3 nm, nitrous oxide acetylene, stoichiometric.

ATOMIC FLUORESCENCE. 313.2 nm, nitrogen shielded air or nitrous oxide acetylene flame.

Nd   NEODYMIUM

PHYSICAL CONSTANTS

Atomic number	Atomic weight	E_i eV
60	144.24	5.45

CHARACTERISTIC LINES

λ	f	Flame AAS			Electrothermal atomization			
		Flame type	Rec. sens.	Det. lim.	Ashing	Atom'n	Rec. sens.	Det. lim.
nm			ppm	ppm	°C	°C	pg	pg
463.4	0.08	N_2O C_2H_2 rich	4.5	0.6	1200	2800	—	—
492.5	0.09		(1/1.5)					
471.9			(1/2)					
489.7			(1/2)					

STANDARD AQUEOUS SOLUTIONS

1000 mg l^{-1}	Wt (g)	Reagent	Dissolution
Nd	1.2218	Nd_2O_3	10 ml conc. HCl

INTERFERENCES

Ionization in flame causing low response is reduced by addition of 0.1% of potassium as chloride. Excesses of silicon, aluminium and fluoride depress the signal, but cerium improves the response.

Praseodymium may interfere in absorption at 492.5 nm due to line overlap.

FLAME EMISSION

492.5 nm, nitrous oxide acetylene flame, lean.

NICKEL   Ni

PHYSICAL CONSTANTS

Atomic number	Atomic weight	E_i eV
28	58.71	7.61

CHARACTERISTIC LINES

λ	f	Flame AAS			Electrothermal atomization			
		Flame type	Rec. sens.	Det. lim.	Ashing	Atom'n	Rec. sens.	Det. lim.
nm			ppm	ppm	°C	°C	pg	pg
232.0	0.095	Air C$_2$H$_2$ lean	0.04	0.002	1100	2850	84	(20)
231.1	—		(1/2)					
234.6	0.051		(1/4)					
341.5	0.30		(1/4)					
346.2	0.16		(1/8)					
352.5	0.12		(1/8)					
339.1	—		(1/40)					
247.7	—		(1/300)					

STANDARD AQUEOUS SOLUTIONS

1000 mg l^{-1}	Wt (g)	Reagent	Dissolution
Ni	1.0000	Metal sheet	50 ml 5 M HNO$_3$

REAGENTS FOR NON-AQUEOUS SOLUTIONS

Nickel cyclohexanebutyrate, B, E, M

EXTRACTIONS

(i) APDC–MIBK, pH 1–10 (pH 2–4 recommended); (ii) Dimethyl glyoxime–MIBK or chloroform (from 1 M hydrochloric acid).

INTERFERENCES

Calibrations with 232.0 nm are usually very curved because of a non-resonance line of nickel at 232.14 nm and are suitable only for lower concentration ranges. 341.5 nm is best for most work. High concentrations of iron and chromium may increase the response. In electrothermal atomization, however, chromium has been found to depress the nickel signal. Stability of nickel compounds up to 1100 °C or higher enables high ashing temperatures to be used with consequent better freedom from interfering species during atomization.

FLAME EMISSION

341.5 nm, nitrous oxide acetylene flame, lean.

ATOMIC FLUORESCENCE

232.0 nm, argon oxygen hydrogen flame.

Nb   NIOBIUM

PHYSICAL CONSTANTS

Atomic number	Atomic weight	D_0 eV	E_i eV
41	92.9064	4.0	6.88

CHARACTERISTIC LINES

λ	f	Flame AAS			Electrothermal atomization			
		Flame type	Rec. sens.	Det. lim.	Ashing	Atom'n	Rec. sens.	Det. lim.
nm			ppm	ppm	$^{\circ}C$	$^{\circ}C$	pg	pg
334.4	—	$N_2O\ C_2H_2$ rich	20	1.5	1500	2900	—	—
405.9	0.19		(1/1.1)					
358.0	0.12		(1/1.1)					
334.9	0.09		(1/1.2)					
408.0	—		(1/1.4)					
412.3	—		(1/2)					
415.3	—		(1/5)					

STANDARD AQUEOUS SOLUTIONS

1000 mg l^{-1}	Wt (g)	Reagent	Dissolution
Nb	1.0000	Metal powder	10 ml HF (40%) + 5 ml conc. HNO_3

EXTRACTIONS

(i) APDC–MIBK, pH 2–4; (ii) 2 M hydrofluoric acid + 4 M sulfuric acid–MIBK.

INTERFERENCES

Ionization causing low response is reduced by addition of 0.1% of potassium as chloride. The niobium signal is increased in hydrofluoric acid in the presence of iron. The presence of iron also prevents minor interferences from nickel, cobalt, chromium and aluminium.

OSMIUM

Os

PHYSICAL CONSTANTS

Atomic number	Atomic weight	E_i eV
76	190.2	8.73

CHARACTERISTIC LINES

λ	f	Flame AAS			Electrothermal atomization			
		Flame type	Rec. sens.	Det. lim.	Ashing	Atom'n	Rec. sens.	Det. lim.
nm			ppm	ppm	°C	°C	pg	pg
290.9	—	$N_2O\ C_2H_2$ rich	1	0.1	1700	2500	(1000)	(2500)
305.9	—							
426.1	—							

STANDARD AQUEOUS SOLUTIONS

1000 mg l^{-1}	Wt (g)	Reagent	Dissolution
Os	1.336	OsO_4	2 M sodium hydroxide solution

EXTRACTIONS

APDC–MIBK, pH 1–10.

INTERFERENCES

Very little work on osmium has ever been reported.

Pd   **PALLADIUM**

PHYSICAL CONSTANTS

Atomic number	Atomic weight	E_i eV
46	106.4	8.33

CHARACTERISTIC LINES

λ	f	Flame AAS			Electrothermal atomization			
		Flame type	Rec. sens.	Det. lim.	Ashing	Atom'n	Rec. sens.	Det. lim.
nm			ppm	ppm	°C	°C	pg	pg
244.8	0.074	Air C$_2$H$_2$ lean	0.1	0.02	600	2750	(20)	(20)
247.6	0.1		(1/1.2)					
276.3	0.071		(1/5)					
340.5	—		(1/7)					

STANDARD AQUEOUS SOLUTIONS

1000 mg l^{-1}	Wt (g)	Reagent	Dissolution
Pd	2.6680	(NH$_4$)$_2$PdCl$_4$	Water

EXTRACTIONS

(i) APDC–MIBK, pH 1–10; (ii) Ethyl isothiocyanate–butyl acetate.

INTERFERENCES

A number of interferences which occur in acid solutions were found not to exist in 2% potassium cyanide solution.[39] Noble metals exhibit a number of mutual interferences, but addition of excess of a base metal, e.g. 1% of copper, or copper and cadmium, has been found to be effective in reducing them.

At 244.8 nm a spectral bandwidth of no more than 0.2 nm should be used because of the proximity of non-absorbing palladium lines. 247.6 nm is more often used, and 340.5 nm has a higher intensity.

FLAME EMISSION

363.5 nm, nitrous oxide acetylene flame, lean.

ATOMIC FLUORESCENCE

340.5 nm, argon oxygen hydrogen flame.

PHOSPHORUS P

PHYSICAL CONSTANTS

Atomic number	Atomic weight	E_i eV
15	30.9738	10.48

CHARACTERISTIC LINES

λ nm	f	Flame AAS Flame type	Flame AAS Rec. sens. ppm	Flame AAS Det. lim. ppm	Electrothermal atomization Ashing °C	Electrothermal atomization Atom'n °C	Electrothermal atomization Rec. sens. pg	Electrothermal atomization Det. lim. pg
177.5	0.15	N_2O C_2H_2 N_2 separated	5 (Ref. 405)	—	—	—	3–4 (Ref. 13)	—
213.6	—		(1/100)	—	300	2750	—	10^4

STANDARD AQUEOUS SOLUTIONS

1000 mg l^{-1}	Wt (g)	Reagent	Dissolution
P	3.7096	$(NH_4)H_2PO_4$	Water

REAGENTS FOR NON-AQUEOUS SOLUTIONS

Triphenyl phosphate, E

INTERFERENCES

Flame determination of phosphorus has not been very successful because of the unavailability of 177.5 nm to most instruments and the poor intensity of phosphorus hollow cathode lamps.

Electrothermal atomization in conjunction with electrodeless discharge lamps is a useful combination for analysis of steel,[797] oils and biological samples (q.v.).

Phosphorus may be determined indirectly by extraction as phosphomolybdate and measurement of molybdenum (see 'chemical amplification', p. 281).

Before electrothermal atomization, phosphorus should be oxidized to phosphate and, to prevent loss during ashing, stabilized by addition of calcium if an element which forms a stable phosphate is not already present.

FLAME EMISSION

526 nm (HPO radical), cool, e.g. argon hydrogen or air hydrogen diffusion flame.

Pt   **PLATINUM**

PHYSICAL CONSTANTS

Atomic number	Atomic weight	E_i eV
78	195.09	9.0

CHARACTERISTIC LINES

λ	f	Flame AAS			Electrothermal atomization			
		Flame type	Rec. sens.	Det. lim.	Ashing	Atom'n	Rec. sens.	Det. lim.
nm			ppm	ppm	$°C$	$°C$	pg	pg
265.9	0.12	Air C_2H_2 lean	1.2	0.05	1600	2800	300	(200)
306.5	—		(1/2)					
262.8	—		(1/2.5)					
273.4	—		(1/4)					
299.8	—		(1/6)					
271.9	—		(1/8)					

STANDARD AQUEOUS SOLUTIONS

1000 mg l^{-1}	Wt (g)	Reagent	Dissolution
Pt	2.2750	$(NH_4)_2PtCl_6$	Water

EXTRACTIONS

(i) APDC–MIBK, pH 1–10; (ii) as $[PtCl_2(SnCl_3)_2]^{2-}$–ethyl acetate[588] (from hydrochloric acid solutions).

INTERFERENCES

Response of platinum is affected by presence of other noble metals, and the effect is reduced by addition of 1% of copper, 0.5% of copper and cadmium or 1% of lanthanum. A number of metals depress the platinum signal, but sodium sulfate (1%) increases it. Matrix matching between samples and standards is particularly important where noble metals are to be determined, and for single samples the method of standard additions is usually to be recommended.

In electrothermal atomization high ashing temperatures are possible. Effects of other elements appear to be less than in the flame.[38]

FLAME EMISSION

265.95 nm, nitrous oxide acetylene flame, lean.

ATOMIC FLUORESCENCE

265.95 nm, argon oxygen hydrogen or nitrous oxide acetylene flames (the latter giving less good sensitivity but better freedom from interference effects).

POTASSIUM   K

PHYSICAL CONSTANTS

Atomic number	Atomic weight	E_i eV
19	39.102	4.34

CHARACTERISTIC LINES

λ	f	Flame AAS			Electrothermal atomization			
		Flame type	Rec. sens.	Det. lim.	Ashing	Atom'n	Rec. sens.	Det. lim.
nm			ppm	ppm	°C	°C	pg	pg
766.5	0.69	Air C_2H_2 lean	0.02	0.002	400	2250	(5)	(1)
769.9	0.34		(1/3)					
404.4	0.11		(1/250)					

STANDARD AQUEOUS SOLUTIONS

1000 mg l^{-1}	Wt (g)	Reagent	Dissolution
K	1.9050	KCl (dry)	Water
K$_2$O	1.5840	KCl (dry)	Water
10 mmol l^{-1} K	0.7455	KCl (dry)	Water

REAGENTS FOR NON-AQUEOUS SOLUTIONS

Potassium cyclohexanebutyrate, E, M

INTERFERENCES

Ionization in air acetylene flame is reduced by addition of caesium, rubidium or sodium, e.g. 1% as chloride, to sample and standard solutions. High mineral acid concentrations also reduce the potassium signal.

In electrothermal atomization, radiation from the graphite atomizer must be prevented from reaching the detector.

FLAME EMISSION

766.5 nm, air acetylene flame, lean.

Pr   PRASEODYMIUM

PHYSICAL CONSTANTS

Atomic number	Atomic weight	E_i eV
59	140.908	5.57

CHARACTERISTIC LINES

λ	f	Flame AAS		
		Flame type	Rec. sens.	Det. lim.
nm			ppm	ppm
495.1		N_2O C_2H_2 rich	20.0	8.0
491.4			(1/1.5)	
513.3			(1/2)	
473.7			(1/2.3)	
504.6			(1/3)	

STANDARD AQUEOUS SOLUTIONS

1000 mg l^{-1}	Wt (g)	Reagent	Dissolution
Pr	1.1709	Pr_2O_3	10 ml conc. HCl

INTERFERENCES

Ionization in flame causing low response is reduced by addition of 0.1% of potassium as chloride.

FLAME EMISSION

495.1 nm, nitrous oxide acetylene flame, stoichiometric.

RHENIUM   Re

PHYSICAL CONSTANTS

Atomic number	Atomic weight	E_i eV
75	186.2	7.87

CHARACTERISTIC LINES

λ	f	Flame AAS			Electrothermal atomization			
		Flame type	Rec. sens.	Det. lim.	Ashing	Atom'n	Rec. sens.	Det. lim.
nm			ppm	ppm	°C	°C	pg	pg
346.0	0.2	N_2O C_2H_2 rich	10.0	0.5	200	2250	(1000)	—
345.2	0.06		(1/1.7)					
346.4	0.13		(1/2.4)					

STANDARD AQUEOUS SOLUTIONS

1000 mg l^{-1}	Wt (g)	Reagent	Dissolution
Re	1.4406	NH_4ReO_4	Water

EXTRACTIONS

(i) APDC–MIBK, pH 1–9; (ii) 'Aliquat 336'–MIBK. (Aliquat 336 is tricapryl monomethyl ammonium chloride.)

INTERFERENCES

Large excesses of aluminium, calcium, iron, lead, manganese, molybdenum and potassium may reduce the response of rhenium. Standards should be matrix matched to samples.

FLAME EMISSION

346.1 nm, nitrous oxide acetylene flame, lean.

Rh

RHODIUM

PHYSICAL CONSTANTS

Atomic number	Atomic weight	E_i eV
45	102.9055	7.45

CHARACTERISTIC LINES

λ	f	Flame AAS			Electrothermal atomization			
		Flame type	Rec. sens.	Det. lim.	Ashing	Atom'n	Rec. sens.	Det. lim.
nm			ppm	ppm	°C	°C	pg	pg
343.5	0.073	Air C_2H_2 lean	0.10	0.004	1000	2750	(50)	(50)
343.5	0.073	N_2O C_2H_2 lean	0.25					
369.2	—		(1/1.7)					
339.7	—		(1/2.5)					
350.2	—		(1/4)					
370.1	—		(1/8)					
328.1	—		(1/100)					

STANDARD AQUEOUS SOLUTIONS

1000 mg l^{-1}	Wt (g)	Reagent	Dissolution
Rh	3.8580	$(NH_4)_3RhCl_6.1.5H_2O$	Water

EXTRACTIONS

APDC–MIBK, pH 1–14.

INTERFERENCES

Few metals, except other noble metals, interfere, but most anions cause depression of the rhodium signal, and this effect is much reduced in presence of 1% of sodium sulfate. Such effects are also much less in the nitrous oxide acetylene flame and calibration linearity is better, though the overall sensitivity is lower.

Slight depression from other noble metals may also be found in electrothermal atomization.

FLAME EMISSION

346.0 nm or 369.2 nm, nitrous oxide acetylene flame, lean.

ATOMIC FLUORESCENCE

343.5 nm, separated air acetylene flame.

RUBIDIUM   Rb

PHYSICAL CONSTANTS

Atomic number	Atomic weight	E_i eV
37	85.468	4.18

CHARACTERISTIC LINES

λ	f	Flame AAS			Electrothermal atomization			
		Flame type	Rec. sens.	Det. lim.	Ashing	Atom'n	Rec. sens.	Det. lim.
nm			ppm	ppm	°C	°C	pg	pg
780.0	0.80	Air C_2H_2 lean	0.04	0.002	800	2100	15	(10)
794.8	0.40		(1/2)					
420.2	—		(1/100)					
421.6	—		(1/250)					

STANDARD AQUEOUS SOLUTIONS

1000 mg l^{-1}	Wt (g)	Reagent	Dissolution
Rb	1.4140	RbCl	Water

INTERFERENCES

Ionization, even in air acetylene flame, causing low response, is reduced by addition of 0.1% of caesium or potassium as chloride, and is also less in fuel rich flame. Aluminium and some mineral acids also reduce the rubidium signal, and matrix matching of samples and standards is important.

In electrothermal atomization, emission from the heated atomizer should be prevented from reaching the detector.

FLAME EMISSION

780.0 nm, air acetylene flame, lean.

Ru   RUTHENIUM

PHYSICAL CONSTANTS

Atomic number	Atomic weight	E_i eV
44	101.07	7.34

CHARACTERISTIC LINES

λ	f	Flame AAS			Electrothermal atomization			
		Flame type	Rec. sens.	Det. lim.	Ashing	Atom'n	Rec. sens.	Det. lim.
nm			ppm	ppm	°C	°C	pg	pg
349.9	0.10	Air C_2H_2 lean	1.0	0.07	500	2700	(100)	—
372.8	0.087		(1/1.5)					
379.9	—		(1/2.3)					
392.6	—		(1/10)					

STANDARD AQUEOUS SOLUTIONS

1000 mg l^{-1}	Wt (g)	Reagent	Dissolution
Ru	2.8390	$[(OH)(NO)Ru(NH_3)_4]Cl_2$	Water

EXTRACTIONS

APDC–MIBK, pH 1–10.

INTERFERENCES

The ruthenium signal is increased in presence of platinum or rhodium, and in presence of lanthanum and hydrochloric acid, but depressed by molybdenum.

Other interferences in air acetylene flame were reduced by addition of 0.5% of each of copper sulfate and cadmium sulfate.[650]

The nitrous oxide acetylene flame can also be used.

In electrothermal atomization the ruthenium signal is depressed by some other noble metals.

FLAME EMISSION

372.8 nm, nitrous oxide acetylene flame, lean.

ATOMIC FLUORESCENCE

372.7, 372.8 and 373.0 nm triplet, separated air acetylene flame.

SAMARIUM   Sm

PHYSICAL CONSTANTS

Atomic number	Atomic weight	E_i eV
62	150.4	5.6

CHARACTERISTIC LINES

λ	f	Flame AAS			Electrothermal atomization			
		Flame type	Rec. sens.	Det. lim.	Ashing	Atom'n	Rec. sens.	Det. lim.
nm			ppm	ppm	°C	°C	pg	pg
429.7	—	N$_2$O C$_2$H$_2$ rich	2.0	0.06	1500	2550	—	—
476.0	—		(1/1.4)					
520.1	—		(1/1.6)					
472.8	—		(1/2)					
528.3	—		(1/2.7)					

STANDARD AQUEOUS SOLUTIONS

1000 mg l^{-1}	Wt (g)	Reagent	Dissolution
Sm	1.0000	Metal	10 ml conc. HCl

INTERFERENCES

Ionization causing low response is reduced by addition of 0.1% of potassium as chloride.

FLAME EMISSION

476.0 nm, nitrous oxide acetylene flame, lean.

Sc   SCANDIUM

PHYSICAL CONSTANTS

Atomic number	Atomic weight	D_0 eV	E_i eV
21	44.9559	6.0	6.54

CHARACTERISTIC LINES

λ	f	Flame AAS				Electrothermal atomization			
		Flame type	Rec. sens.	Det. lim.	Ashing	Atom'n	Rec. sens.	Det. lim.
nm			ppm	ppm	°C	°C	pg	pg
391.2	—	N_2O C_2H_2 rich	0.5	0.05	700	2450	—	—
390.8	—		(1/1.2)					
402.4	—		(1/1.4)					
402.0	—		(1/1.8)					
327.0	—		(1/4)					
408.2	—		(1/8)					
327.4	—		(1/15)					

STANDARD AQUEOUS SOLUTIONS

1000 mg l^{-1}	Wt (g)	Reagent	Dissolution
Sc	1.5336	Sc_2O_3	Water

INTERFERENCES

Ionization causing low response is reduced by addition of 0.1% of potassium as chloride. Sulfide and fluoride also reduce the signal, and standards should be matrix matched to samples.

FLAME EMISSION

402.0 nm, nitrous oxide acetylene flame, lean.

SELENIUM   Se

PHYSICAL CONSTANTS

Atomic number	Atomic weight	D_0 eV	E_i eV
34	78.96	3.5	9.75

CHARACTERISTIC LINES

| λ | f | Flame AAS | | | Electrothermal atomization | | | |
| | | Flame type | Rec. sens. | Det. lim. | Ashing | Atom'n | Rec. sens. | Det. lim. |
nm			ppm	ppm	°C	°C	pg	pg
196.0	0.12	Air C_2H_2 just luminous	—	—	900	2800	44	—
196.0	0.12	Argon hydrogen	0.26	0.05				
196.0	0.12	N_2O C_2H_2 lean	—	—				
204.0	0.26		(1/3)					
206.3	0.30		(1/12)					
207.5	—		(1/50)					

STANDARD AQUEOUS SOLUTIONS

1000 mg l^{-1}	Wt (g)	Reagent	Dissolution
Se	1.0000	Elemental	15 ml conc. HCl + 5 ml conc. HNO_3

EXTRACTIONS

(i) APDC–MIBK, pH 3–6; (ii) acetophenone–chloroform (from hydrochloric/perchloric acid solution).

INTERFERENCES

An air acetylene flame absorbs or scatters more than 50% of incident radiation at 196.1 nm, though with simultaneous background correction, this effect and its associated noise can be largely eliminated. The argon hydrogen flame is almost transparent, though because of its lower temperature, there are certain interelement interferences.

Better sensitivity (detection limit about 5 ng) is obtained by the hydride generation method, using sodium borohydride as reductant.

In electrothermal atomization, loss of selenium during the ashing step at temperatures up to 900 °C can be avoided by stabilizing with solutions of copper, nickel[209] or molybdenum.[316]

Detection limits in all methods are improved by use of an electrodeless discharge lamp as source of resonance radiation.

ATOMIC FLUORESCENCE

196.1 nm, air hydrogen or air acetylene flames.

Si   SILICON

PHYSICAL CONSTANTS

Atomic number	Atomic weight	E_i eV
14	28.086	8.15

CHARACTERISTIC LINES

λ	f	Flame AAS			Electrothermal atomization			
		Flame type	Rec. sens.	Det. lim.	Ashing	Atom'n	Rec. sens.	Det. lim.
nm			ppm	ppm	°C	°C	pg	pg
251.6	0.26	N_2O C_2H_2 rich	0.8	0.02	1100	2850	(50)	(20)
250.7	0.2		(1/2.6)					
251.4	0.54		(1/3.4)					
252.9	—		(1/3.4)					
221.1	—		(1/10)					

STANDARD AQUEOUS SOLUTIONS

1000 mg l^{-1}	Wt (g)	Reagent	Dissolution
Si	$\sim 7.6^a$	$Na_2SiO_3 . 5H_2O$	Water
SiO_2	$\sim 3.53^a$	$Na_2SiO_3 . 5H_2O$	Water

^{a} Should be standardized gravimetrically.

REAGENTS FOR NON-AQUEOUS SOLUTIONS

Octaphenylcyclotetrasiloxane, B, E

EXTRACTIONS

As silicomolybdate into *n*-amyl alcohol.

INTERFERENCES

Ionization causing slight depression of silicon response is overcome by addition of 0.1% sodium or potassium as chloride. Since many samples or even standards may contain unknown or unspecified amounts of sodium through contamination or otherwise, giving rise to this buffering effect to various degrees, the above concentration of sodium or potassium should *always* be added.

In electrothermal atomization, the use of nitrogen instead of argon as the purge gas is said to improve the response of silicon by about 50%.[482]

FLAME EMISSION

251.6 nm, nitrous oxide acetylene flame, stoichiometric.

ATOMIC FLUORESCENCE

251.4 nm and 251.6 nm, fuel-rich, argon shielded, nitrous oxide acetylene flame.

SILVER   Ag

PHYSICAL CONSTANTS

Atomic number	Atomic weight	D_o eV	E_i eV
47	107.868	1.4	7.57

CHARACTERISTIC LINES

λ	f	Flame AAS			Electrothermal atomization			
nm		Flame type	Rec. sens. ppm	Det. lim. ppm	Ashing °C	Atom'n °C	Rec. sens. pg	Det. lim. pg
328.1	0.51	Air C_2H_2 lean	0.02	0.002	500	2750	3.5	(0.1)
338.3	0.25	Air C_2H_2 lean	(1/2)					

STANDARD AQUEOUS SOLUTIONS

1000 mg l^{-1}	Wt (g)	Reagent	Dissolution
Ag	1.5750	$AgNO_3$	Water

REAGENTS FOR NON-AQUEOUS SOLUTIONS

Silver cyclohexanebutyrate, B, E; silver 2-ethylhexanoate, E.

EXTRACTIONS

(i) APDC–MIBK or chloroform; (ii) Dithizone–MIBK, ethyl propionate or benzene (pH 8.5).

INTERFERENCES

Very few interferences reported in air acetylene flame. Even finely suspended silver chloride gives the same response. Large excesses of aluminium, and some oxyacids may depress the silver signal, the latter by viscosity effects.

In electrothermal atomization, the solutions should be about 0.01% in nitric acid, higher concentrations having a depressive effect. Very low concentrations of silver are not stable in the absence of nitric acid.

FLAME EMISSION

328.1 nm, nitrous oxide acetylene flame, lean.

ATOMIC FLUORESCENCE

328.1 nm, air hydrogen or air acetylene.

Na

SODIUM

PHYSICAL CONSTANTS

Atomic number	Atomic weight	E_i eV
11	22.9898	5.14

CHARACTERISTIC LINES

λ	f	Flame AAS			Electrothermal atomization			
		Flame type	Rec. sens.	Det. lim.	Ashing	Atom'n	Rec. sens.	Det. lim.
nm			ppm	ppm	°C	°C	pg	pg
589.0⎫ 589.6⎭	0.76	Air C$_2$H$_2$ lean	0.008	0.0002	300	2250	(1)	(0.2)
330.2⎫ 330.3⎭	0.055		(1/200)					

STANDARD AQUEOUS SOLUTIONS

1000 mg l^{-1}	Wt (g)	Reagent	Dissolution
Na	2.5420	NaCl (dry)	Water
Na$_2$O	1.8860	NaCl (dry)	Water
10 mmol l^{-1} Na	0.5845	NaCl (dry)	Water

REAGENTS FOR NON-AQUEOUS SOLUTIONS

Sodium cyclohexanebutyrate, B, E, M; sodium naphthasulfonate; sodium triphenyl boron.

INTERFERENCES

Ionization in air acetylene flame is reduced by addition of 0.1% caesium or potassium as chloride to both sample and standard solutions. High mineral acid concentrations also depress the sodium signal. In presence of calcium, emission from CaOH bands affects the signal to noise ratio in the 589 nm region.

In electrothermal atomization, radiation from the heated atomizer must be prevented from reaching the detector. Ashing must be carried out at comparatively low temperatures to avoid losses of sodium by vaporization.

The measurement of sodium, by whichever means, is rendered difficult by its great abundance in nature and the consequent difficulty of preparing uncontaminated standards and blanks. A special means of preparing low concentration standards has already been described (p. 193).

FLAME EMISSION. 589.0 nm, air acetylene flame, lean.

ATOMIC FLUORESCENCE

Although one of the earliest elements for which the fluorescence spectrum was observed,[147] interest in atomic fluorescence for analytical purposes has been minimal, the limits of emission and absorption rarely having been achieved because of contamination difficulties.

589.0 nm, air hydrogen flame.

STRONTIUM   Sr

PHYSICAL CONSTANTS

Atomic number	Atomic weight	D_0 eV	E_i eV
38	87.62	4.85	5.69

CHARACTERISTIC LINES

λ	f		Flame AAS			Electrothermal atomization			
		Flame type	Rec. sens.	Det. lim.	Ashing	Atom'n	Rec. sens.	Det. lim.
nm			ppm	ppm	°C	°C	pg	pg
460.7	1.54	N$_2$O C$_2$H$_2$ stoichio-metric			1100	2500	(20)	(5)
460.7	1.54	Air C$_2$H$_2$ stoichio-metric	0.042	0.002				
407.8	0.76							

STANDARD AQUEOUS SOLUTIONS

1000 mg l^{-1}	Wt (g)	Reagent	Dissolution
Sr	1.6840	SrCO$_3$	20 ml M HCl

REAGENTS FOR NON-AQUEOUS SOLUTIONS

Strontium cyclohexanebutyrate, E, M

INTERFERENCES

In the air acetylene flame, depression of the strontium response caused by stable compound formation in presence of phosphate, aluminium, silicon, titanium, etc., is much reduced by addition of 0.5% lanthanum or calcium as chloride. The nitrous oxide acetylene flame also gives virtual freedom from the effect, but when used, 0.1% of potassium as chloride should be added to sample and standard solutions to minimize the effects of ionization, which is said to be about 84% for strontium at the temperature of this flame.[57]

In electrothermal atomization, emission from the heated atomizer must be prevented from reaching the detector. High ashing temperatures are possible in the absence of chloride and the measurement is relatively free of interferences.

FLAME EMISSION

460.7 nm, nitrous oxide acetylene flame, lean.

ATOMIC FLUORESCENCE

460.7 nm, air acetylene flame.

Ta   **TANTALUM**

PHYSICAL CONSTANTS

Atomic number	Atomic weight	E_i eV
73	180.9479	7.88

CHARACTERISTIC LINES

λ	f	Flame AAS		
nm		Flame type	Rec. sens. ppm	Det. lim. ppm
271.5	0.055	N_2O C_2H_2 rich	10.0	1.0
277.6	—			
255.9	—			
265.7	—			
275.8	—			

STANDARD AQUEOUS SOLUTIONS

1000 mg l^{-1}	Wt (g)	Reagent	Dissolution
Ta	1.9798	$TaCl_5$	Water

EXTRACTIONS

2 M hydrofluoric acid + 4 M sulfuric acid–MIBK.

INTERFERENCES

Iron and hydrofluoric acid are said to improve the response. 0.1 M ammonium fluoride should be added to all sample and standard solutions.[109]

FLAME EMISSION

474.0 nm or 481.3 nm, nitrous oxide acetylene flame, lean.

TELLURIUM   Te

PHYSICAL CONSTANTS

Atomic number	Atomic weight	D_0 eV	E_i eV
52	127.60	2.7	9.01

CHARACTERISTIC LINES

| λ | f | Flame AAS | | | Electrothermal atomization | | | |
| | | Flame type | Rec. sens. | Det. lim. | Ashing | Atom'n | Rec. sens. | Det. lim. |
nm			· ppm	ppm	°C	°C	pg	pg
214.3	0.08	Air C_2H_2 lean	0.25	0.05	600	2400	(100)	(50)
225.9			(1/15)					
238.6			(1/50)					

STANDARD AQUEOUS SOLUTIONS

1000 mg l^{-1}	Wt (g)	Reagent	Dissolution
Te	1.0000	Metal chips	15 ml conc. HCl + 5 ml conc. HNO_3

EXTRACTIONS

(i) APDC–MIBK or chloroform, pH 3–5; (ii) NaDDC–chloroform, pH 8–5.

INTERFERENCES

The response is affected by large excesses of calcium, copper, silicon, sodium, zinc and zirconium, and matrix matching of standards is important.

Tellurium is readily determined by the hydride generation technique using sodium borohydride as reductant. The detection limit should be about 2 ng.

In electrothermal atomization, loss of tellurium during the ashing step at temperatures up to about 600 °C can be averted by stabilizing with nickel.

For all techniques, the best source is an electrodeless discharge lamp.

FLAME EMISSION

238.3 nm, nitrous oxide acetylene flame, stoichiometric.

ATOMIC FLUORESCENCE

214.3 nm, air hydrogen flame or separated air acetylene flame.

Tb   **TERBIUM**

PHYSICAL CONSTANTS

Atomic number	Atomic weight	E_i eV
65	158.9254	5.98

CHARACTERISTIC LINES

λ	f	Flame AAS		
		Flame type	Rec. sens.	Det. lim.
nm			ppm	ppm
432.7	—	N_2O C_2H_2 rich	10.0	0.6
431.9	—		(1/1.2)	
390.1	—		(1/1.8)	
406.2	—		(1/2)	
433.9	—		(1/2.2)	

STANDARD AQUEOUS SOLUTIONS

1000 mg l^{-1}	Wt (g)	Reagent	Dissolution
Tb	1.0000	Metal sheet	20 ml 5 M HCl

INTERFERENCES

Ionization causing low response is reduced by addition of 0.1% of potassium as chloride.

FLAME EMISSION

431.9 nm, nitrous oxide acetylene flame, rich.

THALLIUM   Tl

PHYSICAL CONSTANTS

Atomic number	Atomic weight	E_i eV
81	204.37	6.11

CHARACTERISTIC LINES

λ	f	Flame AAS			Electrothermal atomization			
		Flame type	Rec. sens.	Det. lim.	Ashing	Atom'n	Rec. sens.	Det. lim.
nm			ppm	ppm	°C	°C	pg	pg
276.8	0.27	Air C$_2$H$_2$ stoichio-metric	0.35	0.025	500	1600	(50)	(20)
377.6	0.13		(1/3)					
238.0	0.07		(1/6.5)					
258.0	—		(1/25)					

STANDARD AQUEOUS SOLUTIONS

1000 mg l^{-1}	Wt (g)	Reagent	Dissolution
Tl	1.2350	Tl$_2$SO$_4$	Water

EXTRACTIONS

(i) APDC–MIBK, pH 2–12; (ii) tri-octylphosphine oxide (TOPO)–MIBK (from solution containing hydrochloric acid, ascorbic acid and potassium iodide).

INTERFERENCES

No serious interferences reported, though matrix matching is advisable if high concentrations of other species are present.

In electrothermal atomization, low responses for thallium may be found if chlorides are present.

FLAME EMISSION

535.1 nm, nitrous oxide acetylene flame, lean.

ATOMIC FLUORESCENCE

377.6 nm, air hydrogen or nitrous oxide hydrogen flames.

Th   THORIUM

PHYSICAL CONSTANTS

Atomic number	Atomic weight	D_o eV	E_i eV
90	232.0381	8.6	6.2

CHARACTERISTIC LINES

λ	f	Flame AAS	
		Flame type	Rec. sens.
nm			ppm
324.6	—	N_2O C_2H_2 rich	~ 1000

EXTRACTIONS

APDC–MIBK, pH 4–6.

INTERFERENCES

Thorium is one of the least sensitive of elements in atomic absorption, consequently very little information is available on its behaviour. The emission spectrum of thorium is extremely complex and it is probable therefore that band-pass effects in a conventional monochromator would be severe.

THULIUM   Tm

PHYSICAL CONSTANTS

Atomic number	Atomic weight	E_i eV
69	168.9342	6.2

CHARACTERISTIC LINES

λ	f	Flame AAS		
		Flame type	Rec. sens.	Det. lim.
nm			ppm	ppm
371.8	—	N_2O C_2H_2 rich	0.4	0.03
410.6	0.15		(1/1.5)	
409.4	0.16		(1/1.8)	
418.8	0.12		(1/1.8)	
420.4	0.09		(1/3)	
375.2	—		(1/5.5)	
436.0	—		(1/10)	
341.0	—		(1/15)	

STANDARD AQUEOUS SOLUTIONS

1000 mg l^{-1}	Wt (g)	Reagent	Dissolution
Tm	1.0000	Metal	10 ml conc. HCl

INTERFERENCES

Ionization causing low response is reduced by addition of 0.1% of potassium as chloride.

FLAME EMISSION

371.8 nm, nitrous oxide acetylene flame, rich.

Sn

TIN

PHYSICAL CONSTANTS

Atomic number	Atomic weight	D_0 eV	E_i eV
50	118.69	5.7	7.34

CHARACTERISTIC LINES

λ	f	Flame AAS			Electrothermal atomization			
		Flame type	Rec. sens.	Det. lim.	Ashing	Atom'n	Rec. sens.	Det. lim.
nm			ppm	ppm	°C	°C	pg	pg
224.6	0.41	Air hydrogen rich	0.3	0.01	1100	2850	(100)	(100)
224.6	0.41	Air C_2H_2 luminous						
224.6	0.41	N_2O C_2H_2 rich						
286.3	0.23	Air C_2H_2 luminous	(1/1.5)					
286.3	0.23	N_2O C_2H_2 rich						
235.5	—		(1/2)					
270.6	—		(1/3)					
303.4	—		(1/3.5)					
266.1	—		(1/30)					

STANDARD AQUEOUS SOLUTIONS

1000 mg l^{-1}	Wt (g)	Reagent	Dissolution
Sn	1.0000	Metal	200 ml conc. HCl + 5 ml conc. HNO_3

REAGENTS FOR NON-AQUEOUS SOLUTIONS
Dibutyl tin bis(2-ethylhexanoate), E

EXTRACTIONS
(i) APDC–MIBK, pH 4–6; (ii) TOPO–MIBK (from solution containing hydrochloric and ascorbic acids).

INTERFERENCES
The nitrous oxide acetylene flame is recommended for most routine analyses, as it gives best freedom from interferences. Best sensitivity is given in the air hydrogen flame.[129] Mineral acids, ammonium ions, copper, lead, nickel, zinc, etc., affect the response of tin in cooler flames, and matrix matching of standards and samples is essential.

Tin can also be determined by the hydride generation method, using sodium borohydride as reductant. The detection limit should be about 1 ng.

In electrothermal atomization, sample solutions for tin measurement should be neutralized with ammonia to avoid loss of tin.

FLAME EMISSION.   284.00 nm, nitrous oxide acetylene flame, lean.

ATOMIC FLUORESCENCE.   303.4 nm or 317.5 nm, air hydrogen or argon hydrogen flames.

TITANIUM   Ti

PHYSICAL CONSTANTS

Atomic number	Atomic weight	D_0 eV	E_i eV		
22	47.90	6.9	6.82		

CHARACTERISTIC LINES

λ	f	Flame AAS			Electrothermal atomization			
		Flame type	Rec. sens.	Det. lim.	Ashing	Atom'n	Rec. sens.	Det. lim.
nm			ppm	ppm	°C	°C	pg	pg
364.3	0.25	N_2O C_2H_2 rich	1.2	0.07	1500	2900	660	(500)
365.4	0.22		1.2					
320.0	—		(1/1.2)					
337.2	0.20		(1/1.3)					
319.2	—		(1/1.8)					
399.0	—		(1/2.5)					

STANDARD AQUEOUS SOLUTIONS

1000 mg l^{-1}	Wt (g)	Reagent	Dissolution
Ti	7.3939	$K_2TiO(C_2O_4)_2 . 2H_2O$	Water

REAGENTS FOR NON-AQUEOUS SOLUTIONS

Diethanolamine titanate

INTERFERENCES

The response of titanium is improved in presence of excess of fluoride, aluminium or iron. If fluoride is present 0.1 M ammonium fluoride should be added to all sample and standard solutions, and the effects of iron and aluminium are much reduced in perchloric acid medium.

Ionization resulting in low response is reduced by addition of 0.1% of potassium as chloride.

In electrothermal atomization, titanium forms carbide which is decomposed slowly, giving very broad response curves. If nitrogen is used instead of argon as purge gas, the sensitivity is reduced but the response is quicker.

FLAME EMISSION

398.98 nm, nitrous oxide acetylene flame, lean.

ATOMIC FLUORESCENCE

365.4 nm or 319.99 nm, nitrogen-shielded nitrous oxide acetylene flame.

W TUNGSTEN

PHYSICAL CONSTANTS

Atomic number	Atomic weight	D_o eV	E_i eV
74	183.85	7.2	7.98

CHARACTERISTIC LINES

λ	f	Flame AAS		
		Flame type	Rec. sens.	Det. lim.
nm			ppm	ppm
255.1	0.8	N_2O C_2H_2 rich	5.0	1.0
268.1	—		(1/1.8)	
272.4	—		(1/1.8)	
294.7	0.98		(1/2)	
400.9	—		(1/2.5)	
430.2	—		(1/10)	

STANDARD AQUEOUS SOLUTIONS

1000 mg l^{-1}	Wt (g)	Reagent	Dissolution
W	1.4472	$(NH_4)_{10}W_{12}O_{41}.5H_2O$	Water

EXTRACTIONS

APDC–MIBK, pH 1–3.

INTERFERENCES

Large excesses of iron in some mineral acid solutions depress the response of tungsten. Standards should be matrix matched to samples. The line 400.9 nm gives better signal to noise ratios than 255.1 nm and is recommended for routine analysis.

Sensitivity in electrothermal atomization is very low because of carbide formation.

FLAME EMISSION

400.9 nm, nitrous oxide acetylene flame, lean.

URANIUM   U

PHYSICAL CONSTANTS

Atomic number	Atomic weight	E_i eV
92	238.092	6.2

CHARACTERISTIC LINES

λ	f	Flame AAS		
		Flame type	Rec. sens.	Det. lim.
nm			ppm	ppm
358.5	—	N_2O C_2H_2 rich	45.0	8.0
356.7	—		(1/1.5)	
351.5	—		(1/2.5)	
394.4	—		(1/2.5)	
348.9	—		(1/3)	

EXTRACTIONS

(i) APDC–MIBK, pH 3–4; (ii) tributylphosphate–carbon tetrachloride.

INTERFERENCES

The best analytical lines are 356.7 nm and 351.5 nm because 358.5 nm is subject to noise from flame cyanogen band emission. Some elements, when present in excess, e.g. aluminium, cobalt, iron, lead and nickel, cause an increase in uranium response. Standards should be matrix matched to samples for major elements and anions. Ionization causing depression of response is reduced by addition of 0.1% of potassium as chloride to both samples and standards.

FLAME EMISSION

544.8 nm or 591.5 nm, nitrous oxide acetylene flame, stoichiometric.

ISOTOPE RATIOS

The $^{235}U/^{238}U$ ratio was measured by Goleb in a flameless atomizer using a cooled hollow cathode lamp.[283,284]

V   VANADIUM

PHYSICAL CONSTANTS

Atomic number	Atomic weight	D_0 eV	E_i eV
23	50.9414	5.5	6.74

CHARACTERISTIC LINES

| λ | f | Flame AAS | | | Electrothermal atomization | | | |
| | | Flame type | Rec. sens. | Det. lim. | Ashing | Atom'n | Rec. sens. | Det. lim. |
nm			ppm	ppm	°C	°C	pg	pg
318.5	—	N_2O C_2H_2 rich	0.9	0.04	800	2900	200	100
318.3⎱ 318.4⎰	0.66 0.5		(1/1.2)					
306.6	—		(1/2.8)					
437.9	0.20		(1/4.5)					
320.2	—		(1/6.4)					
385.5⎱ 385.6⎰	— 0.98		(1/6.5)					

STANDARD AQUEOUS SOLUTIONS

1000 mg l^{-1}	Wt (g)	Reagent	Dissolution
V	1.7851	V_2O_5	20 ml conc. H_2SO_4
V	2.2960	NH_4VO_3	20 ml 100 vol. H_2O_2

REAGENTS FOR NON-AQUEOUS SOLUTIONS

Bis(1-phenyl-1,3-butanediono)oxovanadium(IV), B, E

EXTRACTIONS. (i) APDC–MIBK, pH 4–6; (ii) Cupferron–MIBK.[658]

INTERFERENCES

Vanadium is sometimes measured by the integrated absorbance of the triplet 318.5, 318.4 and 318.3 nm. Its response is increased in presence of large amounts of aluminium or titanium. Excesses of iron or phosphoric acid also increase the signal and standard solutions should be matrix matched to samples for major constituents. Ionization causing low response is reduced by addition of 0.1% of potassium as chloride.

In electrothermal atomization, a tendency to form carbide causes prolongation of the response curve and may give rise to 'memory' effects if long atomization and/or 'tube clean' steps are not programmed.

FLAME EMISSION. 437.9 nm nitrous oxide acetylene flame, lean.

ATOMIC FLUORESCENCE

The triplet 318.5, 318.4 and 318.3 nm gives highest fluorescence intensity in an argon or nitrogen shielded nitrous oxide acetylene flame.

YTTERBIUM   Yb

PHYSICAL CONSTANTS

Atomic number	Atomic weight	E_i eV
70	173.04	6.22

CHARACTERISTIC LINES

λ	f	Flame AAS			Electrothermal atomization			
nm		Flame type	Rec. sens. ppm	Det. lim. ppm	Ashing °C	Atom'n °C	Rec. sens. pg	Det. lim. pg
398.8	0.38	N_2O C_2H_2 stoichio-	0.08	0.005	1400	2800	—	—
346.4	0.13	metric	(1/3)					
246.5	0.24		(1/7.5)					
267.2	—		(1/50)					

STANDARD AQUEOUS SOLUTIONS

1000 mg l^{-1}	Wt (g)	Reagent	Dissolution
Yb	1.0000	Metal	20 ml 5 M HCl

INTERFERENCES

Ionization causing low response is reduced by addition of 0.1% of potassium as chloride.

FLAME EMISSION

398.8 nm, nitrous oxide acetylene flame, lean.

Y   YTTRIUM

PHYSICAL CONSTANTS

Atomic number	Atomic weight	D_0 eV	E_i eV
39	88.9059	7	6.51

CHARACTERISTIC LINES

λ	f	Flame AAS		
		Flame type	Rec. sens.	Det. lim.
nm			ppm	ppm
410.2	0.21	N_2O C_2H_2 rich	2.0	0.06
412.8	0.18		(1/1.2)	
407.7	0.27		(1/1.2)	
414.3	0.20		(1/2.2)	
362.1	—		(1/2.5)	

INTERFERENCES

Ionization causing low response is reduced by addition of 0.1% of potassium as chloride. Some mineral acids also cause reduced response. Standards and samples should be matrix matched for major components.

FLAME EMISSION

597.2 nm, nitrous oxide acetylene flame, lean.

ZINC   Zn

PHYSICAL CONSTANTS

Atomic number	Atomic weight	D_0 eV	E_i eV
30	65.37	4	9.39

CHARACTERISTIC LINES

λ nm	f	Flame AAS			Electrothermal atomization			
		Flame type	Rec. sens. ppm	Det. lim. ppm	Ashing °C	Atom'n °C	Rec. sens. pg	Det. lim. pg
213.9	1.2	Air C$_2$H$_2$ lean	0.01	0.0008	900	2750	3	(0.05)
307.6	0.00017		(1/7000)					

STANDARD AQUEOUS SOLUTIONS

1000 mg l^{-1}	Wt (g)	Reagent	Dissolution
Zn	1.0000	Metal chips	30 ml 5 M HCl
ZnO	0.8040	Metal chips	30 ml 5 M HCl

REAGENTS FOR NON-AQUEOUS SOLUTIONS

Zinc cyclohexanebutyrate, B, E, M

EXTRACTIONS

APDC–MIBK.

INTERFERENCES

No serious interferences in air acetylene flame.

In electrothermal atomization, main problems are caused by contamination and the preparation of sufficiently pure water to give low 'blanks'. Response is low in presence of chlorides and phosphate. Atomization of zinc is very rapid.

ATOMIC FLUORESCENCE

213.9 nm, separated air acetylene flame.

Zr   ZIRCONIUM

PHYSICAL CONSTANTS

Atomic number	Atomic weight	D_0 eV	E_i eV
40	91.22	7.8	6.84

CHARACTERISTIC LINES

λ	f	Flame AAS		
nm		Flame type	Rec. sens. ppm	Det. lim. ppm
360.1	0.22	N_2O C_2H_2 rich	8.0	0.7
303.1	—		(1/1.5)	
352.0	—		(1/1.6)	
354.8	---		(1/1.6)	
301.2	—		(1/1.8)	
298.5	—		(1/2)	

STANDARD AQUEOUS SOLUTIONS

1000 mg l^{-1}	Wt (g)	Reagent	Dissolution
Zr	1.0000	Metal	50 ml HF (40%)
ZrO_2	0.7403	Metal	50 ml HF (40%)
Zr	1.3499	ZrO_2	50 ml conc. H_2SO_4

EXTRACTIONS

(i) Cupferron–benzene: isopentanol, 1:1;[752] (ii) thenoyltrifluoroacetone–xylene.

INTERFERENCES

An increased response is obtained in presence of ammonium ions, chloride, fluoride or iron. There may be depression by nickel, nitrate or sulfate. Addition of 0.2% of ammonium fluoride to samples and standards reduces most of these effects, and ferric chloride and/or hydrofluoric acid can be added to increase analytical sensitivity.

FLAME EMISSION

360.1 nm, nitrous oxide acetylene flame, lean.

Appendix 1

Manufacturers of Atomic Absorption Spectrometers and Related Equipment

The following information was supplied by the manufacturers to *ARAAS*, Vol. 7 (1977). In addition to the manufacturing addresses given, many of these suppliers have offices or distributors in other countries.

Supplier	*Type of instrument*
Baird-Atomic Ltd Springwood Industrial Estate Braintree Essex CM7 7YL UK	Single beam AAS Graphite rod ETA
Baird-Atomic Inc. 125 Middlesex Turnpike Bedford, Mass. 01730 USA	
Beckman Instruments GmbH 8 Munich 40 Frankfurter Ring 115 West Germany	Double beam AAS Graphite furnace ETA
Beckman-RIIC Ltd Cressex Industrial Estate High Wycombe Bucks. HP12 3NR UK	
GCA/McPherson Instrument 530 Main Street Acton, Mass. 01720 USA	Single beam AAS

Supplier	*Type of instrument*
Hitachi Ltd Nissei Sangyo Co. Ltd Mori 17th Building 26–5 Toranomon 1-chome, Minato-ku Tokyo Japan	Single beam AAS
Nissei Sangyo Instruments Inc. 392 Potorero Avenue Sunnyvale, Calif. 94086 USA	Single and double beam AAS
Nissei Sangyo GmbH 4 Düsseldorf Am Wehrhahn 41 West Germany	
Instrumentation Laboratory Inc. 68 Jonspin Road Wilmington, Mass. 01887 USA	Single and double beam, double channel AAS Graphite furnace ETA
Instrumentation Laboratory (UK) Ltd Station House Stamford New Road Altrincham Cheshire WA14 1BR UK	
Jarrell-Ash Division Fisher Scientific Co. 590 Lincoln Street Waltham, Mass. 02154 USA	Single and double beam, double channel AAS Graphite furnace ETA
S. & J. Juniper & Co. 7 Potter Street Harlow Essex UK	Graphite furnace ETA
Perkin-Elmer Corp. Main Avenue Norwalk, Conn. 06856 USA	Single and double beam AAS Graphite furnace ETA

Supplier   *Type of instrument*

Perkin-Elmer Ltd
Post Office Lane
Beaconsfield
Bucks. HP9 1QA
UK

Perkin-Elmer & Co. GmbH
Postfach 1120
D-7770 Überlingen
West Germany

Pye Unicam Ltd   Single and double beam AAS
York Street   Graphite furnace ETA
Cambridge CB1 2PX
UK

Philips Electronics Industries addresses in many countries

Rank-Hilger   Single beam AAS
Westwood Industrial Estate   Graphite furnace ETA
Ramsgate Road
Margate
Kent CT9 4JL
UK

Shimadzu-Seisakusho Ltd   Single and double beam AAS
14–5 Uchikanda   Graphite furnace ETA
1-chome, Chiyoda-ku
Tokyo 101
Japan

Varian Techtron Pty   Single and double beam AAS
679 Springvale Road   Graphite furnace, threaded graphite furnace,
Mulgrave   graphite cup
Victoria 3170
Australia

Varian Associates Ltd
Instrument Group
28 Manor Road
Walton on Thames
Surrey
UK

Varian Instrument Division
611 Hansen Way
Palo Alto, Calif. 94303
USA

Appendix 2

General Bibliography

BOOKS

1. W. T. Elwell and J. A. F. Gidley, *Atomic Absorption Spectrophotometry*, Pergamon, Oxford, 1st edn, 1961, 2nd edn, 1966.
2. J. W. Robinson, *Atomic Absorption Spectroscopy*, Arnold, London, 1966.
3. E. E. Angino and G. K. Billings, *Atomic Absorption Spectrometry in Geology*, Elsevier, Amsterdam, 1967, 2nd edn, 1972.
4. I. Rubeška and B. Moldan, *Atomic Absorption Spectrophotometry*, SNTL, Prague, 1967, English edition, Iliffe, London, 1969.
5. W. Slavin, *Atomic Absorption Spectroscopy*, Interscience, New York, 1968.
6. J. Ramírez-Muñoz, *Atomic Absorption Spectroscopy*, Elsevier, Amsterdam, 1968.
7. J. A. Dean and T. C. Rains (Eds), *Flame Emission and Atomic Absorption Spectrometry*, Vol. 1: *Theory* (1969), Vol. 2: *Components and Techniques* (1971), Vol. 3: *Elements and Matrices* (1975), Dekker, New York.
8. G. D. Christian and J. J. Feldman, *Atomic Absorption Spectroscopy: Applications in Agriculture, Biology and Medicine*, Wiley-Interscience, New York, 1970.
9. R. J. Reynolds, K. Aldous and K. C. Thompson, *Atomic Absorption Spectroscopy*, Griffin, London, 1970.
10. F. Rousselet, *Spectrophotométrie par Absorption Atomique appliquée à la Biologie*, Sedes, Paris, 1966.
11. W. J. Price, Chapters on atomic absorption and fluorescence in *Spectroscopy*, D. R. Browning (Ed.), McGraw-Hill, London, 1969.
12. B. V. L'vov, *Atomic Absorption Spectrochemical Analysis* (translated from the Russian by J. H. Dixon), Adam Hilger, London, 1970.
13. B. V. L'vov, in *Atomic Absorption Spectroscopy*, R. M. Dagnall and G. F. Kirkbright (Eds), Butterworth, London, 1970.
14. W. J. Price, *Analytical Atomic Absorption Spectrometry*, Heyden, London, 1972 (also translated into Hungarian and Russian).
15. M. Pinta, *Spectrométrie d'Absorption Atomique*, Vol. 1: *Problèmes généraux*, Vol. 2: *Application à l'analyse chimique*, Masson, Paris, 1972 (English translation, Adam Hilger, London, 1974).
16. G. F. Kirkbright and M. Sargent, *Atomic Absorption and Fluorescence Spectroscopy*, Academic Press, London, 1974.
17. V. Sychra, V. Svoboda and I. Rubeska, *Atomic Fluorescence Spectroscopy*, Van Nostrand, London, 1975.
18. B. Welz, *Atom-Absorptions-Spektroskopie*, Verlag Chemie, Weinheim, W. Germany, 1975 (English edition, 1976).
19. C. W. Fuller, *Electrothermal Atomization for Atomic Absorption Spectrometry*, Chemical Society, London, 1977.

363

ABSTRACT AND SPECIALIST JOURNALS AND REVIEWS

20. *Atomic Absorption Newsletter*, Perkin Elmer Corporation, Norwalk, Conn., USA.
21. *Flame Notes*, Beckman Inst. Inc., California, USA.
22. *Analytical Abstracts*, Chemical Society, London.
23. *Chemical Abstracts*, American Chemical Society.
24. *Atomic Absorption and Flame Emission Spectroscopy Abstracts*, Science and Technology Agency, London.
25. P. Platt, in *Annual Reviews in Analytical Chemistry*, Society for Analytical Chemistry, London, 1971.
26. A. Walsh, *Anal. Chem.* **46**, 698A (1974).
27. R. F. Browner, *Analyst (London)*, **99**, 617 (1974) (critical review of atomic fluorescence).
28. T. S. West, *Analyst (London)*, **99**, 886 (1974).
29. J. T. Winefordner and T. J. Vickers, *Anal. Chem.* **46**, 192R (1974); G. M. Hieftje, T. R. Copeland and D. R. de Olivares, *Anal. Chem.* **48**, 142R (1976); G. M. Hieftje and T. R. Copeland, *Anal. Chem.* **50**, 300R (1978), and subsequent biennial reviews in *Anal. Chem.*, even years.
30. S. D. Brown, *Anal. Chem.* **49**, 1269A (1977).
31. R. A. Newstead, W. J. Price and P. J. Whiteside, *Progress in Analytical Atomic Spectroscopy* (Background Correction in Atomic Absorption) **1**, 267 (1978).
32. *Annual Reports on Analytical Atomic Spectroscopy*, Vols. 1–8 (1971–78), Chemical Society, London (published annually).

REFERENCES

33. *ACB Newsletter*, No. 95 (1971).
34. M. J. Adams and G. F. Kirkbright, *Anal. Chim. Acta* **84**, 79 (1976).
35. M. J. Adams, G. F. Kirkbright and P. Rienvatana, *At. Absorp. Newsl.* **14**, 105 (1975).
36. M. J. Adams, G. F. Kirkbright and T. S. West, *Talanta* **21**, 573 (1974).
37. P. B. Adams and W. O. Passmore, *Anal. Chem.* **38**, 630 (1966).
38. E. Adriaenssens and P. Knoop, *Anal. Chim. Acta* **68**, 37 (1973).
39. E. Adriaenssens and F. Verbeek, *At. Absorp. Newsl.* **12**, 57 (1973).
40. H. Agemian and A. S. Y. Chau, *Analyst (London)* **101**, 761 (1976).
41. J. Aggett and A. J. Sprott, *Anal. Chim. Acta* **72**, 49 (1974).
42. J. Aggett and T. S. West, *Anal. Chim. Acta* **57**, 15 (1971).
43. J. F. Alder and T. S. West, *Anal. Chim. Acta* **51**, 365 (1970).
44. K. M. Aldous, B. W. Bailey and J. M. Rankin, *Anal. Chem.* **44**, 191 (1972).
45. K. M. Aldous, D. G. Mitchell and F. J. Ray, *Anal. Chem.* **45**, 1990 (1973).
46. C. Th. J. Alkemade, *Appl. Opt.* **7**, 1261 (1968).
47. P. Allain, *Clin. Chim. Acta* **64**, 281 (1975).
48. J. E. Allan, *Analyst (London)* **86**, 530 (1961).
49. J. E. Allan, *Spectrochim. Acta* **17**, 459 (1961).
50. J. E. Allan, *Spectrochim. Acta* **24B**, 13 (1969).
51. S. E. Allen, private communication. See also Atomic Absorption Method, *Metallic Elements in Soils*, Pye Unicam Ltd., Cambridge, 1975.
52. S. E. Allen and J. A. Parkinson, *Spectrovision* **22**, 2 (1969).
53. A. D. Ambrose, 27th BISRA/BSC Conference, Scarborough, May 1974.
54. F. Amore, *Anal. Chem.* **46**, 1597 (1974).
55. M. D. Amos and P. A. Bennett, FACSS, 2nd National Meeting, Indianapolis, USA (1975).

56. M. D. Amos and P. E. Thomas, *Anal. Chim. Acta* **32**, 139 (1965).

57. M. D. Amos and J. B. Willis, *Spectrochim. Acta* **22**, 1325 (1966).

58. Analytical Methods Committee (Analytical Division of the Chemical Society), *Analyst* (*London*) **102**, 769 (1977).

59. V. D. Anand and M. C. Lancaster, *26th National Meeting of the American Association of Clinical Chemists*, Las Vegas (1974).

60. U. Anders and G. Hailer, *Z. Anal. Chem.* **278**, 203 (1976).

61. G. R. Anderson, I. S. Mains and T. S. West, *Anal. Chim. Acta* **51**, 355 (1970).

62. T. R. Andrew and P. N. R. Nichols, *Analyst* (*London*) **87**, 25 (1962).

63. D. G. Andrews and J. B. Headridge, *Analyst* (*London*) **102**, 436 (1977).

64. A. Antic-Jovanovic, V. Bojovic and M. Marinkovic, *Spectrochim. Acta* **25B**, 405 (1970).

65. M. A. Ashy and J. B. Headridge, *Analyst* (*London*), **99**, 285 (1974).

66. M. A. Ashy, J. B. Headridge and A. Sowerbutts, *Talanta* **21**, 649 (1974).

67. ASTM Procedure D3237.

68. H. J. Auslitz, *Arch. Lebensmittelhyg.* **27**, 68 (1976).

69. E. T. Backer, *Clin. Chim. Acta* **24**, 233 (1969).

70. H. Bader and H. Brandenburger, *At. Absorp. Newsl.* **7**, 1 (1968).

71. R. B. Baird and S. M. Gabrielian, *Appl. Spectrosc.* **28**, 273 (1974).

72. C. A. Baker and F. W. J. Garton, *UK At. Energy Auth. Rep.* R3490, HM Stationery Office, London (1961).

73. S. J. Baker, K. J. Mills and D. S. Widmer, *5th International Conference on Atomic Spectroscopy*, Melbourne (1975).

74. N. D. H. Balazs, D. J. Pole and J. R. Masarei, *Clin. Chim. Acta* **40**, 213 (1972).

75. K. Z. Balla, E. G. Harsanyi, L. Polos and E. Pungor, *Mikrochim. Acta* 107 (1975).

76. A. E. Ballard, D. W. Stewart, W. O. Kamm and C. W. Zuelke, *Anal. Chem.* **26**, 921 (1954).

77. A. E. Ballard and C. D. W. Thornton, *Ind. Eng. Chem. Anal. Ed.* **13**, 893 (1941).

78. L. Barnes, Jr., *Anal. Chem.* **38**, 1083 (1966).

79. W. B. Barnett, *At. Absorp. Newsl.* **12**, 142 (1973).

80. W. B. Barnett and H. L. Kahn, *Anal. Chem.* **44**, 935 (1972).

81. W. B. Barnett and H. L. Kahn, *Clin. Chem.* **18**, 923 (1972).

82. R. C. Barras and J. D. Helwig, *Proc. Amer. Petrol. Inst.* Section III **43**, 223 (1963).

83. G. Baudin, J. Normand and J. Fijalkowski, *Spectrochim. Acta* **23B**, 587 (1968).

84. A. S. Bazhov, *Zh. Anal. Khim.* **23**, 1640 (1968).

85. M. Bedard and J. D. Kerbyson, *Anal. Chem.* **47**, 1441 (1975).

86. M. Bedard and J. D. Kerbyson, *Can. J. Spectrosc.* **21**, 64 (1976).

87. D. Behne, P. Bratter and W. Wolters, *Z. Anal. Chem.* **277**, 355 (1976).

88. F. Bek, J. Janouskova and B. Moldan, *At. Absorp. Newsl.* **13**, 47 (1974).

89. C. B. Belcher and H. M. Bray, *Anal. Chim. Acta* **26**, 322 (1962).

90. C. B. Belcher and K. Kinson, *Anal. Chim. Acta* **30**, 483 (1964).

91. G. F. Bell, *At. Absorp. Newsl.* **5**, 73 (1966).

92. H. F. Bell, *Anal. Chem.* **45**, 2296 (1973).

93. C. B. Belt, Jr., *Anal. Chem.* **39**, 676 (1967).

94. O. Beniot and G. Geiger, *Bull. Liaison Lab. Ponts et Chaussées* **79**, 83 (1975).

95. R. Berman, *Appl. Spectrosc.* **29**, 1 (1975).

96. D. G. Berge, R. T. Pflaum, D. A. Lehman and C. W. Frank, *Anal. Lett.* **1**, 613 (1968).

97. B. Bernas, *Anal. Chem.* **40**, 1682 (1968).

98. M. E. Beyer and A. M. Bond, *Anal. Chim. Acta* **75**, 409 (1975).

99. D. G. Biechler, *Anal. Chem.* **37**, 1054 (1965).

100. G. K. Billings and P. C. Ragland, *Can. Spectrosc.* **14**, 8 (1969).

101. C. A. Black *et al.* (Eds), *Methods of Soil Analysis*, American Society of Agronomy, 1965.

102. L. C. Blakemore, *N.Z. J. Agric. Res.* **11**, 515 (1968).

103. J. Blomfield, J. McPherson and C. R. P. George, *Brit. Med. J.* **2**, 141 (1969).

104. P. L. Boar and L. K. Ingram, *Analyst (London)* **95**, 124 (1970).
105. E. A. Boettner and F. I. Grunder, in *Trace Inorganics in Water: Advances in Chemistry Series* No. 73, American Chemical Society, Washington D.C., 1968.
106. D. L. Bokowski, *Amer. Ind. Hyg. Assoc. J.* **29**, 474 (1968).
107. E. A. Boling, *Spectrochim. Acta* **22**, 425 (1966).
108. E. Bonilla, *Clin. Chem.* **24**, 471 (1978).
109. A. M. Bond, *Anal. Chem.* **42**, 932 (1970).
110. A. M. Bond and T. A. O'Donnell, *Anal. Chem.* **40**, 560 (1968).
111. A. M. Bond and J. B. Willis, *Anal. Chem.* **40**, 2087 (1968).
112. H. Bosch, *Mikrochim. Acta* **I**, 49 (1976).
113. J. A. Bowman and J. B. Willis, *Anal. Chem.* **39**, 1210 (1967).
114. E. G. Bradfield and D. Spencer, *J. Sci. Food Agric.* **16**, 33 (1965).
115. H. von Brandenberger and H. Bader, *Helv. Chim. Acta* **50**, 1409 (1967).
116. L. A. Brandvold and S. J. Marson, *At. Absorp. Newsl.* **13**, 125 (1974).
117. M. P. Bratzel, R. M. Dagnall and J. D. Winefordner, *Anal. Chem.* **41**, 1527 (1969).
118. M. P. Bratzel, R. M. Dagnall and J. D. Winefordner, *Anal. Chim. Acta* **48**, 197 (1969).
119. M. Briska, *Fresenius' Z. Anal. Chem.* **273**, 283 (1975).
120. R. F. Browner, B. M. Patel, T. H. Glenn, M. E. Rietta and J. D. Winefordner, *Spectrosc. Lett.* **5**, 331 (1972).
121. R. F. Browner and J. D. Winefordner, *Spectrochim. Acta* **28B**, 263 (1973).
122. J. A. Burrows, J. C. Heerdt and J. B. Willis, *Anal. Chem.* **37**, 579 (1965).
123. L. R. P. Butler and A. Fulton, *Appl. Opt.* **7**, 2131 (1968).
124. L. R. P. Butler and A. Strasheim, *Spectrochim. Acta* **21**, 1207 (1965).
125. W. C. Campbell, Ph.D. Thesis, University of Strathclyde, Glasgow (1975).
126. W. C. Campbell and J. M. Ottaway, *Talanta* **21**, 837 (1974).
127. W. C. Campbell and J. M. Ottaway, *Analyst (London)* **102**, 495 (1977).
128. L. Capacho-Delgado and D. C. Manning, *At. Absorp. Newsl.* **5**, 1 (1966).
129. L. Capacho-Delgado and D. C. Manning, *Spectrochim. Acta* **22**, 1505 (1966).
130. J. B. Carlsen, *Anal. Biochem.* **64**, 53 (1975).
131. G. Castellari and P. G. Fiorentini, *G. Clin. Med. (Bologna)* **54**, 116 (1973).
132. J. E. Caupeil, P. W. Hendrikse and J. S. Bongers, *Anal. Chim. Acta* **81**, 53 (1976).
133. I. Cavalleri, C. Minoia, L. Pozzoli and A. Baruttini, *Brit. J. Ind. Med.* **35**, 21 (1978).
134. P. Cavalli and G. Rossi, *Analyst (London)* **101**, 272 (1976).
135. A. A. Cernik, *Brit. J. Ind. Med.* **26**, 44 (1969).
136. A. A. Cernik, *At. Absorp. Newsl.* **12**, 42 (1973).
137. A. A. Cernik, *At. Absorp. Newsl.* **12**, 163 (1973).
138. A. A. Cernik and M. H. P. Sayers, *Brit. J. Ind. Med.* **32**, 155 (1975).
139. Y. S. Chae, J. P. Vacik and W. H. Shelver, *J. Pharm. Sci.* **62**, 1838 (1973).
140. T. T. Chao, M. J. Fishman and J. W. Ball, *Anal. Chim. Acta* **47**, 189 (1969).
141. J. F. Chapman and L. S. Dale, *Anal. Chim. Acta* **87**, 91 (1976).
142. Y. K. Chau and K. Lum-Shue-Chan, *Anal. Chim. Acta* **48**, 205 (1969).
143. Y. K. Chau, P. T. S. Wong and P. D. Goulden, *Anal. Chim. Acta* **85**, 421 (1976).
144. J. E. Chester, R. M. Dagnall and M. R. G. Taylor, *Analyst (London)* **95**, 702 (1970).
145. G. D. Christian, *Clin. Chem.* **11**, 459 (1965).
146. G. D. Christian and F. J. Feldman, *Anal. Chim. Acta* **40**, 173 (1968).
147. C. J. Christiensen, and G. K. Rollefsen, *Phys. Rev.* **34**, 1157 (1929).
148. R. C. Chu, G. P. Barron and P. A. W. Baumgarner, *Anal. Chem.* **44**, 1480 (1972).
149. N. C. Clampit and G. M. Hieftje, *Anal. Chem.* **46**, 382 (1974).
150. D. Clark, R. M. Dagnall and T. S. West, *Anal. Chim. Acta* **63**, 11 (1973).
151. O. E. Clinton, *Analyst (London)* **102**, 187 (1977).
152. S. A. Clyburn, T. Kantor and C. Veillon, *Anal. Chem.* **46**, 2213 (1974).
153. W. D. Cobb, W. W. Foster and T. S. Harrison, *Lab. Pract.* **24**, 123 (1975).
154. W. J. Collinson and D. F. Boltz, *Anal. Chem.* **40**, 1896 (1968).

155. P. A. Cooke and W. J. Price, *Spectrovision* **16**, 7 (1966).
156. T. B. Cooper, G. M. Simpson and D. Allen, *At. Absorp. Newsl.* **13**, 119 (1974).
157. D. H. Cotton and D. R. Jenkins, *Spectrochim. Acta* **25B**, 283 (1970).
158. L. E. Cox, *U.S.A.E.C. LA5791* (1974).
159. L. E. Cox, *Appl. Spectrosc.* **30**, 225 (1976).
160. L. R. Crawford, Jr. and T. Greweling, *Appl. Spectrosc.* **22**, 793 (1968).
160a. M. S. Cresser, *Solvent Extraction in Flame Spectroscopic Analysis*, Butterworth, London, 1978.
161. M. S. Cresser and P. N. Keliher, *Am. Lab.* p. 8 (Aug. 1970).
162. M. S. Cresser and P. N. Keliher, *Int. Lab.* p. 17 (Jan./Feb. 1971).
163. M. S. Cresser and T. S. West, *Spectrochim. Acta* **25B**, 61 (1970).
164. B. R. Culver and T. Surles, *Anal. Chem.* **47**, 920 (1975).
165. A. F. Cunningham, *At. Absorp. Newsl.* **8**, 70 (1969).
166. D. H. Curnow, D. H. Gutteridge and E. D. Horgan, *At. Absorp. Newsl.* **7**, 45 (1968).
167. A. S. Curry, J. F. Read and A. R. Knott, *Analyst (London)* **94**, 744 (1969).
168. R. M. Dagnall, K. C. Thompson and T. S. West, *Talanta* **14**, 551 (1967).
169. R. M. Dagnall, K. C. Thompson and T. S. West, *Talanta* **14**, 557 (1967).
170. R. M. Dagnall, K. C. Thompson and T. S. West, *Talanta* **14**, 1151 (1967).
171. R. M. Dagnall, K. C. Thompson and T. S. West, *Talanta* **14**, 1467 (1967).
172. R. M. Dagnall and T. S. West, *Appl. Opt.* **7**, 1287 (1968).
173. R. M. Dagnall, T. S. West and P. Young, *Anal. Chem.* **38**, 358 (1966).
174. E. F. Dalton and A. J. Melanoski, *J. Assoc. Off. Anal. Chem.* **52**, 1035 (1969).
175. D. J. D'Amico and H. L. Klawans, *Anal. Chem.* **48**, 1469 (1976).
176. D. J. David, *Analyst (London)* **89**, 747 (1964).
177. D. J. David, *Analyst (London)* **87**, 576 (1962).
178. D. J. David, *Analyst (London)* **86**, 730 (1961).
179. D. J. David, *Analyst (London)* **85**, 495 (1960).
180. D. J. David, *Analyst (London)* **84**, 536 (1959).
181. D. J. David, *Analyst (London)* **83**, 655 (1958).
182. J. B. Dawson, E. Grassam, D. J. Ellis and M. J. Keir, *Analyst (London)* **101**, 315 (1976).
183. J. B. Dawson and F. W. Heaton, *Spectrochemical Analysis of Clinical Materials*, Thomas, Springfield, USA, 1967.
184. J. B. Dawson and F. W. Heaton, *Biochem. J.* **80**, 99 (1961).
185. J. R. Deily, *At. Absorp. Newsl.* **5**, 119 (1966).
186. B. Delaughter, *At. Absorp. Newsl.* **4**, 273 (1965).
187. H. T. Delves, *Analyst (London),* **95**, 431 (1970).
188. H. T. Delves, *Biochem. J.* **112**, 34P (1969).
189. H. T. Delves, *Essays in Medical Biochemistry*, Biochem. Soc. and Assoc. Clin. Biochem. **2**, 37 (1976).
190. H. T. Delves, *Clin. Chim. Acta* **71**, 495 (1976).
191. H. T. Delves, *XX CSI/7th ICAS*, Prague, 1977.
192. H. T. Delves and P. J. Aggett, *Fourth SAC Conference on Analytical Chemistry*, Birmingham, UK, 1977.
193. D. R. Demers and D. C. Mitchell, *Technicon International Congress*, 2–4 Nov. 1970.
194. M. B. Denton and D. B. Swartz, *Rev. Sci. Inst.* **45**, 81 (1974).
195. R. Djudzman, E. van der Eeckhout and P. de Moerloose, *Analyst (London)* **102**, 688 (1977).
196. F. Dolinsek and J. Stupar, *Analyst (London),* **98**, 841 (1973).
197. H. M. Donega and T. E. Burgess, *Anal. Chem.* **42**, 1521 (1970).
198. R. R. Dorsch, US Patent 3937577.
199. J. E. Drinkwater, *Analyst (London)* **101**, 672 (1976).
200. D. J. Driscoll, D. A. Clay and R. H. Jungers, *1977 Pittsburgh Conference on Analytical Chemistry and Applied Spectroscopy*, Cleveland, USA (1977).

201. J. Dumanski and K. Sosin, *Vth Polish Spectroanalytical Conference*, Władisławów, Oct. 1976.
202. G. Duncan and R. J. Herridge, *Talanta* **17**, 766 (1970).
203. J. V. Dunkley, *Clin. Chem.* **19**, 1081 (1973).
204. R. Dyck, *At. Absorp. Newsl.* **4**, 170 (1965).
205. U. Ebbestad, N. Gundersen and T. Torgrimsen, *At. Absorp. Newsl.* **14**, 142 (1975).
206. L. Ebdon, D. P. Hubbard and R. G. Michel, *Anal. Chim. Acta* **74**, 281 (1975).
207. L. Ebdon, *Proc. Anal. Div. Chem. Soc.*, to be published.
208. J. Ebert and H. Jungmann, *Fresenius' Z. Anal. Chem.* **272**, 287 (1974).
209. R. D. Ediger, *At. Absorp. Newsl.* **14**, 127 (1975).
210. R. D. Ediger, G. E. Peterson and J. D. Kerber, *At. Absorp. Newsl.* **13**, 61 (1974).
211. D. W. Eggiman and P. R. Betzer, *Anal. Chem.* **48**, 886 (1976).
212. W. Elenbaas and J. Riemans, *Philips Tech. Rev.* **11**, 299 (1950).
213. W. T. Elwell and I. R. Scholes, *Analysis of Copper and its Alloys*, Pergamon, Oxford, 1967.
214. W. T. Elwell and D. F. Wood, *Analysis of the New Metals*, Pergamon, Oxford, 1966.
215. K. Eng, *Skand. Tidskr. Faerg Lack* **21**, 7 (1975).
216. M. S. Epstein, T. C. Rains and O. Menis, *Can. J. Spectrosc.* **20**, 22 (1975).
217. G. Erinc and R. J. Magee, *Anal. Chim. Acta* **31**, 197 (1964).
218. A. Eskamani, M. S. Vigler, H. A. Strecker and N. R. Anthony, *4th International Conference on Atomic Spectroscopy*, Toronto, 1973.
219. G. L. Everett, *Analyst (London)* **101**, 348 (1976).
220. R. J. Everson and H. E. Parker, *Anal. Chem.* **46**, 1966 (1974).
221. H. Falk, *XIX CSI/6th ICAS*, Philadelphia, 1976.
222. B. Farrar, *At. Absorp. Newsl.* **4**, 325 (1965).
223. R. O. Farrelly and J. Pybus, *J. Clin. Chem.* **15**, 566 (1969).
224. V. A. Fassel, J. A. Rasmuson and T. G. Cowley, *Spectrochim. Acta* **23B**, 579 (1968).
225. F. J. Feldman, *Res. Dev.* p. 22 (Oct. 1969).
226. F. J. Feldman, *Anal. Chem.* **42**, 719 (1970).
227. F. J. Feldman, J. A. Blasi and S. B. Smith, Jr., *Anal. Chem.* **41**, 1095 (1969).
228. F. J. Feldman, E. C. Knoblock and W. C. Purdy, *Anal. Chim. Acta* **38**, 489 (1967).
229. F. J. Feldman and W. C. Purdy, *Anal. Chim. Acta* **33**, 273 (1965).
230. G. S. Fell, D. C. Hough, F. E. R. Hussein and J. M. Ottaway, *Proc. Anal. Div. Chem. Soc.* **13**, 271 (1976).
231. G. S. Fell and R. T. Peaston, *Proc. Soc. Anal. Chem.* **10**, 256 (1973).
232. J. A. Fiorino, J. W. Jones and S. G. Caper, *Anal. Chem.* **48**, 120 (1976).
233. M. J. Fishman, *At. Absorp. Newsl.* **11**, 46 (1972).
234. M. J. Fishman and M. E. Midgett, in *Trace Inorganics in Water: Advances in Chemistry Series* No. 73. American Chemical Society, Washington D.C., 1968.
235. A. W. Fitchett, E. H. Daughtrey and P. Mushak, *Anal. Chim. Acta* **79**, 93 (1975).
236. D. E. Fleming and G. A. Taylor, *Analyst (London)* **103**, 101 (1978).
237. H. D. Fleming and R. G. Ide, *Anal. Chim. Acta* **83**, 67 (1976).
238. M. Floyd and L. E. Sommers, *J. Environ. Qual.* **4**, 323 (1975).
239. T. R. Folson, N. Hansen, G. T. Parks and W. E. Weitz, *Appl. Spectrosc.* **28**, 345 (1974).
240. M. A. Ford, *Photoelectr. Spectrom. Group, Bull.* **18**, 554 (1968).
241. R. Frache and A. Mazzucotelli, *Talanta* **23**, 389 (1976).
242. C. W. Frank, W. G. Schrenk and C. E. Meloan, *Anal. Chem.* **38**, 1005 (1966).
243. W. Frech and A. Cedergren, *Anal. Chim. Acta* **82**, 83, 93 (1976).
244. W. Frech, G. Lundgren and S. E. Lunner, *At. Absorp. Newsl.* **15**, 57 (1976).
245. H. Freeman, J. F. Uthe and B. Flemming, *At. Absorp. Newsl.* **15**, 49 (1976).
246. S. W. Frey, W. G. Dewitt and B. R. Bellomy, *Am. Soc. Brew. Chem. Proc.* 172 (1966).
247. S. W. Frey, W. G. Dewitt and B. R. Bellomy, *Am. Soc. Brew. Chem. Proc.* 199 (1967).
248. M. T. Friend, C. A. Smith and D. Wishart, *At. Absorp. Newsl.* **16**, 46 (1977).
249. P. Friggieri and R. Trucco, *XX CSI/7th ICAS*, Prague, 1977.

250. C. Fuchs, M. Brashe, K. Paschen, A. Nordbeck, E. Quellhorst and U. Peek, *Clin. Chim. Acta* **52**, 71 (1974).

251. D. L. Fuhrman, *At. Absorp. Newsl.* **8**, 105 (1969).

252. C. W. Fuller, *Anal. Chim. Acta* **62**, 44 (1972).

253. C. W. Fuller, *At. Absorp. Newsl.* **11**, 65 (1972).

254. C. W. Fuller, *Analyst (London)* **99**, 739 (1974).

255. C. W. Fuller, *At. Absorp. Newsl.* **14**, 73 (1975).

256. C. W. Fuller, *Oil Col. Chem. Assoc. Meeting*, Newcastle upon Tyne, March 1975.

257. C. W. Fuller, *Analyst (London)* **101**, 798 (1976).

258. C. W. Fuller, *Anal. Chim. Acta* **81**, 199 (1976).

259. C. W. Fuller, *At. Absorp. Newsl.* **15**, 73 (1976).

260. C. W. Fuller and I. Thompson, *Analyst (London)* **102**, 143 (1977).

261. A. Fulton, K. C. Thompson and T. S. West, *Anal. Chim. Acta* **51**, 373 (1970).

262. K. Fuwa and B. Vallee, *Anal. Chem.* **35**, 942 (1963).

263. J. C. Gage and J. M. Warren, *Ann. Occup. Hyg.* **13**, 115 (1970).

264. L. de Galan, *Spectrochim. Acta* **28B**, 157 (1973).

265. L. de Galan, *XX CSI/7th ICAS*, Prague, 1977.

266. W. M. G. T. van den Broek and L. de Galan, *Anal. Chem.* **49**, 2176 (1977).

267. L. de Galan and G. F. Samaey, *Spectrochim. Acta* **25B**, 245 (1970).

268. E. Gamot, J. Philibert and Y. Vialette, *Colloq. Nat. C.N.R.S.*, Nancy, 4–6 December 1968, p. 287.

269. B. W. Gandrud and R. K. Skogerboe, *Appl. Spectrosc.* **25**, 243 (1971).

270. D. J. Gardner, J. A. Pritchard and M. A. Sadler, *Analyst (London)* **101**, 278 (1976).

271. R. D. Gardner, A. L. Henicksman and W. H. Ashley, *U.S.A.E.C. LA5539* (1974).

272. K. Garmestani, A. J. Blotcky and E. P. Rock, *Anal. Chem.* **50**, 144 (1978).

273. V. P. Garnys and J. P. Matousek, *Clin. Chem.* **21**, 891 (1975).

274. V. P. Garnys and L. E. Smythe, *Talanta* **22**, 881 (1975).

275. B. M. Gatehouse and A. Walsh, *Spectrochim. Acta* **16**, 602 (1960).

276. A. G. Gaydon and H. G. Wolfard, *Flames, their Structure, Radiation and Temperature*. Chapman and Hall, London, p. 304, 1960.

277. R. A. George, *Pye Unicam 3rd Analytical Conference*, Birmingham, 1976.

278. D. M. Gershman and V. B. Glushanok, *An. Metody Geokhim. Issled.*, Mater. Geokhim. Konf. 4th 100 (1970).

279. E. G. Gimblet, A. G. Marney, and R. W. Bonsnes, *Clin. Chem.* **13**, 204 (1967).

280. V. L. Ginzburg, D. M. Livshits and G. I. Satarina, *Zh. Anal. Khim.* **19**, 1089 (1964).

281. P. Girgis-Takla and I. Chroneos, *Analyst (London)* **103**, 122 (1978).

282. C. E. Gleit and W. D. Holland, *Anal. Chem.* **34**, 1454 (1962).

283. J. A. Goleb, *Anal. Chem.* **35**, 1978 (1963).

284. J. A. Goleb, *Anal. Chim. Acta* **34**, 135 (1966).

285. J. A. Goleb and J. K. Brody, *Anal. Chim. Acta* **28**, 457 (1963).

286. J. A. Goleb and Y. Yokoyama, *Anal. Chim. Acta* **30**, 213 (1964).

287. S. Gomiscek, Z. Lengar, J. Cernetic and V. Hudnik, *Anal. Chim. Acta* **73**, 97 (1974).

288. A. Gomez Coedo and M. T. Dorado Lopez, *Rev. Metal (Madrid)* **12**, 88 (1976).

289. T. T. Gorsuch, *Analyst (London)*, **84**, 135 (1959).

290. D. S. Gough, *Anal. Chem.* **48**, 1926 (1976).

291. D. S. Gough, P. Hannaford and A. Walsh, *9th Australian Spectroscopy Conference*, Canberra, 1973.

292. D. S. Gough, P. Hannaford and A. Walsh, *Spectrochim. Acta* **28B**, 197 (1973).

293. C. L. Grant, *Atomic Absorption Spectroscopy*, A.S.T.M. Special Technical Publication 443, p. 37, 1969.

294. E. Grassam, J. B. Dawson and D. J. Ellis, *Analyst (London)*, **102**, 804 (1977).

295. H. C. Green, *Analyst (London)* **100**, 640 (1975).

296. R. B. Green, J. C. Travis and R. A. Keller, *Anal. Chem.* **48**, 1954 (1976).

297. D. S. Grewal and F. X. Kearns, *At. Absorp. Newsl.* **16**, 131 (1977).
298. W. H. Gries and E. Norval, *Anal. Chim. Acta* **75**, 289 (1975).
299. H. K. L. Gupta, *1977 Pittsburgh Conference on Analytical Chemistry and Applied Spectroscopy*, Cleveland, Ohio, USA, 1977.
300. B. Gutsche, K. Rudiger and R. Herrmann, *Spectrochim. Acta* **30B**, 441 (1975).
301. J. P. S. Haarsma, J. Vlogtman and J. Agterdenbos, *Spectrochim. Acta* **31B**, 129 (1976).
302. T. Hadeishi and R. D. McLaughlin, *Am. Lab.* p. 57, Oct. 1975.
303. B. J. Hale, *J. Agr. Sci.* **37**, 236 (1947).
304. E. H. Halstead, S. A. Barber, D. D. Warncke and J. B. Bole, *Proc. Am. Soil Sci. Soc.* **32**, 69 (1968).
305. R. L. Halstead, B. J. Finn and A. J. MacLean, *Can. J. Soil* **49**, 335 (1969).
306. D. J. Hankinson, *J. Dairy Sci.* **58**, 326 (1975).
307. F. R. Hartley and A. S. Inglis, *Analyst (London)* **92**, 622 (1967).
308. W. R. Hatch and W. L. Ott, *Anal. Chem.* **40**, 2085 (1968).
309. J. E. Hawley and J. D. Ingle, *Anal. Chim. Acta* **77**, 71 (1975).
310. J. B. Headridge and D. P. Hubbard, *Anal. Chim. Acta* **37**, 151 (1967).
311. B. J. Heffernan, R. O. Archibold and T. J. Vickers, *Australas. Inst. Min. Metall. Proc.* **223**, 65 (1967).
312. G. Heinemann, *Z. Klin. Chem. Klin. Biochem.* **11**, 197 (1973).
313. M. R. Hendzel and D. M. Jamieson, *Anal. Chem.* **48**, 926 (1976).
314. E. L. Henn, *Paint Varn. Prod.* **63**, 29 (1973).
315. E. L. Henn, *Anal. Chim. Acta* **73**, 273 (1974).
316. E. L. Henn, *Anal. Chem.* **47**, 428 (1975).
317. L. G. Hickey, *Anal. Chim. Acta* **41**, 546 (1968).
318. G. M. Hieftje, *Appl. Spectrosc.* **25**, 653 (1971).
319. D. N. Hingle, G. F. Kirkbright and T. S. West, *Talanta* **15**, 199 (1968).
320. M. Hiraide and A. Mizuike, *Talanta* **22**, 539 (1975).
321. S. Hirano, Y. Tofuku and T. Fujii, *Jpn. Analyst* **18**, 574 (1969).
322. P. Hocquellet and N. Labeyrie, *At. Absorp. Newsl.* **16**, 124 (1977).
323. A. Hofer, *Z. Anal. Chem.* **249**, 115 (1970).
324. W. Holak, *Anal. Chem.* **41**, 1712 (1969).
325. W. Holak, *Anal. Chim. Acta* **74**, 216 (1975).
326. S. T. Holding and P. H. D. Matthews, *Analyst (London)* **97**, 189 (1972).
327. S. T. Holding and J. J. Rowson, *Analyst (London)* **100**, 465 (1975).
328. W. L. Hoover, J. C. Reagor and J. C. Garner, *J. Assoc. Off. Anal. Chem.* **52**, 708 (1969).
329. E. H. W. Hornbrook, *Can. Mining J.* **90**, 107 (1969).
330. P. Hornick, *U.S.N.T.I.S. AD Rep.* 757015 (1973).
331. R. Horton and J. J. Lynch, *Geol. Surv. Can.*, Paper No. 75–1 part A, 213 (1975).
332. D. P. Hubbard and R. G. Michel, *Spectrovision* **27**, 15 (1972).
333. D. P. Hubbard and R. G. Michel, *Anal. Chim. Acta* **67**, 55 (1973).
334. D. P. Hubbard and R. G. Michel, *Anal. Chim. Acta* **72**, 285 (1974).
335. D. P. Hubbard and H. H. Monks, *Anal. Chim. Acta* **47**, 197 (1969).
336. J. W. Huckabee, C. Feldman and Y. Talmi, *Anal. Chim. Acta* **70**, 41 (1974).
337. R. P. Hullin, M. Kapel and J. A. Drinkall, *J. Food Technol.* **4**, 235 (1969).
338. J. R. Humphrey, *Anal. Chem.* **37**, 1604 (1965).
339. T. R. Hurford and D. F. Boltz, *Anal. Chem.* **40**, 379 (1968).
340. J. A. Hurlbut, *U.S.A.E.C. Rep.* RFP 2151 (1974).
341. J. A. Hurlbut and D. L. Bokowski, *U.S. Dep. Commer. Rep.* RFP 2442 (1977).
342. R. J. Hurtubise, *J. Pharm. Sci.* **63**, 1128 (1974).
343. R. J. Hurtubise, *J. Pharm. Sci.* **63**, 1131 (1974).
344. J. Husler, *At. Absorp. Newsl.* **9**, 31 (1970).
345. J. Y. Hwang and L. M. Sandonato, *Anal. Chem.* **42**, 744 (1970).
346. J. Inczedy, *Analytical Applications of Ion Exchangers*, Pergamon, New York, 1966.

347. M. Ihnat, *Anal. Chim. Acta* **82**, 293 (1976).
348. M. Ihnat and R. J. Westerby, *Anal. Lett.* **7**, 257 (1974).
349. E. S. Ioffe, V. S. Prisenko and I. I. Romazonova, *Zavod. Lab.* **40**, 358 (1974).
350. *I.P. Standards for Petroleum and its Products*, 37th edn, Part I, Volume 2, p. 288.1, Heyden, London, 1978.
351. *I.P. Standards for Petroleum and its Products*, 37th edn, Part I, Volume 2, p. 308.1, Heyden, London, 1978.
352. W. Ishibashi, M. Sato and M. Hashimoto, *Bunseki Kagaku* **23**, 597 (1974).
353. H. J. Issaq and L. P. Morgenthaler, *Anal. Chem.* **47**, 1661, 1668, 1748 (1975).
354. I.U.P.A.C., *Nomenclature, Symbols, Units and their Usage in Spectrochemical Analysis*, Parts II and III, Pergamon, Oxford, 1975.
355. N. P. Ivanov, M. N. Gusinsky and A. D. Jesikov, *Zh. Anal. Khim.* **20**, 11, 33 (1965).
356. R. J. Jakubiec and D. F. Boltz, *Anal. Lett.* **1**, 347 (1968).
357. A. Janssen and F. Umland, *Z. Anal. Chem.* **251**, 101 (1970).
358. J. L. Jimenez Seco and A. G. Coldo, *Rev. Met.* **4**, 621 (1968).
359. P. Johns, *Spectrovision* **24**, 6 (1970).
360. P. Johns, *Spectrovision* **26**, 15 (1971).
361. P. Johns and W. J. Price, *Metallurgia* **81**, 75 (1970).
362. P. Johns and W. J. Price, *Pittsburgh Conference*, March 1970, Abstract 108, Cleveland, Ohio, 1970.
363. C. A. Johnson, *Anal. Chim. Acta* **81**, 69 (1976).
364. D. J. Johnson and J. D. Winefordner, *Anal. Chem.* **48**, 341 (1976).
365. S. C. Jolly (Ed.) (1) *Official, Standardised and Recommended Methods of Analysis.* London, 1963; (2) *Supplement to Official, Standardised and Recommended Methods of Analysis.* Society for Analytical Chemistry, London, 1967.
366. A. H. Jones, *Anal. Chem.* **48**, 1472 (1976).
367. D. R. Jones and S. E. Manahan, *Anal. Lett.* **8**, 569 (1975).
368. M. Jones, G. F. Kirkbright, L. Ranson and T. S. West, *Anal. Chim. Acta* **63**, 210 (1973).
369. J. Jordan, *At. Absorp. Newsl.* **7**, 48 (1968).
370. M. Josephson and K. Dixon, *Natl. Inst. Metall. Repub. S. Afr. Rep.* 1665 (1974).
371. K. Julshamn and O. R. Braekkan, *At. Absorp. Newsl.* **12**, 139 (1973).
372. E. Jungreis and S. Kraus, *Mikrochim. Acta* **I**, 43 (1976).
373. M. L. Jursik, *At. Absorp. Newsl.* **6**, 21 (1967).
374. M. Kahl, D. G. Mitchell, G. L. Kaufman and K. M. Aldous, *Anal. Chim. Acta* **87**, 215 (1976).
375. H. Kahn, *XX CSI/7th ICAS*, Prague, 1977.
376. H. L. Kahn, G. E. Peterson and J. E. Schallis, *At. Absorp. Newsl.* **7**, 35 (1968).
377. H. L. Kahn and J. E. Schallis, *At. Absorp. Newsl.* **7**, 5 (1968).
378. S. Kallman and E. W. Hobart, *Anal. Chim. Acta* **51**, 120 (1970).
379. H. Kamel, D. H. Brown, J. M. Ottaway and W. E. Smith, *Analyst (London)* **102**, 645 (1977).
380. T. Kantor, L. Polos, P. Fodor and E. Pungor, *Talanta* **23**, 585 (1976).
381. J. P. Kapetan, *Atomic Absorption Spectroscopy*, A.S.T.M. Special Publication 443, 1969, p. 78.
382. M. Kh. Karapet'yanto and M. K. Karapet'yanto, *Handbook of Thermodynamic Constants of Inorganic and Organic Compounds*, Ann Arbor–Humphrey Science Publishers, Ann Arbor and London, 1970.
383. M. Kashiki, S. Yamazoe and S. Oshima, *Anal. Chim. Acta* **53**, 95 (1971).
384. D. A. Katsova, L. P. Kruglikova and B. V. L'vov, *Zh. Anal. Khim.* **30**, 238 (1975).
385. J. R. Kelson and R. J. Shamberger, *Clin. Chem.* **24**, 240 (1978).
386. J. D. Kerber, *Appl. Spectrosc.* **20**, 212 (1966).
387. J. D. Kerber, A. Koch and G. E. Peterson, *At. Absorp. Newsl.* **12**, 104 (1973).
388. J. D. Kerber, D. C. Manning and S. Slavin, *XX CSI/7th ICAS*, Prague, 1977.

389. P. P. Kharkhanis and J. R. Anfinsen, *J. Ass. Off. Anal. Chem.* **56**, 358 (1973).
390. Y. Kidani, K. Inagaki, T. Saotome and H. Koike, *Bunseki Kagaku* **22**, 896 (1973).
391. Y. Kidani and E. Ito, *Bunseki Kagaku* **25**, 57 (1976).
392. Y. Kidani, K. Nakamura, K. Inagaki and H. Koike, *Bunseki Kagaku* **24**, 742 (1975).
393. Y. Kidani, T. Saotome, M. Kato and H. Koike, *Bunseki Kagaku* **23**, 265 (1974).
394. Y. Kidani, T. Saotome, K. Inagaki and H. Koike, *Bunseki Kagaku* **24**, 463 (1975).
395. A. S. King, *Astrophys. J.* **27**, 353 (1908).
396. R. B. King and D. C. Stockbarger, *Astrophys. J.* **91**, 488 (1940).
397. R. B. King, *Astrophys. J.* **95**, 78 (1942).
398. J. Kinnunen and O. Lindsjö, *Chemist-Analyst* **56**, 25 and 76 (1967).
399. J. D. Kinrade and J. C. van Loon, *Anal. Chem.* **46**, 1894 (1974).
400. K. Kinson and C. B. Belcher, *Anal. Chim. Acta* **30**, 64 (1964).
401. K. Kinson and C. B. Belcher, *Anal. Chim. Acta* **31**, 180 (1964).
402. K. Kinson, R. J. Hodges and C. B. Belcher, *Anal. Chim. Acta* **29**, 134 (1963).
403. G. Kirchhoff and R. Bunsen, *Pogg. Ann.* **110**, 161 (1860); **113**, 337 (1861); *Philos. Mag.* **22**, 329 (1861).
404. M. Kirk, E. G. Perry and J. M. Arrett, *Anal. Chim. Acta* **80**, 163 (1975).
405. G. F. Kirkbright and M. Marshall, *Anal. Chem.* **45**, 1610 (1973).
406. G. F. Kirkbright, M. K. Peters and T. S. West, *Talanta* **14**, 789 (1967).
407. G. F. Kirkbright, M. K. Peters and T. S. West, *Analyst (London)* **91**, 411 (1966).
408. G. F. Kirkbright, A. Semb and T. S. West, *Talanta* **15**, 441 (1968).
409. G. F. Kirkbright, A. M. Smith and T. S. West, *Analyst (London)* **91**, 700 (1966).
410. G. F. Kirkbright, A. M. Smith and T. S. West, *Analyst (London)* **92**, 411 (1967).
411. G. F. Kirkbright, A. M. Smith and T. S. West, *Analyst (London)* **93**, 292 (1968).
412. G. F. Kirkbright, A. M. Smith, T. S. West and R. Wood, *Analyst (London)* **94**, 754 (1969).
413. G. F. Kirkbright and T. S. West, *Appl. Opt.* **7**, 1305 (1968).
414. G. F. Kirkbright, T. S. West and P. J. Wilson, *Analyst (London)* **98**, 49 (1973).
415. G. F. Kirkbright and P. J. Wilson, *Anal. Chim. Acta* **68**, 462 (1974).
416. J. R. Knechtel and J. L. Fraser, *Analyst (London)* **103**, 105 (1978).
417. D. M. Knight and M. K. Pyzyna, *At. Absorp. Newsl.* **8**, 129 (1969).
418. D. W. Kohlenberger, *At. Absorp. Newsl.* **8**, 108 (1969).
419. H. Koizumi and K. Yasuda, *Bunko Kenkyu* **23**, 290 (1974). See also *Anal. Chem.* **47**, 1679 (1975).
420. S. R. Koirtyohann and C. A. Hopkins, *Analyst (London)* **101**, 870 (1976).
421. T. Y. Kometani, *Plating* 1251 (Nov. 1969).
422. K. H. Konig and P. Neumann, *Z. Anal. Chem.* **279**, 337 (1976).
423. L. E. Kopito, M. A. Davis and H. Schwachman, *Clin. Chem.* **20**, 205 (1974).
424. L. Kopito and H. Schwachman, *J. Lab. Clin. Med.* **70**, 326 (1967). See also L. Kopito, H. Schwachman and L. A. Williams, *Stand. Methods Clin. Chem.* **7**, 151 (1972).
425. J. Korkisch and H. Gross, *Mikrochim. Acta* **II**, 413 (1975).
426. J. Korkisch and A. Sorio, *Anal. Chim. Acta* **82**, 311 (1976).
427. J. Korkisch, A. Sorio and I. Steffan, *Talanta* **23**, 289 (1976).
428. N. E. Korte, J. L. Moyes and M. B. Denton, *Anal. Chem.* **45**, 530 (1973).
429. H. Kraemer, *Galvanotechnik* **64**, 289 (1973).
430. S. S. Krishnan, S. Quittkat and D. R. Crapper, *Can. J. Spectrosc.* **21**, 25 (1976).
431. N. P. Kubasik, H. E. Sine and M. T. Volosin, *Clin. Chem.* **18**, 1326 (1972).
432. T. Kumamaru, *Anal. Chim. Acta* **43**, 19 (1968).
433. Y. Kuwae, T. Hasegawa and T. Shono, *Anal. Chim. Acta* **84**, 185 (1976).
434. Y. Laflamme, *At. Absorp. Newsl* **7**, 101 (1968).
435. I. Lang, G. Sebor, V. Sychra, D. Kolihova and O. Weisser, *Anal. Chim. Acta* **84**, 299 (1976).
436. F. J. Langmyhr, *Talanta* **24**, 277 (1977).
437. F. J. Langmyhr and P. R. Graff, *Anal. Chim. Acta* **21**, 334 (1959).

438. F. J. Langmyhr and J. T. Hakedal, *Anal. Chim. Acta* **83**, 127 (1976).
439. F. J. Langmyhr and T. Lind, *Anal. Chim. Acta* **80**, 297 (1975).
440. F. J. Langmyhr and P. E. Paus, *Anal. Chim. Acta* **43**, 397 (1968).
441. F. J. Langmyhr and P. E. Paus, *Anal. Chim. Acta* **43**, 506 (1968).
442. F. J. Langmyhr and P. E. Paus, *Anal. Chim. Acta* **43**, 508 (1968).
443. F. J. Langmyhr and P. E .Paus, *Anal. Chim. Acta* **45**, 173 (1969).
444. F. J. Langmyhr and P. E. Paus, *Anal. Chim. Acta* **45**, 176 (1969).
445. F. J. Langmyhr and P. E. Paus, *Anal. Chim. Acta* **45**, 157 (1969).
446. F. J. Langmyhr and P. E. Paus, *Anal. Chim. Acta* **50**, 515 (1970).
447. F. J. Langmyhr and S. Sveen, *Anal. Chim. Acta* **32**, 1 (1965).
448. F. J. Langmyhr and D. L. Tsalev, *Anal. Chim. Acta* **92**, 79 (1977).
449. O. W. Lau and K. L. Li, *Analyst (London)* **100**, 430 (1975).
450. S. L. Law and T. E. Green, *Anal. Chem.* **41**, 1008 (1969).
451. C. J. Least, A. Rejent and H. Lees, *At. Absorp. Newsl.* **13**, 4 (1974).
452. J. R. Leaton, *J. Assoc. Off. Anal. Chem.* **53**, 237 (1970).
453. E. F. Legg, *Clin. Chim. Acta* **50**, 157 (1974).
454. V. Lehmann, *Clin. Chim. Acta* **20**, 523 (1968).
455. C. L. Lewis, W. L. Ott and N. M. Sine, *The Analysis of Nickel*. Pergamon, Oxford, 1966.
456. L. L. Lewis, *Atomic Absorption Spectroscopy*, A.S.T.M. Special Technical Publication 443, 1969, p. 47.
457. G. Lindstedt, *Analyst (London)* **95**, 264 (1970).
458. D. L. Littlejohn, G. S. Fell and J. M. Ottaway, *Clin. Chem.* **22**, 1719 (1976).
459. A. Lorber, R. L. Cohen, C. C. Chang and H. E. Anderson, *Arthritis Rheum.* **11**, 170 (1968).
460. R. J. Lovett, D. L. Welch and M. L. Parsons, *Appl. Spectrosc.* **29**, 470 (1975).
461. R. J. Lukasiewiez and B. E. Buell, *169th A.C.S. National Meeting*, Philadelphia, USA, 1975.
462. G. Lundgren, *Talanta* **23**, 309 (1976).
463. G. Lundgren and G. Johanssen, *Talanta* **21**, 257 (1974).
464. G. Lundgren, L. Lundmark and G. Johannson, *Anal. Chem.* **46**, 1028 (1974).
465. B. V. L'vov, *Spectrochim. Acta* **17**, 761 (1961).
466. B. V. L'vov, *Spectrochim. Acta* **24B**, 53 (1969).
467. B. V. L'vov, *Spectrochim. Acta* **24B**, 53 (1970).
468. B. V. L'vov, *Zh. Anal. Khim.* **30**, 1870 (1975).
469. B. V. L'vov and A. D. Khartsyzov, *Zh. Anal. Khim.* **24**, 799 (1969).
470. B. V. L'vov and A. D. Khartsyzov, *Zh. Anal. Khim.* **25**, 1824 (1970).
471. B. V. L'vov and N. A. Orloff, *Zh. Anal. Khim.* **30**, 1661 (1973).
472. R. Machiroux and D. T. K. Anh, *Anal. Chim. Acta* **86**, 35 (1976).
473. D. Mack, *Dtsch. Lebensm.-Rundsch.* **71**, 71 (1975).
474. W. S. G. MacPhee and D. F. Ball, *J. Sci. Food Agr.* **18**, 376 (1967).
475. J. P. Mahoney, K. Sargent, M. Greland and W. Small, *Clin. Chem.* **15**, 312 (1969).
476. I. S. Maines and C. L. Chakrabarti, *21st Canadian Spectroscopy Symposium*, Ottawa, October 1974.
477. H. Malissa and E. Schöffmann, *Mikrochim. Acta* **I**, 187 (1955).
478. S. E. Manahan and R. Kunkel, *Anal. Lett.* **6**, 547 (1973).
479. D. C. Manning, *At. Absorp. Newsl.* **10**, 86 (1971).
480. D. C. Manning and R. D. Eddiger, *At. Absorp. Newsl.* **15**, 42 (1976).
481. D. C. Manning and F. Fernandez, *At. Absorp. Newsl.* **7**, 24 (1968).
482. D. C. Manning and F. Fernandez, *At. Absorp. Newsl.* **9**, 65 (1970).
483. D. C. Manning and W. Slavin, *At. Absorp. Newsl.* No. 8 (November 1962).
484. R. E. Mansell, H. W. Emmel and E. L. McLaughlin, *Appl. Spectrosc.* **20**, 231 (1966).
485. M. Mantel, A. Aladjem and R. Nothmann, *Anal. Lett.* **9**, 671 (1976).
486. M. V. Marcec, K. Kinson and C. B. Belcher, *Anal. Chim. Acta* **41**, 447 (1968).
487. J. Y. Marks and G. G. Welcher, *Anal. Chem.* **42**, 1033 (1970).

488. J. Y. Marks, G. G. Welcher and R. J. Spellman, *Appl. Spectrosc.* **31**, 9 (1977).
489. G. B. Marshall and T. S. West, *Analyst (London)* **95**, 343 (1970).
490. H. Massmann, *Méthodes Phys. Anal.* **4**, 193 (1968).
491. H. Massmann, *Spectrochim. Acta* **23B**, 215 (1968).
492. H. Massmann and S. Gucer, *Spectrochim. Acta* **29B**, 283 (1974).
493. J. Matousek, *Talanta* **24**, 315 (1977).
494. J. P. Matousek and B. J. Orr, *5th International Conference on Atomic Spectroscopy*, Melbourne, Australia, 1975.
495. J. Matousek and V. Sychra, *Anal. Chem.* **41**, 518 (1969).
496. J. Matousek and V. Sychra, *Anal. Chim. Acta* **49**, 175 (1970).
497. T. Matsuo, J. Shida and M. Motoki, *Jpn Analyst* **18**, 521 (1969).
498. T. Matsuo, J. Shida, M. Abiko and K. Konno, *Bunseki Kagaku* **24**, 723 (1975).
499. K. Matsusaki, *Bunseki Kagaku* **24**, 442 (1975).
500. R. Mavrodineanu and H. Boiteux, *Flame Spectroscopy*. Wiley, New York, 1965.
501. R. Mavrodineanu and R. C. Hughes, *Appl. Opt.* **7**, 1281 (1968).
502. J. R. McDermott and I. Whitehill, *Anal. Chim. Acta* **85**, 195 (1976).
503. J. D. McCrackan, M. C. Vecchione and S. L. Longo, *At. Absorp. Newsl.* **8**, 102 (1969).
504. C. L. McIsaac, *Eng. Min. J.* **170**, 55 (1969).
505. A. J. McLean, R. L. Halstead and B. J. Finn, *Can. J. Soil Sci.* **49**, 327 (1969).
506. E. A. Means and D. Ratcliffe, *At. Absorp. Newsl.* **4**, 174 (1975).
507. J. M. L. Mee and H. W. Hilton, *J. Agr. Food Chem.* **17**, 1398 (1969).
508. J. R. Melton, W. L. Hoover and P. A. Howard, *J. Assoc. Off. Anal. Chem.* **52**, 950 (1969).
509. R. Menache, *Clin. Chem.* **20**, 1444 (1974).
510. O. Menis and T. C. Rains, *Anal. Chem.* **41**, 952 (1969).
511. J. C. Meranger, *J. Assoc. Off. Anal. Chem.* **58**, 1143 (1975).
512. J. C. Meranger and E. Somers, *Analyst (London)* **93**, 799 (1968).
513. M. K. Meredith, S. Baldwin and A. A. Andreasen, *J. Assoc. Off. Anal. Chem.* **53**, 12 (1970).
514. T. H. Miller and W. H. Edwards, *At. Absorp. Newsl.* **15**, 75 (1976).
515. B. A. Milner, P. J. Whiteside and W. J. Price, unpublished.
516. T. Minamikawa and K. Matsumura, *Yakugaku Zasshi* **96**, 440 (1976).
517. T. Minamikawa, K. Sakai, N. Hashitani, E. Fukushima and N. Yamagishi, *Chem. Pharm. Bull. Tokyo* **21**, 1632 (1973).
518. P. Minkkinen, *Kem.-Kemi* **3**, 282 (1976).
519. K. Mitani, *Eisei Kagaku* **22**, 65 (1976).
520. A. C. G. Mitchell and M. W. Zemansky, *Resonance Radiation and Excited Atoms*. Cambridge University Press, London, 1934, reprinted 1961.
521. D. G. Mitchell, *Technicon International Congress*, 2–4 November 1970.
522. D. G. Mitchell, K. M. Aldous and A. F. Ward, *At. Absorp. Newsl.* **13**, 121 (1974).
523. D. G. Mitchell and A. Johansson, *Spectrochim. Acta* **25B**, 175 (1970); **26B**, 677 (1971).
524. J. W. Mitchell, *Anal. Chem.* **45**, 492A (1973).
525. J. W. Mitchell and D. L. Nash, *Anal. Chem.* **46**, 326 (1974).
526. T. Mitsui and Y. Fujimura, *Bunseki Kagaku* **24**, 575 (1975).
527. T. Mizuno, A. Harada, Y. Kudo and N. Hasegawa, *Jpn Analyst* **19**, 251 (1970).
528. A. Montaser, S. R. Goode and S. R. Crouch, *Anal. Chem.* **46**, 599 (1974).
529. B. Montford and S. C. Cribbs, *At. Absorp. Newsl.* **8**, 77 (1969).
530. E. J. Moore, O. I. Milner and J. R. Glass, *Microchem. J.* **10**, 148 (1966).
531. N. M. Morris, M. A. Clarke, V. W. Tripp and F. G. Carpenter, *J. Agric. Food Chem.* **24**, 45 (1976).
532. G. H. Morrison and H. Freiser, *Solvent Extraction in Analytical Chemistry*. Wiley, New York, 1957.
533. V. G. Mossotti, K. Lacqua and W. D. Hagenah, *Spectrochim. Acta* **23B**, 197 (1967).
534. R. A. Mostyn and A. F. Cunningham, *Anal. Chem.* **38**, 121 (1966).

535. R. A. Mostyn and A. F. Cunningham, *At. Absorp. Newsl.* **6**, 86 (1967).
536. R. A. Mostyn and A. F. Cunningham, *Anal. Chem.* **39**, 433 (1967).
537. R. A. Mostyn and A. F. Cunningham, *J. Inst. Petrol.* **53**, 101 (1967).
538. J. D. Mullen, *Talanta* **23**, 846 (1976).
539. C. Mullins, *5th Working Seminar on AAS*, Røros, Norway, November 1976.
540. G. M. Murphy, K. M. Ryan and G. W. Blight, *4th Australian Symposium on Analytical Chemistry*, Brisbane, 1977.
541. V. I. Muscat, T. J. Vickers and A. Andren, *Anal. Chim. Acta* **44**, 218 (1972).
542. R. A. A. Muzzarelli and R. Rocchetti, *Anal. Chem.* **70**, 283 (1974).
543. R. A. A. Muzzarelli and R. Rocchetti, *Talanta* **22**, 683 (1975).
544. N. Nadirshaw and A. H. Cornfield, *Analyst* **93**, 475 (1968).
545. T. Nakahara, M. Munemori and S. Musha, *Anal. Chim. Acta* **50**, 51 (1970).
546. W. R. Nall, D. Brumhead and R. Witham, *Analyst* **100**, 555 (1975).
547. R. A. Newstead, British Patent 55340/72.
548. R. A. Newstead, W. J. Price, P. J. Whiteside, T. C. Dymott and I, Bowater, *29th Pittsburgh Conference on Analytical Chemistry and Applied Spectroscopy*, 1978.
549. R. A. Newstead and P. J. Whiteside, private communication.
550. C. Van Nieuwenhauyzen, *Chem. Weekbl.* m 106 (1976).
551. G. I. Nikolayev, *Zh. Anal. Khim.* **20**, 44 (1965).
552. R. J. Noga, *Anal. Chem.* **47**, 332 (1975).
553. J. D. Norris and T. S. West, *Anal. Chem.* **45**, 2148 (1973).
554. J. D. Norris and T. S. West, *Anal. Chim. Acta* **71**, 289 (1974).
555. J. D. Norris and T. S. West, *Anal. Chem.* **46**, 1423 (1974).
556. J. D. Norris and T. S. West, *Anal. Chim. Acta* **71**, 458 (1976).
557. J. W. Novak and R. F. Browner, *Anal. Chem.* **50**, 407 (1978).
558. H. Oguro, *Bunseki Kagaku* **25**, 785 (1976).
559. K. Ohta and M. Suzuki, *Anal. Chim. Acta* **77**, 288 (1975).
560. K. Ohta and M. Suzuki, *Talanta* **22**, 465 (1975).
561. K. Ohta and M. Suzuki, *Anal. Chim. Acta* **85**, 83 (1976).
562. M. Olivier, *Z. Anal. Chem.* **248**, 145 (1969).
563. R. D. Olsen and M. R. Sommerfeld, *At. Absorp. Newsl.* **12**, 165 (1973).
564. S. H. Omang, *Anal. Chim. Acta* **53**, 415 (1971).
565. N. Omenetto and G. Rossi, *Anal. Chim. Acta* **40**, 195 (1968).
566. T. Ono and S. Odagiri, *Rakuno Kagaku No Kenkyu* **24**, A133 (1975).
567. R. M. Orheim and H. H. Bovee, *Anal. Chem.* **46**, 921 (1974).
568. J. M. Ottaway, *27th BISRA/BSC Conference*, Scarborough, May 1974.
569. J. M. Ottaway and W. C. Campbell, *Int. J. Environ. Anal. Chem.* **4**, 233 (1976).
570. J. M. Ottaway, D. T. Coker and J. A. Davies, *Anal. Lett.* **3**, 385 (1970).
571. J. M. Ottaway and F. Shaw, *Analyst (London)* **100**, 438 (1975).
572. V. Ovnic and S. Gomiscek, *Erzmetall* **26**, 179 (1973).
573. T. Owa, K. Hiro and T. Tanaka, *Jpn Analyst* **21**, 878 (1972).
574. J. W. Owens and E. S. Gladney, *At. Absorp. Newsl.* **15**, 95 (1976).
575. J. E. Parker, *Spectrovision* **24**, 10 (1970).
576. M. W. Parker, F. L. Humoller and D. J. Mahler, *Clin. Chem.* **13**, 40 (1967).
577. Parr Instrument Co., *Bulletin* No. 4745, 1973.
578. J. A. Parsson, W. Frech and A. Cedergren, *Anal. Chim. Acta* **89**, 119 (1977).
579. P. E. Paus, *At. Absorp. Newsl.* **11**, 129 (1972).
580. P. E. Paus, *Interan Conference*, Prague, 1976.
581. S. L. Paveri-Fontana, G. Tessari and G. Torsi, *Anal. Chem.* **46**, 1032 (1974).
582. I. D. Pearce, R. R. Brooks and R. D. Reeves, *J. Assoc. Off. Anal. Chem.* **59**, 655 (1976).
583. B. Perry, *Spectrovision* **25**, 8 (1971).
584. E. F. Perry, S. R. Koirtyohann and H. M. Perry, *Clin. Chem.* **21**, 626 (1975).
585. E. A. Peterson, *At. Absorp. Newsl.* **8**, 53 (1969).

586. C. J. Pickford and G. Rossi, *At. Absorp. Newsl.* **14**, 78 (1975).
587. M. Pinta, *Method Phys. Anal.* **6**, 268 (1970).
588. A. E. Pitts and F. E. Beamish, *Anal. Chim. Acta* **52**, 405 (1970).
589. A. E. Pitts, J. C. van Loon and F. E. Beamish, *Anal. Chim. Acta* **50**, 181 (1970).
590. A. E. Pitts, J. C. van Loon and F. E. Beamish, *Anal. Chim. Acta* **50**, 195 (1970).
591. J. A. Platte, in *Trace Inorganics in Water: Advances in Chemistry Series* No. 73, American Chemical Society, Washington D.C., 1968.
592. J. A. Platte and V. M. Marcy, *At. Absorp. Newsl.* **4**, 289 (1965).
593. E. N. Pollock, *At. Absorp. Newsl.* **9**, 47 (1970).
594. E. N. Pollock and E. J. West, *At. Absorp. Newsl.* **11**, 104 (1972).
595. E. N. Pollock and E. J. West, *At. Absorp. Newsl.* **12**, 6 (1973).
596. W. K. Porter, *J. Assoc. Off. Anal. Chem.* **57**, 614 (1974).
597. A. L. Potter, E. D. Ducay and R. M. McCready, *J. Assoc. Off. Anal. Chem.* **51**, 748 (1968).
598. P. R. Premi and A. H. Cornfield, *Spectrovision* **19**, 15 (1968).
599. A. Prévot and M. Gente-Jauniaux, *At. Absorp. Newsl.* **17**, 1 (1978).
600. W. J. Price, *Chem. Br.* **14**, 140 (1978).
601. W. J. Price, *Effluent Water Treatment J.*, April 1967.
602. W. J. Price, *Medical Laboratory World* **2**, 135 (1978).
603. W. J. Price, *Paint, Oil, Colour J.* 3 (21 August 1970).
604. W. J. Price and P. A. Cooke, *Spectrovision* **18**, 2 (1967).
605. W. J. Price, T. C. Dymott and M. Wassall, *XX CSI/7th ICAS*, Prague, 1977.
606. W. J. Price and J. T. H. Roos, *Analyst (London)* **93**, 709 (1968).
607. W. J. Price and J. T. H. Roos, *J. Sci. Food Agr.* **20**, 437 (1969).
608. W. J. Price, J. T. H. Roos and A. F. Clay, *Analyst (London)* **95**, 760 (1970).
609. W. J. Price and P. J. Whitseside, *Analusis* **5**, 275 (1977).
610. W. J. Price and P. J. Whiteside, *Analyst (London)* **102**, 664 (1977).
611. J. Pybus, *Clin. Chim. Acta* **23**, 309 (1969).
612. J. Pybus and G. N. Bowers, *Clin. Chem.* **16**, 139 (1970).
613. J. Pybus, F. J. Feldman and G. N. Bowers, *Clin. Chem.* **16**, 998 (1970).
614. T. M. Quarrell, R. J. W. Powell and H. J. Cluley, *Analyst (London)*, **98**, 443 (1973).
615. A. Quentmeier, W. D. Hagenah and K. Lacqua, *5th International Conference on Atomic Spectroscopy*, Melbourne, Australia, 1975.
616. B. D. Radcliffe, C. S. Byford and P. B. Osman, *Anal. Chim. Acta* **75**, 457 (1975).
617. T. C. Rains, M. S. Epstein and O. Menis, *Anal. Chem.* **46**, 207 (1974).
618. T. V. Ramakrishna, J. W. Robinson and P. W. West, *Anal. Chim. Acta* **45**, 43 (1969).
619. T. V. Ramakrishna, P. W. West and J. W. Robinson, *Anal. Chim. Acta* **44**, 437 (1969).
620. J. Ramirez-Muñoz, *Beckman Flame Notes* **7**, 49 (1975).
621. J. Ramirez-Muñoz, *Beckman Flame Notes* **2**, 77 (1967).
622. H. Rampon and R. Cuvelier, *Anal. Chim. Acta* **60**, 226 (1972).
623. A. Rattonetti, *Anal. Chem.* **46**, 739 (1974).
624. R. A. G. Rawson, *Analyst (London)* **91**, 630 (1966).
625. J. R. Reeves, *Econ. Geol.* **62**, 426 (1967).
626. J. G. T. Regan and J. Warren, *Analyst (London)* **101**, 220 (1976).
627. W. Reichel and B. G. Bleakeley, *Anal. Chem.* **46**, 59 (1974).
628. J. Reid, J. M. Galloway, J. MacDonald and B. B. Bach, *Metallurgia* **81**, 243 (1970).
629. I. Reif, V. A. Fassel and R. N. Kniseley, *Spectrochim. Acta* **28B** 105 (1973).
630. I. Reif, V. A. Fassel and R. N. Kniseley, *Spectrochim. Acta* **29B**, 79 (1974).
631. G. D. Renshaw, *At. Absorp. Newsl.* **12**, 158 (1973).
632. R. A. Richardson, *Clin. Chem.* **22**, 1604 (1976).
633. J. P. Riley and D. Taylor, *Anal. Chim. Acta* **40**, 479 (1968).
634. A. G. Roach, P. Sanderson and D. R. Williams, *Analyst (London)* **63**, 42 (1968).
635. W. K. Robbins, J. H. Runnels and R. Merryfield, *Anal. Chem.* **47**, 2095 (1975).

636. R. V. D. Robert, *IUPAC Sponsored International Symposium on Analytical Chemistry etc.*, Johannesburg, South Africa (1976).
637. G. R. Rogers, *J. Assoc. Off. Anal. Chem.* **51**, 1042 (1968).
638. R. C. Rooney, *Analyst (London)* **100**, 471 (1975).
639. R. C. Rooney, *Analyst (London)* **101**, 678 (1976).
640. R. C. Rooney, *Analyst (London)* **101**, 749 (1976).
641. J. T. H. Roos, *Spectrochim. Acta* **24B**, 255 (1969).
642. J. T. H. Roos and W. J. Price, *Analyst (London)* **94**, 89 (1969).
643. J. T. H. Roos and W. J. Price, *J. Sci. Food Agr.* **21**, 51 (1970).
644. J. T. H. Roos and W. J. Price, *Spectrochim. Acta* **26B**, 279 (1971).
645. J. T. H. Roos and W. J. Price, *Spectrochim. Acta* **26B**, 441 (1971).
646. R. Roosels and J. V. Vanderkeel, *At. Absorp. Newsl.* **7**, 9 (1968).
647. S. A. Rose and D. F. Boltz, *Anal. Chim. Acta* **44**, 239 (1969).
648. R. T. Ross and J. G. Gonzalez, *Bull. Environ. Contam. Toxicol.* **12**, 470 (1974).
649. R. T. Ross, J. G. Gonzalez and D. A. Segar, *Anal. Chim. Acta* **63**, 205 (1973).
650. W. B. Rowston and J. M. Ottaway, *Anal. Lett.* **3**, 411 (1970).
651. I. Rubeska, J. Koreckova and D. Weiss, *Interan Conference*, Prague, August 1976.
652. I. Rubeska and M. Miksovsky, *Collect. Czech. Chem. Commun.* **39**, 3485 (1974).
653. I. Rubeska, M. Miksovsky and M. Huka, *At. Absorp. Newsl.* **14**, 26 (1975).
654. I. Rubeska and B. Moldan, *Appl. Opt.* **7**, 1341 (1968).
655. I. Rubeska and B. Moldan, *Analyst (London)* **93**, 148 (1968).
656. I. Rubeska and J. Stupar, *At. Absorp. Newsl.* **5**, 69 (1966).
657. I. Rubeska and V. Svoboda, *Anal. Chim. Acta* **32**, 253 (1965).
658. S. L. Sachdev, J. W. Robinson and P. W. West, *Anal. Chim. Acta* **37**, 12 (1967).
659. M. N. Saha and H. K. Saha, *A Treatise on Modern Physics*, Vol. 1. Allahabad, Calcutta, 1934.
660. A. M. Saltykova, N. K. Davidovitch and S. G. Melamed, *J. Anal. Chem. USSR* **27**, 1091 (1972).
661. A. Sapek, *Biol. Nauki (Moscow)* **19**, 131 (1976).
662. V. S. Sastri, C. L. Chakrabarti and D. E. Willis, *Can. J. Chem.* **47**, 587 (1969).
663. V. S. Sastri, C. L. Chakrabarti and D. E. Willis, *Talanta* **16**, 1093 (1969).
664. M. Satake, T. Asano, Y. Tagaki and T. Yonekubo, *Nippon Kagaku Kaishi* **8**, 762 (1976).
665. J. M. Scarborough, *Anal. Chem.* **41**, 250 (1969).
666. J. M. Scarborough, C. D. Bingham and P. F. DeVries, *Anal. Chem.* **39**, 1394 (1967).
667. M. Schattenkirchner and Z. Grobenski, *At. Absorp. Newsl.* **16**, 84 (1977).
667a. P. S. Schmidt, *Instrum. Technol.* **22**, 35 (1975).
668. M. M. Schnepfe and F. S. Grimaldi, *Talanta* **16**, 1461 (1969).
669. P. H. Scholes, *Analyst (London)* **93**, 197 (1968).
670. J. Scott and F. C. A. Killer, *Proc. Soc. Anal. Chem.* **7**, 18 (January 1970).
671. E. Sebastiani, K. Ohis and G. Riemer, *Z. Anal. Chem.* 105 (1973).
672. G. Sebor, I. Lang, P. Vavrecka, V. Sychra and O. Weisser, *Anal. Chim. Acta* **78**, 99 (1975).
673. D. A. Segar, *Anal. Lett.* **7**, 89 (1974).
674. S. Selander and K. Cramer, *Brit. J. Ind. Med.* **25**, 139 (1968).
675. R. G. Shafto, *Prod. Finishing (Cincinnati)* **28**, 138 (1964).
676. F. Shaw and J. M. Ottaway, *Analyst (London)* **100**, 217 (1975).
677. B. J. Shelton and E. B. T. Cook, *Nat. Inst. Metall. Repub. S. Afr. Rep.* 1723 (1975).
678. J. E. Sheridan, *Spectrovision* **25**, 10 (1971).
679. I. L. Shrista and T. S. West, *Bull. Soc. Chim. Belg.* **84**, 549 (1975).
680. L. H. Siertsema, *Clin. Chim. Acta* **69**, 533 (1976).
681. G. P. Sighinolfi and A. M. Santos, *Mikrochim. Acta* **I**, 477 (1976).
682. G. P. Sighinolfi and A. M. Santos, *Mikrochim. Acta* **II**, 33 (1976).
683. S. J. Simon and D. F. Boltz, *Microchem. J.* **20**, 468 (1975).

684. J. V. Simonian, *At. Absorp. Newsl.* **7**, 63 (1968).
685. A. Simonson, *Anal. Chim. Acta* **49**, 368 (1970).
686. W. R. Simpson and G. Nickless, *Analyst (London)* **102**, 86 (1977).
687. K. C. Singhal, A. C. Banerji and B. K. Banerjee, *Technology (Sindri, India)* **5**, 117 (1968).
688. S. Slavin and T. W. Sattur, *At. Absorp. Newsl.* **7**, 99 (1968).
689. W. Slavin, *At. Absorp. Newsl.* **4**, 243 (1965).
690. H. T. Smart and D. J. Campbell, *Can. J. Pharm. Sci.* **4**, 73 (1969).
691. A. E. Smith, *Analyst (London)* **100**, 300 (1975).
692. D. C. Smith, J. R. Johnson and G. C. Soth, *Appl. Spectrosc.* **24**, 576 (1970).
693. J. C. Smith and J. E. Kench, *Brit. J. Ind. Med.* **14**, 240 (1957).
694. S. B. Smith, J. A. Blasi and F. J. Feldman, *Anal. Chem.* **40**, 1525 (1968).
695. A. Sohler, P. Wolcott and C. C. Pfeiffer, *Clin. Chim. Acta* **70**, 391 (1976).
696. S. D. Soman, V. K. Panday and K. T. Joseph, *Am. Ind. Hyg. Assoc. J.* **30**, 527 (1969).
697. G. I. Spielholtz and G. C. Toralballa, *Analyst (London)* **94**, 1072 (1969).
698. B. Y. Spivakov, V. I. Lebedev, V. M. Shikinev, N. P. Krivenkova, T. S. Plotnikova, I. P. Kharlamov and Y. A. Zolotov, *Zh. Anal. Chim.* **31**, 757 (1976).
699. S. Sprague, D. C. Manning and W. Slavin, *At. Absorp. Newsl.* No. 20, p. 1 (May 1964).
700. S. Sprague and W. Slavin, *At. Absorp. Newsl.* **3**, 160 (1964).
701. S. Sprague and W. Slavin, *At. Absorp. Newsl.* **4**, 367 (1965).
702. *Standard Methods of Testing Paint, Varnish, Lacquer and Related Products.* H.M. Stationery Office, London, 1964.
703. J. Stary, *The Solvent Extraction of Metal Chelates.* Pergamon, Oxford, 1964.
704. T. Stiefel, K. Schulze, G. Tolg and H. Zorn, *Anal. Chim. Acta* **87**, 67 (1976).
705. A. Strasheim, F. W. E. Strelow and L. R. P. Butler, *J.S.A. Chem. Inst.* **13**, 73 (1960).
706. A. Strasheim and G. J. Wessels, *Appl. Spectrosc.* **17**, 65 (1963).
707. A. Strecker, A. Varnes and M. S. Vigler, *1977 Pittsburgh Conference on Analytical Chemistry and Applied Spectroscopy*, Cleveland, USA, 1977.
708. F. W. E. Strelow, E. C. Feast, P. M. Mathews, C. J. C. Bothma and C. R. Van Zyl, *Anal. Chem.* **38**, 115 (1966).
709. D. H. Strunk and A. A. Andreasen, *J. Assoc. Off. Anal. Chem.* **50**, 339 (1967).
710. J. Stupar and J. B. Dawson, *Appl. Opt.* **7**, 1351 (1968).
711. J. Stupar and J. B. Dawson, *At. Absorp. Newsl.* **8**, 38 (1969).
712. R. E. Sturgeon and C. L. Chakrabarti, *Spectrochim. Acta* **32B**, 231 (1977).
713. R. E. Sturgeon, C. L. Chakrabarti and P. C. Bertels, *Anal. Chem.* **47**, 1240, 1250 (1975).
714. R. E. Sturgeon, C. L. Chakrabarti and C. H. Langford, *Anal. Chem.* **48**, 1792 (1976).
715. R. F. Suddendorf, D. E. Gutzler and M. B. Denton, *Spectrochim. Acta* **31B**, 281 (1976).
716. J. E. Sueiras, K. S. Subramanian, C. L. Chakrabarti, D. J. Young and I. S. Maines, *2nd Int. Conf. Chem. Inst. Can. and Am. Chem. Soc.*, Montreal, 1977.
717. Y. Sugawara, J. Miuru, F. Shirato and K. Sakai, *Bunseki Kagaku* **25**, 210 (1976).
718. S. Sukiman, *Anal. Chim. Acta* **84**, 419 (1976).
719. J. V. Sullivan, *5th International Conference of Atomic Spectroscopy*, Melbourne, Australia, 1975.
720. J. V. Sullivan and A. Walsh, *Spectrochim. Acta* **21**, 721 (1965).
721. F. W. Sunderman, *Hum. Pathol.* **4**, 549 (1973).
722. H. J. Svec and A. R. Anderson Jr, *Geochim. Cosmochim. Acta* **29**, 633 (1965).
723. R. T. Swider, *At. Absorp. Newsl.* **7**, 111 (1968).
724. V. Sychra and J. Matousek, *Anal. Chim. Acta* **52**, 376 (1970); see also V. Sychra, P. T. Slevin, J. Matousek and F. Bek, *Anal. Chim. Acta* **52**, 259 (1970).
725. T. Takeuchi and M. Yanagisawa, *Jpn Analyst* **15**, 1059 (1966).
726. M. S. Taylor, *Proc. Anal. Div. Chem. Soc.* **13**, 342 (1976).
727. N. J. Teclu, *J. Prakt. Chem.* **44**, 246 (1891).
728. A. M. Tenny, *Perkin-Elmer Instr. News* **18**, 1 and 14 (1967).
729. S. Terashima, *Bunseki Kagaku* **24**, 319 (1975).

730. Y. Thomassen, B. V. Larsen, F. J. Langmyhr and W. Lund, *Anal. Chim. Acta* **83**, 103 (1976).
731. D. R. Thomerson, *Spectrovision* **26**, 13 (1972).
732. D. R. Thomerson and W. J. Price, *Analyst (London)* **96**, 321 (1971).
733. D. R. Thomerson and W. J. Price, *Analyst (London)* **96**, 825 (1971).
734. D. R. Thomerson and W. J. Price, *Anal. Chim. Acta* **72**, 188 (1974).
735. A. J. Thompson, *Spectrovision* **20**, 7 (1968).
736. A. J. Thompson and P. A. Thoresby, *Analyst (London)* **102**, 9 (1977).
737. K. C. Thompson, *Analyst (London)* **95**, 1043 (1970).
738. K. C. Thompson and R. G. Godden, *Analyst (London)* **100**, 544 (1975).
739. K. C. Thompson and R. G. Godden, *Analyst (London)* **101**, 96 (1976).
740. K. C. Thompson, R. G. Godden and D. R. Thomerson, *Anal. Chim. Acta* **74**, 289 (1975).
741. K. C. Thompson and D. R. Thomerson, *Analyst (London)* **99**, 595 (1974).
742. J. B. Tillery and D. E. Johnson, *Environ. Health Perspect.* **12**, 19 (1975).
743. F. M. Tindall, *At. Absorp. Newsl.* **4**, 339 (1965).
744. F. M. Tindall, *At. Absorp. Newsl.* **5**, 140 (1966).
745. J. Toffaletti and J. Savory, *Anal. Chem.* **47**, 2091 (1975).
746. S. Tolansky, *High Resolution Spectroscopy*. Methuen, London, 1947.
747. S. L. Tompsett, *Proc. SSOC. Clin. Biochem.* **5**, 125 (1968).
748. F. Torres, *Nat. Bur. Stand. (U.S.) Reprint* SC-RR-69-784, December 1969.
749. T. Toyoguchi and H. Shimizu, *Jpn Analyst* **16**, 565 (1967).
750. D. Trent and W. Slavin, *At. Absorp. Newsl.* No. 19, March 1964, p. 1.
751. D. L. Trudeau and E. F. Freier, *Clin. Chem.* **13**, 101 (1967).
752. J. B. Tyler, *At. Absorp. Newsl.* **6**, 14 (1967).
753. T. Uchida, M. Nagase and C. Iida, *Anal. Lett.* **8**, 825 (1975).
754. U. Ulfvarson, *Acta Chem. Scand.* **21**, 641 (1967).
755. A. M. Ure, *Anal. Chim. Acta* **76**, 1 (1975).
756. A. M. Ure and M. C. Mitchell, *Anal. Chim. Acta* **87**, 283 (1976).
757. A. M. Ure and R. L. Mitchell, *Spectrochim. Acta* **23B**, 79 (1967).
758. J. F. Uthe, F. A. J. Armstrong and M. P. Stainton, *J. Fisheries Res. Board Can.* **27**, 805 (1970).
759. O. W. Van Assendelft, W. G. Zijlstra, A. Buursma, E. J. Van Kampen and W. Hoek, *Clin. Chim. Acta* **22**, 281 (1968).
760. J. C. van Loon, *Z. Anal. Chem.* **246**, 122 (1969).
761. J. C. van Loon, J. H. Galbraith and H. M. Aarden, *Analyst (London)* **96**, 47 (1971).
762. J. C. van Loon and C. M. Parissis, *Analyst (London)* **94**, 1057 (1969).
763. M. Varju, *Vth Polish Spectroanalytical Conference*, Władislawow, 1976.
764. C. Veillon, J. M. Mansfield, M. L. Parsons and J. D. Winefordner, *Anal. Chem.* **38**, 204 (1966).
765. G. Velghe, M. Verloo and A. Cottenie, *Z. Lebensm.-Unters. Forsch.* **156**, 77 (1974).
766. O. Vesterberg and T. Bergstrom, *Clin. Chem.* **23**, 555 (1977).
767. M. S. Vigler and V. F. Gaylor, *Appl. Spectrosc.* **28**, 342 (1974).
768. V. A. Vilenkin, L. L. Kalinin and Y. V. Mikulin, *Khim. Tekhnol. Topl. Masel.* **12**, 54 (1975).
769. A. Walsh, *Spectrochim. Acta* **7**, 108 (1955).
770. A. Walsh, *Proceedings of the X Colloquium Spectroscopicum Internationale* 27 (1962).
771. A. Walsh, *Proceedings of the XIII Colloquium Spectroscopicum Internationale*, Ottawa, Canada, 1967. Hilger, London, 1968, p. 257.
772. A. Walsh, *Atomic Absorption Spectroscopy*. Plenary Lectures of International Conference, Sheffield, July 1969. IUPAC/Butterworth, London, 1970, p. 1.
773. A. Walsh, *Appl. Spectrosc.* **27**, 335 (1973).
774. D. A. Ward and D. G. Biechler, *At. Absorp. Newsl.* **14**, 29 (1975).
775. G. M. Ward and M. J. Miller, *Can. J. Plant Sci.* **49**, 53 (1969).

776. J. Warren and M. P. Harrison, *Proc. Anal. Div. Chem. Soc.* **13**, 287 (1976).
777. B. Watne and R. Woodriff, *Appl. Spectrosc.* **30**, 71 (1976).
778. C. A. Watson, *Ammonium Pyrrolidine Dithiocarbamate*, Monograph 74. Hopkin and Williams, Chadwell Heath, England, October 1971.
779. S. J. Weger Jr, L. R. Hossner and L. W. Ferrar, *J. Agr. Food Chem.* **17**, 1276 (1969).
780. J. P. Weiner and L. Taylor, *J. Inst. Brewing* **75**, 195 (1969).
781. D. R. Weir and R. P. Kofluk, *At. Absorp. Newsl.* **6**, 24 (1967).
782. G. G. Welcher and O. H. Kriege, *At. Absorp. Newsl.* **9**, 61 (1970).
783. G. G. Welcher, O. H. Kriege and J. Y. Marks, *Anal. Chem.* **46**, 1227 (1974).
784. G. G. Welcher, O. H. Kriege and H. Owen, *At. Absorp. Newsl.* **8**, 97 (1969).
785. B. Welz, *Joint Symposium on Detection of Major Components*, Society for Analytical Chemistry and Royal Dutch Chemical Society, London, 1970.
786. B. Welz *et al.*, *Interan Conference*, Prague, 1976.
787. R. H. Wendt and V. A. Fassel, *Anal. Chem.* **38**, 337 (1966).
788. T. S. West, *Atomic Absorption Spectroscopy*. Plenary Lectures of International Conference Sheffield, July 1969. IUPAC/Butterworth, London, 1970, p. 99.
789. T. S. West, *Proc. Anal. Div. Chem. Soc.* **13**, 266 (1976).
790. T. S. West and X. K. Williams, *Anal. Chem.* **40**, 335 (1968).
791. T. S. West and X. K. Williams, *Anal. Chim. Acta* **42**, 29 (1968).
792. T. S. West and X. K. Williams, *Anal. Chim. Acta* **45**, 27 (1969).
793. J. A. White, W. L. Harper, A. P. Friedman and V. E. Banas, *Appl. Spectrosc.* **28**, 192 (1974).
794. R. A. White, *International Atomic Absorption Conference*, Sheffield, 1969. Abstract G5.
795. P. J. Whiteside (Ed.), *Atomic Absorption with Electrothermal Atomization*, Pye Unicam, Cambridge, 1977.
796. P. J. Whiteside and W. J. Price, *XVIII CSI*, Grenoble, 1975.
797. P. J. Whiteside and W. J. Price, *Analyst (London)* **102**, 618 (1977).
798. L. M. Whitlock, J. R. Milton and T. J. Billings, *J. Assoc. Off. Anal. Chem.* **59**, 580 (1976).
799. C. M. Whittington and J. B. Willis, *Plating* **51**, 767 (1964).
800. C. H. Williams, D. J. David and O. Iismaa, *J. Agr. Sci.* **59**, 381 (1962).
801. J. B. Willis, *Nature* **184**, 186 (1959).
802. J. B. Willis, *Spectrochim. Acta* **16**, 273 (1960).
803. J. B. Willis, *Nature* **186**, 249 (1960).
804. J. B. Willis, *Spectrochim. Acta* **16**, 259 (1960).
805. J. B. Willis, *Spectrochim. Acta* **16**, 551 (1960).
806. J. B. Willis, *Nature* **191**, 381 (1961).
807. J. B. Willis, *Anal. Chem.* **33**, 556 (1961).
808. J. B. Willis, *Anal. Chem.* **34**, 614 (1962).
809. J. B. Willis, *Methods of Biochemical Analysis*. Interscience, New York, 1963, Vol. XI, p. 1.
810. J. B. Willis, *Nature* **207**, 715 (1965).
811. J. B. Willis, *Appl. Opt.* **7**, 1295 (1968); see also chapter in *Analytical Flame Spectroscopy* (Ed. R. Mavrodineanu), Macmillan, London, 1970.
812. J. B. Willis, *Anal. Chem.* **47**, 1752 (1975).
813. J. B. Willis, J. O. Rasmuson, J. R. Kniseley and V. A. Fassel, *Spectrochim. Acta* **23B**, 725 (1968).
814. A. L. Wilson, *Chem. Ind.* **36**, 1253 (1969).
815. A. L. Wilson, *The Chemical Analysis of Water*, Analytical Sciences Monograph No. 2, Analytical Division of the Chemical Society, London, 1974.
816. L. Wilson, *Anal. Chim. Acta* **30**, 377 (1964).
817. L. Wilson, *Metallurgy Note 42*. Aeronautical Research Laboratories. Dept. of Supply, Melbourne, Australia, 1966.
818. L. Wilson, *Anal. Chim. Acta* **40**, 503 (1968).
819. H. Windemann and U. Mueller, *Mitt. Geb. Lebensmittelunters. Hyg.* **66**, 64 (1975).

820. J. D. Winefordner, *Atomic Absorption Spectroscopy*. Plenary Lectures of International Conference, Sheffield, July 1969. IUPAC/Butterworth, London, 1970, p. 35.
821. J. D. Winefordner and R. A. Staab, *Anal. Chem.* **36**, 1367 (1964).
822. J. D. Winefordner and R. A. Staab, *Anal. Chem.* **36**, 165, 1367 (1964).
823. J. D. Winefordner and T. J. Vickers, *Anal. Chem.* **36**, 161 (1964).
824. H. Woidich and W. Pfannhauser, *Euroanalysis II*, Budapest, 1975.
825. H. Woidich and W. Pfannhauser, *Fresenius' Z. Anal. Chem.* **276**, 61 (1975).
826. W. H. Wollaston, *Phil. Trans. Roy. Soc. London Ser. A* **92**, 365 (1802).
827. R. Woodriff, *XX CSI/7th ICAS*, Prague, 1977.
828. R. Woodriff and G. Ramelow, *Spectrochim. Acta* **23B**, 665 (1968).
829. T. T. Woodson, *Rev. Sci. Instrum.* **10**, 308 (1939).
830. J. W. Woollen and M. G. Wells, *Ann. Clin. Biochem.* **10**, 85 (1973).
831. J. F. Woolley, *Spectrovision* **22**, 7 (1969).
832. J. F. Woolley, *Analyst (London)* **100**, 896 (1975).
833. Y. Yamamoto, T. Kumamaru, Y. Hayashi and Y. Otani, *Jpn Analyst* **17**, 92 (1968).
834. Y. Yamamoto, T. Kumamaru, Y. Hayashi and Y. Otani, *Anal. Lett.* **1**, 955 (1968).
835. Y. Yamamoto, T. Kumamaru, Y. Hayashi and M. Kanke, *Gisei Kagaku* **17**, 251 (1971).
836. M. Yanagisawa, M. Suzuki and T. Takeuchi, *Anal. Chim. Acta* **46**, 152 (1969).
837. H. Zachariasen, I. Andersen, C. Kostol and R. Barton, *Artzl. Lab.* **22**, 172 (1976); see also *Clin. Chem.* **21**, 562 (1975).
838. W. S. Zaugg and R. J. Knox, *Anal. Chem.* **38**, 1759 (1966).
839. P. B. Zeeman and L. R. P. Butler, *Appl. Spectrosc.* **16**, 120 (1962).
840. A. Zettner and D. Seligson, *Clin. Chem.* **10**, 869 (1964).
841. A. Zlatkis, W. Bruening and E. Bayer, *Anal. Chem.* **41**, 1692 (1969).
842. Y. A. Zolotov and Petrikova (Eds), *Uspekhi Analiticheskoi Khimii*, Nauka, Moscow, 1974, pp. 54–58.
843. E. G. Zook, J. J. Powell, B. M. Hackley, J. A. Emerson, J. R. Brooker and G. M. Knobl, *J. Agric. Food Chem.* **24**, 47 (1976).

Index